THE POLYTECHF

SYMPOSIUM ADVISORY COMMITTEE

Professor Sir Hugh Ford (Chairman)	Sir Hugh Ford and Associates
Professor S. F. Brown	University of Nottingham
Mr J. W. Eastwood	Science and Engineering Research Council
Dr J. P. Giroud	GeoServices Consulting Engineers, U.S.A.
Professor R. Haas	University of Waterloo, Canada
Dr C. J. F. P. Jones	West Yorkshire Metropolitan County Council
Dr G. J. A. Kennepohl	The Tensar Corporation, U.S.A.
Dr A. McGown	University of Strathclyde
Mr V. Milligan	Golder Associates, Canada
Mr R. T. Murray	Transport & Road Research Laboratory
Mr N. Paine	Binnie & Partners
Professor I. M. Ward	University of Leeds
Dr A. J. Watson	University of Sheffield
Professor C. P. Wroth	University of Oxford

SYMPOSIUM ORGANISING COMMITTEE

Mr J. E. Templeman (Chairman)	Netlon Limited
Professor S. F. Brown	University of Nottingham
Mr J. W. Eastwood	Science and Engineering Research Council
Dr A. McGown	University of Strathclyde
Mr N. Paine	Binnie & Partners
R. F. Gibson	Netlon Limited
D. DuBois	Netlon Limited

Telford Ltd, Telford House, PO Box 101, 26-34 Old Street, London EC1P 1JH, England

g in Publication Data:

nt: proceedings of a
y the Science and
uncil and Netlon Ltd
and 23 March 1984

tion
Research Council

ord, Norfolk

POLYMER GRID REINFORCEMENT

Proceedings of a conference sponsored by the
Science and Engineering Research Council
and Netlon Ltd and held in London on
22 and 23 March 1984

Published by Thomas

First published 1985

British Library Cataloguin

Polymer grid reinforceme
conference sponsored
Engineering Research Co
and held in London on 2

1. Polymers and polymeriz
I. Science and Engineering
II. Netlon Ltd

620.1'92   TA455.P58

ISBN: 0 7277 0242 4

© authors of papers, 1984

All rights, including translation, reserved.
mitted in any form or by any means, ele
Managing Editor, Publications Division, Th
Printed by The Thetford Press Limited, Theth

CONTENTS

Polymer grid reinforcement. Thomas Telford Ltd, London, 1985

ERRATA

p.31. The following names should be added to the list of authors of the paper 'Use of geogrid properties in limit equilibrium analysis'

K. Z. Andrawes, *University of Strathclyde*
R. A. Jewell, *Binnie and Partners*

p.45. The third sentence of Dr R. B. Singh's discussion should read 'Towards this end the contributor has proposed a system of using TRRL z-type patented anchors (Murray & Irwin, 1981) alternating with geogrids for almost all types of soils, and especially so for embankment situations.'

Opening address

Sir Alan Muir-Wood, *Sir William Halcrow & Partners*

I am very pleased to be here on several counts. First of all, I sit on the council of SERC, which is usually backing success, and I am delighted therefore to see that you have an overflow for this meeting as an example of this degree of discrimination. Secondly, on behalf of the Institution of Civil Engineers, I am very pleased to see this conference being held here, and I would remind those who do not know that quite a lot of the innovation in the applications to reinforced earth took place in the bailiwick of Mr Tony Gaffney, the present President of this Institution. Thirdly, this Institution always has seen innovation as being at the centre of its activities, and particularly innovation which cuts across disciplines, which this occasion clearly exemplifies.

I was brought up on the myth of isotropic soils. It was then gradually appreciated that natural soils, by the way in which they are laid down, are almost certain to be anisotropic. We then went in for artificial soils used as fills, and these tended to be rather more isotropic than in nature. Now we reinforce the soils and we put them back to being anisotropic. But just to complete the story I see that paper 8.1 for this occasion is dealing with random reinforcement, so presumably from that we return full circle to isotropic soils of a sort.

Reinforced earth in concept is quite literally as old as the hills, and anybody who has looked at mesas and seen why they stand up will appreciate this. The concepts of reinforced earth were widely used for military engineering through the centuries. I am indebted to Professor Tom Hanna for enlightening me on the building of ziggurats by the Babylonians, reinforced by layers of papyrus in bitumen. So we are not talking about anything particularly new in fundamentals. But on the other hand what we are always talking about as civil engineers is finding how to make use of the forces of nature towards our benefit rather than our destruction.

The military uses I alluded to were for making ravelins and all the other things that Gilbert's modern major-general was so expert upon - in, as I recall, a purely theoretical sense. But nevertheless model tests were undertaken early in the 19th century on the use of reinforced earth. More recently we had the ideas of Vidal which came over in <u>Terre Armée</u>.

The subject area for this symposium is that of net fabrics. It is a pity that there is no account of the extreme ingenuity which went into the origin of the method of forming these nets, but I hope the omission is going to be repaired during the next two days.

In the general process of innovation commonly seen in civil engineering the innovators first step forward out of line, taking a certain amount of risk in so doing. This then results in trials based upon a rudimentary analysis. There is a period of development of the theory, and this leads to acceptance by the few who are in the van. Following this there is great activity in research where it becomes fashionable throughout the academic world. Subsequently this leads to production of codes of practice with their wider acceptance.

There are two particular problems in this area that are worth mentioning. The first is the necessity for those who are familiar with the application of plastics in one form or another to understand the long time-scale of civil engineers. I noticed, for instance, that Professor Ward mentions in his paper that it is not practicable to measure strain rates below 10^{-9} s^{-1}. Now, 10^9 s is a period of about 30 years, which is well below the normal design life for a civil engineer, so if the strains occur at a rate of 10^{-9} s^{-1}, that may mean that he is going to have rather a pregnant-looking embankment after 30 years.

The second problem is the question of durability. I see that there is no specific paper on durability, which I would have thought needs to be covered fairly fully, because as civil engineers we are well aware of the various forms of attack that we have to contend with, from the aggressive soils of one sort and another. It would be nice to know that what we are talking about in general can also be applied in particular; or, if not, to know what the cautionary tales are so that there is not a history of defect which then tends towards a more suspicious attitude to the use of nets. This is particularly so because of the behaviour of certain polymers in the open air - particularly in sunshine - and the fact that nets have been sold for purposes which should have been accompanied by a government health warning. Those who are gardeners will know that nets, after 2-5 years depending upon which particular material they are made of, tend to look rather like the white suit that Alec Guinness wore in the film of that name. Hence it is necessary to give assurances. Underground you do not have these particular problems, but there may be others which affect longevity.

With those cautionary words, I nevertheless wish to share a feeling of great enthusiasm towards this symposium. I am extremely pleased to see this coming-together of those of totally different disciplines, and I look forward to this as being a thoroughly constructive discussion.

Polymer grid reinforcement. Thomas Telford Limited, London, 1985

The research and development programmes for polymer grid reinforcement in civil engineering

Sir Hugh Ford, *Sir Hugh Ford and Associates Ltd*

The development of the properties of polymers to allow them to be used as load bearing structural materials has only slowly invaded the more conventional engineering fields. Much ingenuity has been needed to develop high strength and long term stability. Tensar geogrids are now available to the Civil Engineer for soil stabilisation, reinforcement, road pavements etc. but it is by extensive and careful test work that design guidelines can evolve.

To provide the necessary knowledge of engineering properties, Netlon Ltd is working with four University Civil Engineering Departments through a SERC co-operative award.

The work has progressed sufficiently to mount this Symposium to discuss the results and present examples of applications worldwide.

Although polymeric materials have been around for many years, the development of their properties to allow them to be used as load-bearing structures has been slow to invade the more conventional engineering fields. With their low elastic modulus relative to metals their elasto plastic behaviour and strain-rate dependency, it has been difficult to get acceptance among engineers in general and in heavy engineering situations in particular.

Nor is this surprising. So much of engineering depends upon a sufficient body of practising engineers at all levels gaining sufficient experience and mutual awareness of a new development that behaviour in a range of conditions and environments is appropriately established for reasonable confidence to be generated. Nowhere is this more important than with a new material especially a material like a polymer that has penetrated very dramatically into everyday life as a packaging material or as a substitute for a conventional, well established material.

With Civil and Structural Engineering applications a new material has very considerable obstacles to surmount - perhaps greater than in other branches of engineering. Most long term structures are required to be designed for a life of 120 years. Few polymeric materials have been in large scale production for longer than 50 years and, with their known creep behaviour, predicting acceptable lives of 120 years from short term tests has provided formidable problems to researchers and designers alike.

There could be two ways to overcome these barriers to progress. One that is too often used is make a frontal attack with a measure of risk that some applications will fail and need to be remedied - or compensated for - at a later date. Enough experience, it is hoped, will have been gained meanwhile to avoid the difficulties in future applications without the new product's reputation being irreparably tarnished in the process. The other way is to approach the potential market cautiously while undertaking the necessary development and application trials to ensure satisfactory performance to be confidently anticipated: a slower method, perhaps, but one that inevitably has to be followed with a technically sophisticated material such as highly oriented polymer.

It is only by careful and exhaustive experimental studies that it has been possible to evolve grid structures that have the high strength and durability necessary to be used as geogrids in real engineering applications. To take heavy gauge polypropylene or high density polyethylene sheet, to punch clean, regular holes and then to extend the sheet either uniaxially or biaxially under controlled temperatures and strain rates, demanded very great ingenuity in mechanical and control engineering. It will well be appreciated that, while the strain rate, and orientation strengthening characteristics of polymers compensate for the reduction in area effect of the stretching there is inherent in the process the "plastic instability" phenomenon familiar to all in the tensile test.

It is important to draw attention to this aspect of the manufacture of geogrids because upon the knowledge of the behaviour of polymers in such a process and the precision and quality control at each stage depends the reliability and mechanical properties of the final product. The present stage of realisation of a practical engineering material has not been reached without very sophisticated and carefully controlled processes and continuing research into polymeric materials and the optimisation of their orientation.

2

Polymer grid reinforcement. Thomas Telford Limited, London, 1985

One of the most important aspects and certainly the most critical consequence of orientation is the development of stable long term properties. While high strength and an enhancement of the pseudo-elastic properties of the polymers are major objectives that have been achieved by orientation, these properties in themselves are not enough: the creep rates under steady load, of an order to be of practical value, are of paramount importance, particularly in Civil Engineering applications, as has already been said, to be able to predict satisfactory performance over periods of 120 years. Such a requirement is of particular difficulty because, not only are the potential applications extremely varied, but the manufacturers and processors of polymers are not standing still and better materials and controlled processes are constantly coming forward. Moreover, Civil Engineering depends upon a great variety of practitioners, from the highly discerning to the reverse. While the applications of inclusions, whether geotextiles or geogrids, are now established to the extent that there is a growing appreciation of the methods of their applications for soil stabilisation, reinforcement of soils, road and loaded areas, asphalt structures and the like, there is still much debate as to the best design methods and guidelines to be followed to ensure the most economic and long-term performance of such materials. Their use in cement and concrete type composites is less well defined.

With a view to providing both the background knowledge and the necessary design data on the engineering properties of Tensar geogrids, the Netlon Company sought the assistance of the Science and Engineering Research Council in enlisting the services of University Departments to carry out a comprehensive programme of cooperative research. The SERC's scheme of cooperative awards with industry has enabled an extensive programme of objective studies to be undertaken and prosecuted with much greater speed and effectiveness than any other means.

There are four Universities in the scheme and each has support from the SERC and Netlon for at least a three year programme. The Civil Engineering Departments of Nottingham, Oxford, Sheffield and Strathclyde are involved in appropriate parts of the overall programme, while Professor Ward of Leeds University has given considerable help in advice on orientation in relation to the physical performance of the geogrids. More recently the Teaching Company scheme of SERC is being invoked in association with Bradford and Strathclyde in the applicational work.

As Chairman of the SERC Steering Committee the members of which cover the main engineering interests both in the UK and North America, it appeared to me that the work had progressed sufficiently rapidly to enable a Symposium to be mounted, both as a means of reporting on the work done to date and its promise for the future, and also to provide a forum for a

wider discussion and dissemination of the actual and potential applications of these important new civil engineering materials. Although there are now many installations around the world - and several of these are being described and assessed in later papers to this Symposium - the objective views and considerations of practising Civil Engineers is most warmly to be welcomed. It is for this reason that we have arranged for as much discussion from the floor as possible.

In the foregoing, the importance of determining the creep behaviour of oriented polymers has been emphasised. The problem from the viewpoint of the application of geogrids as inclusions in soils, unmade roads and pavements or embankments is that while it is necessary to study the creep behaviour of the grids in isolation, creep data in itself is only part of the whole behaviour of the grid and the matrix, since the latter has its own characteristics which interact in varying degrees and in different ways with the geogrid. Much of the work of the Group has been directed towards an understanding of these phenomena and to develop design guides to assist those who wish to employ Tensar grids. At Strathclyde, a major programme of creep testing of significant test pieces of the various grid types has been interlocked with the extensive laboratory facilities built up at Blackburn. Creep tests have been going on continuously now for several years and the whole programme, together with some specific testing at Nottingham is probably the most comprehensive test facility for oriented polymers on a practical basis to be found anywhere.

The interpretation of these data when the grid is included in a matrix, be it for an embankment, a retaining wall, an unpaved road, an asphalt layer or a cement composite, is the crux of our work and the papers presented in the Symposium will, it is confidently hoped, demonstrate that, while much remains to be done, reliable guidelines for the designer can now be proposed.

The application of new materials in situations requiring fitness for purpose for periods up to a hundred years or more presents the researcher and the practising engineer with redoubtable problems. There are no short cuts. Only continuous testing and development, feedback from the field and large scale trials and a professional team of experienced people can ensure the full economic advantages of new technology to be realised. Until there is a sufficient crowd of witness from the field, it is advisable to be cautious in proposing guidelines for load and strain limits when using geogrids. Yet it is my considered view that the thorough and wide-ranging programme of in-house and University research and development is of a quality and extent to justify its presentation at a Symposium on Tensar alone. It is a good example of what can be achieved by vigorous collaboration of industry with Universities, a Research Council and Consulting Engineers on a broad front.

The orientation of polymers to produce high performance materials

Recent research at Leeds University has shown that the physical properties of polymers can be improved dramatically by stretching processes whose principal aim is to produce a high degree of molecular orientation. In polyethylene and a few other polymers, very high degrees of stretch have led to spectacular improvements in stiffness and strength.

In polyethylene, particular attention has been given to the creep behaviour under sustained load. Guidelines have been established for the sensitivity of the creep behaviour to such variables as polymer molecular weight, degree of copolymerisation and draw ratio. With this information, it is possible to optimise the situation with regard to the possibility of ultimate failure.

I. M. Ward, *University of Leeds*

INTRODUCTION

The discovery of methods for the preparation of ultra high modulus polyethylene by tensile drawing to very high draw ratios[1] raised the possibility that such materials might be used for the reinforcement of brittle matrices such as cement[2], concrete[3] or polymeric resins[4]. This requires that the oriented polymer should be subjected to continuous loading without this leading to failure. During recent years extensive studies of the creep and recovery behaviour of oriented polyethylene[5] and polypropylene[6] have been carried out at Leeds University. The aim of the work was two fold (1) to establish criteria for creep failure (2) to identify structural factors which lead to improvement in creep performance. The results of these studies have been of direct relevance to Tensar and have provided guidelines for the development of Tensar for engineering applications.

THE PRODUCTION AND PROPERTIES OF DRAWN POLYETHYLENE

The discovery of high modulus polyethylene stemmed from the recognition that the Young's modulus of fibres and tapes produced by tensile drawing relates to the draw ratio, and increases steadily with increasing draw ratio[1]. The limitation of a natural draw ratio was replaced by the concept of effective drawing i.e. drawing at a temperature and strain rate which avoids fracture on the one hand, and "flow drawing" where the molecules relax to isotropy, on the other.

It is therefore useful to consider the properties of oriented polymers in terms of the draw ratio as a key variable, alongside other variables such as molecular weight, molecular weight distribution, and degree of branching or cross-linking. Fig. 1 shows the creep compliance (extensional strain/applied stress) for linear polyethylene (Rigidex 50 grade) of draw ratios λ = 10 and 30, each sample being measured at three levels of stress. It can be seen that there is a marked reduction in creep compliance with draw ratio. Although in all cases the creep compliance is not independent of stress level (the behaviour is non-linear) the response is more nearly linear at the highest draw ratio.

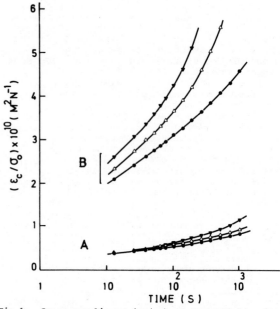

Fig.1. Creep compliance (ε_c/σ_o) of drawn Rigidex 50 samples A, λ=30 and C, λ=10 at 0.1 ($\bullet$), 0.15 ($\triangle$) and 0.2 GPa ($\blacktriangledown$) applied stress σ_o as a function of time. (Reproduced from Polymer 19, 969 (1978) by permission of the publishers, Butterworth & Co. (Publishers) Ltd. (C).

Although Fig. 1 shows the effect of draw ratio on the creep behaviour, it is not so revealing as the plots of creep rate versus total creep strain. Sherby and Dorn[6] used such plots to represent their data for the creep of isotropic polymethylmethacrylate (PMMA). This was because they adopted the approach of considering that the total creep behaviour can be represented by a thermally activated process, so that it should be possible to construct a master curve for data over a range of temperatures. Although the case of oriented polyethylene is not quite as tractable as PMMA, we have

Polymer grid reinforcement. Thomas Telford Limited, London, 1985

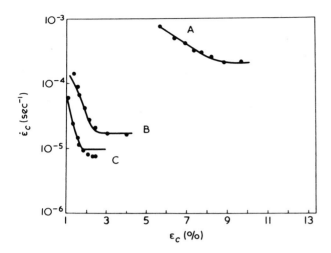

Fig.2. Creep strain rate $\dot{\varepsilon}_c$ as a function of ε_c at 0.2 GPa applied stress for drawn Rigidex 50 samples A, λ=10, B, λ=20, C, λ=30. (Reproduced from Polymer 19, 969 (1978) by permission of the publishers, Butterworth & Co. (Publishers) Ltd. (C).

used "Sherby-Dorn Plots" to describe the results to some advantage.

Fig.2 shows these plots for the same samples as those described by Fig. 1. In contrast to Sherby and Dorn's results, the creep rates fall to a constant rate which is independent of strain and it was established that there was no simplistic stress/temperature super-position rule for the plots as a whole. However, the constant creep rates (which we have termed the plateau creep rates $\dot{\varepsilon}_p$) were shown to depend on stress σ and temperature T is a manner expected for a single thermally activated process, leading to the so-called Eyring equation

$$\dot{\varepsilon}_p = \dot{\varepsilon}_o \exp - \frac{\Delta H}{kT} \sinh \frac{\sigma v}{kT} \qquad (1)$$

where ΔH, v are the activation energy and activation volume respectively, $\dot{\varepsilon}_o$ is a constant pre-exponential factor and k is Boltzmann's constant.

It was concluded that there is an initial viscoelastic region, where the creep rate falls with time (and hence creep strain) followed by a second region where the creep rate is constant, and corresponds to a permanent flow plastic deformation process which is irrecoverable and will eventually lead to failure of the sample.

This is the behaviour of the first sample of oriented polyethylene to be examined, a linear polymer of low molecular weight. It could be very well modelled by the arrangement of springs and dashpots shown in Fig.3(b), which is essentially a linear Voigt element in series with a Maxwell element which contains an Eyring dashpot. Materials of this kind would be unsuitable for load-bearing applications as there will always be an element of permanent plastic deformation, even at very low stresses, which could lead to failure, albeit at very long times.

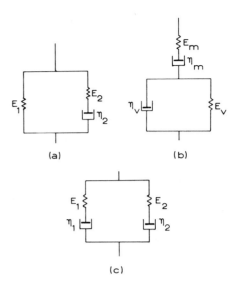

Fig.3. Schematic representation of the four element mechanical model for creep and recovery. (Reproduced from Polymer 22, 870 (1981) by permission of the publishers, Butterworth & Co. (Publishers) Ltd. (C).

It was therefore considered essential to produce oriented polyethylenes of greatly improved creep response esponse, and this has been achieved in three ways

(1) by increasing molecular weight
(2) by cross-linking the polymer
(3) by use of copolymers.

Details of samples studied are given in Table 1.

Table 1.

Polymer Grade†	Sample details		Chemical Composition
	$\bar{M}_n$	$\bar{M}_w$	
Rigidex 50	6,180	101,450	Homopolymer
Rigidex 006-60	6,000	130,000	Homopolymer
Rigidex HO20-54P	33,000	312,000	Homopolymer
Rigidex 002-55	16,900	155,000	Ethylene/Hexene Copolymer ca 1.5 butyl/ 10^3 carbon atoms
Rigidex 50	6,180	101,450	Homopolymer γ-irradiated (2Mr) before drawing

† BP Chemicals International Ltd.

In Fig. 4, Sherby-Dorn plots for a higher molecular weight polyethylene homopolymer are compared with those for the lower molecular weight material shown in Figs. 1 and 2. It can be seen that at the lowest stress level, the Sherby-Dorn plot for the higher molecular weight polymer shows a continuously decreasing strain-rate with increasing strain. This implies that at low stress levels this material reaches an equilibrium

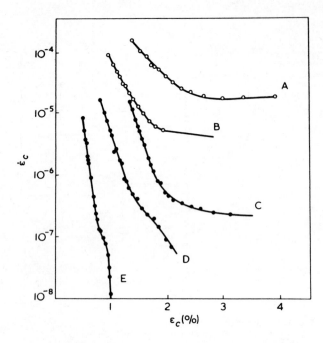

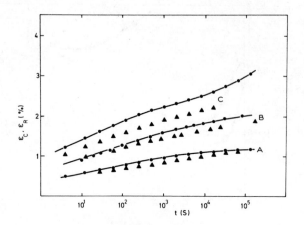

Fig. 6.   Creep ε_c and recovery ε_R curves for Rigidex 002-55 copolymer, λ=20 ($\bar{M}_w$ = 155,000, ca 1.5 butyl branches per 10^3 carbon atoms) ($\bullet$) creep ($\blacktriangle$) recovery. A, 0.1 GPa, B, 0.15 GPa, C, 0.2 GPa. (Reproduced from Polymer 22, 870 (1981) by permission of the publisher, Butterworth & Co. (Publishers) Ltd. (C)).

Fig. 4.   Creep strain rate $\dot{\varepsilon}_c$ as a function of strain for Rigidex 50, λ=20 (o) ($\bar{M}_w \sim$ 100,000) and Rigidex HO20, λ=20 ($\bullet$) ($\bar{M}_w \sim$ 300,000) A, 0.2; B, 0.15; C, 0.2; D, 0.15, E, 0.1 GPa. (Reproduced from Polymer 19, 969 (1978) by permission of the publishers, Butterworth & Co. (Publishers) Ltd. (C)).

safe critical stress level for long term loading.   In Fig. 6, similar results are shown for a copolymer, and it can be seen that the presence of a very small number of side branches ($\sim$ 1 per 1,000 carbon atoms) is extremely effective in altering the creep behaviour.

The concept of a critical stress for permanent flow creep, although valuable for engineering design is somewhat obscure in terms of polymer physics.   It is therefore necessary to pursue the analysis of the data further to bring the behaviour within the general scheme of our present understanding of deformation processes in polymers.

It has been proposed that a more appropriate first-order representation for the viscoelastic behaviour of these materials is by two Maxwell elements in parallel, the two dashpots each corresponding to thermally activated processes, (Figure 3(c)).   The equilibrium creep behaviour is then described by the situation where the two springs have reached their equilibrium extension and the total creep rate is constant, with the stress divided across the two dashpots.   If the two thermally activated dashpots are very different in kind, one having a large activation volume v_1 and the other a small activation volume v_2, we have

$$\frac{\sigma}{T} = \frac{k}{v_1}\left[\frac{\Delta H_1}{kT} + \ln \frac{2\dot{\varepsilon}_p}{[\dot{\varepsilon}_o]}\right]_1 + \frac{k}{v_2}\sinh^{-1}\frac{\dot{\varepsilon}_p}{[\dot{\varepsilon}_o]}\exp\frac{\Delta H_2}{kT}$$

(2)

where the two activated processes are denoted by the subscript symbols 1 and 2 respectively. σ is the applied tensile stress at a temperature T°K, $\dot{\varepsilon}_p$ the creep rate, $[\dot{\varepsilon}_o]$ the pre-exponential factor, ΔH the activation energy and k is Boltzmann's constant.

strain, as shown in Fig. 5, so that failure does not occur.   In practical terms, this means that there is a

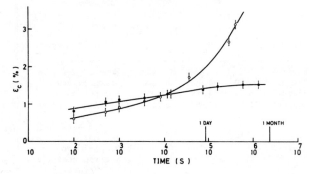

Fig.5.   A comparison of the long term creep response of Rigidex 50, λ=20 (o) and Rigidex HO20, λ=20 ($\bullet$) at 0.1 GPa applied stress (——) are visual fits and the error bars correspond to the accuracy of the measurements. (Reproduced from Polymer 19, 969 (1978) by permission of the publishers, Butterworth & Co.(Publishers) Ltd. (C).

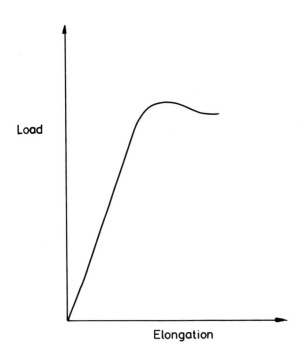

Fig. 7. The yield point in a constant strain rate test.

RELATIONSHIP OF CREEP BEHAVIOUR TO STRUCTURE

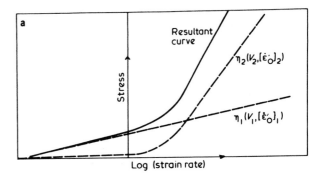

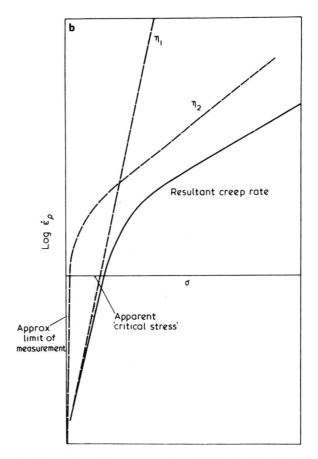

This representation is familiar to polymer scientists in respect of the yield behaviour of isotropic polymers[7,8]. In general, a yield point is observed in a constant strain rate test (Figure 7). The point of maximum load represents the point at which the plastic flow rate $\dot{\varepsilon}_p$ matches the imposed strain rate of the testing machine. The relationship between the maximum stress (the yield or flow stress) and the imposed strain rate can generally be represented by equation (2), as shown in Fig. 8(a). In Fig. 8(b) an exactly equivalent plot is presented which is the more useful form for considering creep data. Here the creep rate (exactly equivalent to the strain rate) is plotted on the ordinate and stress level is plotted on the abscissa.

At low stress levels, the viscosity of the dashpot η_1 in Fig. 3(c) is much larger than that of the dashpot η_2. Hence η_1 can be regarded as infinite and model 3(c) reduces to model 3(a), the Eyring formulation of a standard linear solid[9]. The apparent critical stress can now also be explained in a more satisfactory fashion. Because of limitations imposed by the creep test itself, it is not practicable to measure strain rates below $\sim 10^{-9} s^{-1}$. There is hence a practical cut-off at very low strain rates, where the plateau creep rates are very small, and plateau creep behaviour would only be observed at very long times ($\sim$ several years). The apparent critical stress corresponding to these very low strain rates has been indicated schematically in Fig. 8(b). At these low stress levels we would also expect to find almost total recovery in terms of the modelling proposed here, and this expectation has been borne out by experiment[5].

Fig. 8. Schematic representation of plastic flow for the 2-process model; (a) yield stress vs applied strain rate; (b) creep strain rate $\dot{\varepsilon}_p$ vs applied stress. (Reproduced from Polymer 22, 870 (1981) by permission of the publishers, Butterworth & Co. (Publishers) Ltd. (C)).

RELATIONSHIP OF CREEP BEHAVIOUR TO STRUCTURE

At high stress levels, equation (2) gives the relationship between the plateau creep rate and stress in terms of an apparent single activated process as

$$\frac{\sigma}{T} = \frac{k}{v_{eff}} \quad \frac{\Delta H_{eff}}{kT} + \ln \frac{2\dot{\varepsilon}_p}{\left[\dot{\varepsilon}_o\right]_{eff}} \quad (3)$$

where v_{eff} and ΔH_{eff} are the effective activation volume and activation energy respectively.

Because the term σv_{eff} is small, small variations in v_{eff} with temperature can be neglected and plateau creep rates determined at a constant stress level can be described by the equation

$$(\log \dot{\varepsilon}_p)_\sigma = \log \left[\dot{\varepsilon}_o\right]_{eff/2} - \frac{\Delta G_{eff}}{2.3kT} \quad (4)$$

where ΔG_{eff} is an activation free energy.

Results for two samples (Rigidex 50, λ=20 and Rigidex HO20, λ=20) are shown in Fig. 9. There is a

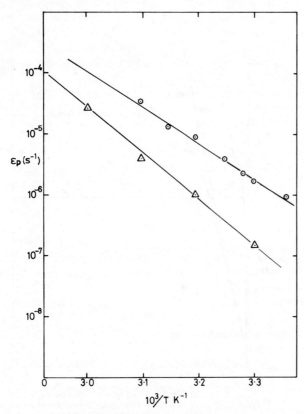

Fig.9. Temperature dependence of plateau creep rates $\dot{\varepsilon}_p$ for two samples Δ Rigidex 50, λ=20, stress level 0.1 GPa; Θ Rigidex HO20, λ=20, stress level 0.25 GPa. (Reproduced from J.Polym.Sci., Polym. Phys. Edn., (in press).

good linear relationship between $\log \dot{\varepsilon}_p$ and $1/T$, giving a value for ΔG_{eff} of 28 kcals/mole. Assuming a constant activation volume v_{eff} of 100Å^3 gives a value of 30 kcals/mole for ΔH_{eff}. This is in the range quoted for the α-relaxation in polyethylene by previous workers on the basis of dynamic mechanical experiments[10-12]. Now

$$\Delta H_{eff} = \frac{v_2 \Delta H_1}{v_1+v_2} + \frac{v_1 \Delta H_2}{v_1+v_2} \quad (5)$$

so that ΔH_{eff} is primarily determined by ΔH_2 (unless ΔH_1 is very much larger than ΔH_2). It is therefore reasonable to conclude that the smaller activation volume process relates to the α-relaxation process.

To examine the changes in behaviour which occur on changing the structure of the oriented polymer, plateau creep data at 20°C for a number of different samples were fitted to the constant temperature form of equation (2)

$$\frac{\sigma}{T} = \frac{k}{v_1} \quad \ln \dot{\varepsilon}_p - \ln \left[\frac{\dot{\varepsilon}_o'}{2}\right]_1 + \frac{k}{v_2} \quad \sinh^{-1} \frac{\dot{\varepsilon}_p}{\left[\dot{\varepsilon}_o'\right]_2} \quad (6)$$

where $\left[\dot{\varepsilon}_o'\right]_1$ and $\left[\dot{\varepsilon}_o'\right]_2$ include the temperature dependence $\exp \Delta H/kT$ term.

It was shown that the data could be equally well fitted to either a variable value for v_1 or a constant average value for v_1 and that the latter constant affected the values obtained v_2 to a very small extent only[13]. The results of these fitting procedures for a fixed value of v_1 are shown in Table 2, and some of data

TABLE 2

Activation parameters for the 2-process model

Sample	v_1 (Å^3)	$\left[\dot{\varepsilon}_o'\right]_1$ (sec^{-1})	v_2 (Å^3)	$\left[\dot{\varepsilon}_o'\right]_2$ (sec^{-1})
Rigidex 50, λ=20	456	2.2×10^{-8}	78	1.0×10^{-6}
006-60, λ=10	456	2.1×10^{-9}	720	2.6×10^{-11}
006-60, λ=20	456	1.7×10^{-14}	126	7.8×10^{-7}
006-60, λ=30	456	2.1×10^{-16}	78	6.6×10^{-7}
HO20-54P, λ=10	456	5×10^{-14}	98	1.3×10^{-6}
HO20-54P, λ=20	456	8.4×10^{-18}	95	1.6×10^{-7}
γ-irradiated Rigidex 50, λ=20	456	1.1×10^{-16}	103	1.2×10^{-6}
002-55, λ=20	456	1.6×10^{-18}	88	4.7×10^{-7}

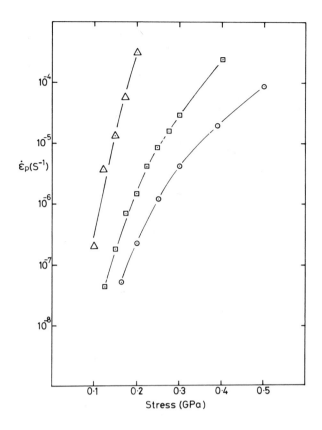

Fig. 10.   Effect of draw ratio on plateau creep
            behaviour at 20°C for 006-60 samples
            Δ λ=10,   □ λ=20,   Θ λ=30
            (Reproduced from J.Polym. Sci., Polym.
            Phys. Edn., (in press).

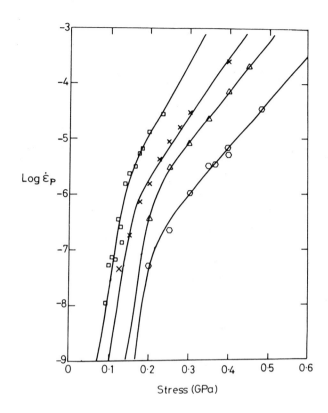

Fig. 11.   Fitted curves to plateau creep data on the
            basis of the two process model, assuming that
            process 1 has an activation volume of 456Å³
            □ Rigidex 50, λ=20, x 006-60, λ=20,
            Δ γ-irradiated Rigidex 50, λ=20,
            Θ H020, λ=20
            (Reproduced from J. Polym. Sci., Polym. Phys.
            Edn., (in press).

(there is considerable overlap between results for several samples) in Figs. 10 and 11.

The differences between different materials are primarily reflected in the relative contribution of process 1, the large activation volume process. We note from Table 2 that the pre-exponential factor $[\dot{\varepsilon}_o']_1$ falls substantially with increasing draw ratio and there are large differences between samples of different chemical composition. In particular both γ-irradiated low molecular weight Rigidex 50 and the 002-55 copolymer behave like the high molecular weight H020-55P. The activation volume of ∿ 500 Å³ for process 1 is consistent with that for an oriented non-crystalline polymer, and it has been suggested that this process is associated with the non-crystalline regions[5]. It is then attractive to speculate that γ-irradiation of low molecular weight polymer produces a comparable molecular network by incorporating chemical cross-links to that obtained in high molecular weight polymer due to physical entanglements. It is surprising that introducing a small proportion of branch points as in

the 002-55 copolymer has the same effect. However, these results are consistent with observations of the tensile drawing of these materials. Increasing molecular weight, γ-irradiation prior to drawing and the introduction of side branches have all been shown to give rise to increased strain hardening[14-16].

With regard to the practical consequences of this work, it can be seen from Figs. 10 and 11 and Table 2 that although there is also merit in increasing the draw ratio, either increasing the molecular weight or γ-irradiation prior to drawing or introducing side branches,   are equally effective in increasing the critical stress below which plateau creep cannot be observed within a sensible experimental time-scale. In the case of Tensar, a copolymer with a somewhat larger branch content than the Rigidex 002-55 grade was selected, to ensure a maximum critical stress compatible with achieving a good degree of molecular orientation by drawing.

CONCLUSIONS

The steady state creep behaviour of oriented poly-ethylene can be satisfactorily represented by two thermally activated processes acting in parallel. One process has a comparatively small activation volume and is primarily affected by the draw ratio, decreasing with increasing draw ratio. Its activation energy is close to that for the α-relaxation process, consistent with its identification with a deformation process in the crystalline regions of the polymer.

The second process has a comparatively high activation volume which is not significantly affected by changes in structure or chemical composition. The contribution of this process is markedly affected by the molecular weight of the polymer, by γ-irradiation prior to drawing and by branch content. It has been tentatively associated with the molecular network.

A result of practical importance is that for each drawn polymer, a critical stress can be defined, below which permanent flow is negligible and on unloading, recovery is virtually complete. This concept has been used to select a suitable polymer for Tensar and can also be applied to practical performance of Tensar grids, as described in a subsequent contribution to this symposium.

REFERENCES

1. Capaccio, G., Ward, I.M. (1973). Properties of ultra-high modulus linear polyethylenes. Nature Phys. Sci., 243, 130, 143; (1974) Preparation of ultra-high modulus linear polyethylenes; effect of molecular weight and molecular weight distribution on drawing behaviour and mechanical properties. Polymer, 15, 233.

2. Hannant, D.J. Symposium 'Polymers in Civil Engineering', Plastics & Rubber Institute, London 20th May, 1980.

3. Kamal, M. (1983). Ph.D. thesis Leeds University.

4. Ladizesky, N.H., Ward, I.M., Phillips, L.N. (1983). Novel fibre/resin interface for improved mechanical properties of composites reinforced with high modulus polyethylene filaments. Progress in Science & Engineering Composites, T. Hayashi, K. Kawata & S. Umekawa, Ed. ICCM-IV, Tokyo, 203-210.

5. Wilding, M.A., Ward, I.M. (1978). Tensile creep and recovery in ultra-high modulus linear polyethylenes. Polymer, 19, 969; (1981). Creep and recovery of ultra-high modulus polyethylene. Polymer, 22, 870. (1981). Routes to improved creep behaviour in drawn linear polyethylene. Plastics & Rubber Processing & Applications, 1, 167.

6. Sherby, O.D., Dorn, J.E. (1956). Anelastic creep of poly methyl-methacrylate. J. Mech. Phys. Solids, 6, 145.

7. Roetling, J.A. (1965). Yield stress behaviour of polymethylmethacrylate. Polymer, 6, 311.

8. Bauwens-Crowet, C., Bauwens, J.C., Homes, G. (1969). Tensile yield-stress behavior of glassy polymers. J. Polym. Sci., A2, 7, 735.

9. Halsey, G., White, H.J., Eyring, H. (1945). Mechanical properties of textiles, 1. Text. Res. J. 15, 295.

10. Sandiford, D.J.H., Wilbourn, A.H. (1960). "Polyethylene" (A. Renfrew & P. Morgan, eds) Iliffe, London, Ch.8.

11. Kajiyama, T., Okada, T., Sakoda, A., Takayanagi, M. (1973). Analysis of the α-relaxation process of bulk crystallized polyethylene based on that of single crystal mat. J. Macromol. Sci. Phys., B7, 583.

12. Nakayasu, H., Markowitz, H., Plazek, D.J. (1961). The frequency and temperature dependence of the dynamic mechanical properties of a high density polyethylene. Trans. Soc. Rheol. 5, 261.

13. Wilding, M.A., Ward, I.M. (1984). Creep behaviour of ultra high modulus polyethylene: Influence of draw ratio and polymer composition. J. Polym. Sci., Polym. Phys. Edn., (in press).

14. Capaccio, G., Crompton, T.A., Ward, I.M. (1976). The drawing behaviour of linear polyethylene. I. Rate of drawing as a function of polymer molecular weight and initial thermal treatment. J. Polym. Sci., Polym. Phys. Edn. 14, 1641. (1980). Drawing behaviour of linear poly-ethylene. II. Effect of draw temperature and molecular weight on draw ratio and modulus. J. Polym. Sci., Polym. Phys. Edn., 18, 301.

15. Capaccio, G., Ward, I.M., Wilding, M.A. (1978). The plastic deformation of γ-irradiated linear polyethylene. J. Polym. Sci., Polym. Phys. Edn., 16, 2083.

16. Capaccio, G., Ward, I.M. (1984). The drawing behaviour of polyethylene copolymers. J. Polym. Sci., Polym. Phys. Edn., (in press).

The load-strain-time behaviour of Tensar geogrids

A. McGown, K. Z. Andrawes and K. C. Yeo,
University of Strathclyde, and D. DuBois,
Netlon Ltd

TENSAR geogrids are a new generation of soil reinforcing materials manufactured by heat stretching a perforated sheet of co-polymer to form a grid-like structure. In order to use them to their maximum efficiency it is necessary to establish the loads which they may carry for any period up to 120 years, the design lifetime of many civil engineering structures. In this paper the basic definitions that should be used to describe the geometry of the grids are given and then employed to identify the representative sizes and shapes of test specimens. Test methods which have been developed for quality control purposes and for performance data acquisition are detailed. Following this the method of analysing and presenting test data are given together with typical data for both uniaxial and biaxial grids.

INTRODUCTION

Soils have little or no tensile resistance; therefore inclusion of tension-resistant materials within the soils is an effective means of reinforcing them. To optimise the efficiency of these inclusions, they should be placed in the zones of largest tensile strains and in the directions of principal tensile strains. The mechanism of stress transfer from the soil to the inclusion depends primarily on the geometrical properties of the inclusions. For rods, strips and sheets, the stress transfer mechanism is one of surface friction; but for grids, nets and meshes, it depends upon the more efficient interlock principle. When stressed, the inclusions must of course be capable of sustaining the load without rupture and without generating unacceptably large deformations during the design lifetime of the structure. Thus the two properties of tension resistant soil inclusions which must be established in order to allow the design of reinforced soil structures are their surface friction or interlock properties and their load-strain-time behaviour over the lifetime of the structure.

Tensar geogrids are molecularly oriented polymeric grid structures specifically developed for use as tension-resistant inclusions in soils. Jewell et al (1984) deal with the measurement of the interlock that develops between Tensar geogrids and various soil types. In this paper the methods of measuring and presenting the load-strain-time behaviour of the geogrids are detailed and their applicability to civil engineering specifications and designs is fully identified.

MATERIALS TESTED

In order to develop and prove the methods of testing and the analysis of test data for geogrids, control batches of two geogrids, Tensar SR2 and SS2, were used throughout. Tensar SR2 is a uniaxial geogrid manufactured from co-polymer grade high density polyethylene (H.D.P.E.). Tensar SS2 is a biaxial geogrid manufactured from homo-polymer polypropylene. Prior to the production of both these grids, 2.5% of carbon black was added to provide ultraviolet protection.

The terminology used to describe the sizes and shapes of Tensar SR2 and SS2 is as shown in Fig. 1 and the mean dimensions of the materials in the control batches are given in Table 1, together with the mass per unit area of each product.

CLAMPING ARRANGEMENTS AND TEST SPECIMEN SIZES

In order to determine the load-strain-time relationship properties of the geogrids, specialised end clamps were developed, as shown in Fig. 2. The top and bottom bars of uniaxial grid test specimens are held directly by these clamps but the cross ribs of biaxial grid test specimens are cast into the clamps using a low melt point alloy "Ostalloy 158", manufactured by Fry's Metal Limited, Glasgow. The composition and properties of the alloy are given in Table 2.

By a programme of constant rate of strain testing at 2% strain per minute, 20°±2°C, and 65%±2% relative humidity, the minimum sizes of test specimens at which the measured load-strain properties of the grids are independent of sample size were established to be as shown in Table 3. These test specimen sizes are used in all subsequent testing.

CONSTANT RATE OF STRAIN TESTING

A constant rate of strain of 2% per minute and standard test conditions of 20°±2°C and 65%±2% relative humidity were used to establish reference load-strain properties for the geogrids. These properties are termed the "Index Properties". The effect of humidity on the properties of Tensar geogrids is negligible, but as with all polymeric materials, their load-strain properties vary with strain rates and temperature.

Polymer grid reinforcement. Thomas Telford Limited, London, 1985

11

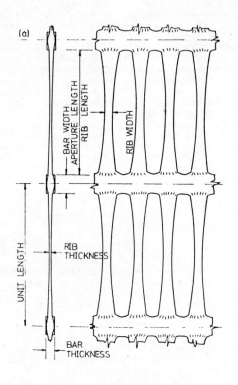

FIG. 1(a). TERMINOLOGY USED FOR TENSAR
GEOGRID STRUCTURE, SR2.

TABLE 1.   PHYSICAL PROPERTIES OF TENSAR SR2
AND TENSAR SS2

Dimensional Properties	Mean
Product Width	1000 mm
Number of Ribs	44 per metre
Aperture Length	90.73 mm
Transverse Bar Width	12.69 mm
Transverse Bar Thickness (max)	4.56 mm
(min)	4.36 mm
Rib Thickness	1.34 mm
Rib Width	5.72 mm
Product Mass	
Mass per unit area	971.6 g/m²

(a) TENSAR SR2

Dimensional Properties	Mean
Product Width	3000 mm
Number of Ribs (primary)	37 per metre
Number of Ribs (secondary)	27 per metre
Max. Aperture Length	34.42 mm
Max. Aperture Width	24.73 mm
Min. Rib Width (primary)	2.68 mm
Min. Rib Width (secondary)	2.95 mm
Grid Pitch (primary)	36.17 mm
Grid Pitch (secondary)	27.23 mm
Min. Rib Thickness (primary)	1.54 mm
Min. Rib Thickness (secondary)	1.74 mm
Max. Junction Thickness	4.17 mm
Product Mass	
Mass per unit area	202.6 g/m²

(b) TENSAR SS2

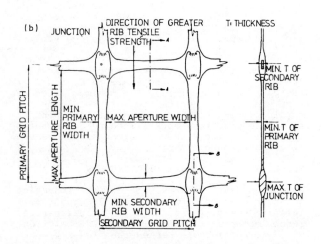

FIG. 1(b). TERMINOLOGY USED FOR TENSAR
GEOGRID STRUCTURE, SS2.

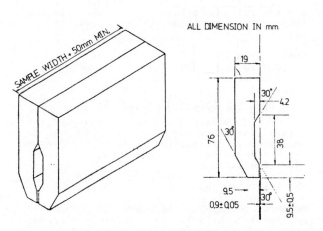

FIG. 2.   ISOMETRIC VIEW AND HALF SECTION OF
THE TENSAR JAWS.

The Index properties of Tensar SR2 and SS2 are as shown in Fig. 3. Corrections must be applied to these Index properties to obtain the constant rate of strain, load-strain properties at any other test conditions. The temperature and strain rate corrections for Tensar SR2 and SS2 are as given in Figs. 4 and 5 respectively.

TABLE 2. PHYSICAL PROPERTIES OF OSTALLOY 158

Melting temperature °C	70
Density g/cm	9.38
Specific heat liquid j/g°C	0.167
solid j/g°C	0.167
Latent heat of fusion KU/kg	32.564
Brinell hardness number	9.2
Tensile strength mpa	41.3
% elongation in slow loading	200
Loading mpa	
Max. load 30 secs.	68.9
Max. load 5 mins.	27.6
Safe sustained load	2.1

Composition %	
Bismuth	50.0
Lead	26.7
Tin	13.3
Cadmium	10.0

TABLE 3. TEST SPECIMEN SIZES FOR GEOGRIDS

Grid	Specimen Size	
Uniaxial Product SR2	5 bars x 15 ribs	
Biaxial Product	Vertical Ribs x Horizontal Ribs	
SS2	4 x 8 (secondary direction)	
	5 x 6 (primary direction)	

Even with the slowest practical rates of strain of 10^{-3} to 10^{-4}% per minute in constant rate of strain testing, the time to reach strain levels operating in grids within soil structures is often no more than 1-2 months. Thus, there is a practical limit to the usefulness of constant rate of strain testing for the prediction of the long-term load-strain behaviour of geogrids. A more suitable means of determining this is creep testing.

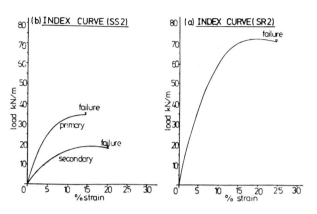

FIG. 3. INDEX TEST CURVES (a) SR2; (b) SS2.

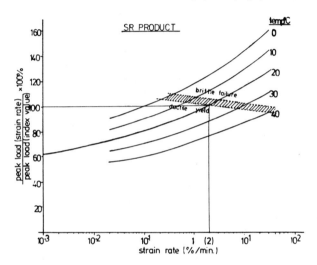

FIG. 4. TEMP/STRAIN RATE CORRECTION FOR SR2.

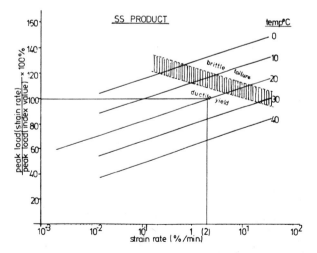

FIG. 5. TEMP/STRAIN RATE CORRECTION FOR SS2.

RAPID LOADING CREEP TESTING

(a) Test Method

To determine the long-term load-strain properties of Tensar geogrids, rapid loading creep
tests were performed. Temperature and relative
humidity conditions are controlled throughout
the tests ±2°C and ±2% respectively. In each
test the load is applied smoothly in less than
5 seconds, either by direct dead loading or
through an adjustable 5 to 1 level arm system.
The loads are maintained for the duration of
the test at ±1% of full load, with continuous
electronic monitoring undertaken. The deformation of the test specimen is measured electron-
ically over the initial rapidly varying part of
the test using two LVDT's, one each end of the
specimen, but thereafter measurements are made
by mechanical dial gauges at the same locations.
To record the variations in load and deformations during the test, an electronic data log-
ging device is used which records the output of
the load cell and LVDT's in a pre-set pattern
varying from every 0.2 seconds for the first
10 seconds to once per day after 24 hours.

The data obtained from each test is plotted as
load against time and strain against time, with
at least five levels of load applied at each
test temperature. The range of test temperatures presently used is from 0°C to +40°C. The
duration of loading is usually in excess of
1000 hours, unless rupture of the test specimen
occurs before the allotted period of testing.
Where rupture does not occur, tests are terminated when the daily creep strain rate either
reaches a minimum value or becomes insignificant
(i.e. when the $\log_{10}$ value of the strain rate
is less than the order of -8 . The unit of
strain rate is % strain/min.). At the end of
each test period, the test specimen is partially
unloaded and left under reduced sustained loading for 100 hours, to establish the elastic
rebound of the material. During the unloading
and reduced loading period, the load and deformation of the test specimen is measured in the
same manner as during loading. Normally a
number of short term load tests of five hours
duration with partial unloading sustained for a
half hour are also conducted to improve the
interpretation of the initial loading and
unloading behaviour of these materials.

(b) Presentation of Test Data

The load-strain-time behaviour of geogrids at
any temperature (T°C) can be represented by a
rheological model which is as shown in Fig. 6.
This comprises both recoverable (elastic) and
irrecoverable (plastic) instantaneous strains
and linear and non-linear time-dependent creep
strains. In order to assess these various components of the total strain it is, however,
necessary to perform a correction to the initial
part of the rapid loading creep test data to
establish an equivalent instantaneous loading
strain-time curve, as shown in Fig. 7(a). In
the same manner, the rapidly unloaded test data
is corrected to establish an equivalent instantaneous unloading strain-time curve, as shown in

Fig. 7(b). Having corrected the strain-time
test data in this manner, equivalent instant-

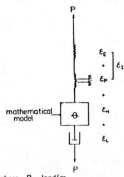

where: P = load/m
ε_E = elastic strain (recoverable)
ε_P = initial plastic strain (irrecoverable)
ε_I = $\varepsilon_E + \varepsilon_P$ = instantaneous strain
ε_N = non-linear time dependent strain
ε_L = linear time dependent strain

FIG. 6. RHEOLOGICAL MODEL.

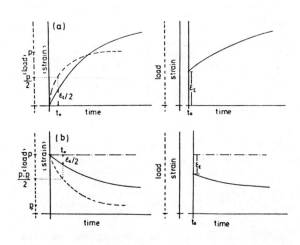

FIG. 7. INSTANTANEOUS CREEP DATA CORRECTION
FOR (a) LOADING and (b) UNLOADING.

aneous strain-time curves may be produced, as is shown in Fig. 8 for Tensar SR2. Similar unloading curves may also be produced after appropriate corrections to the unloading test data.

A further method of presenting the test data at any temperature is to plot the daily creep strain rate on a logarithmic scale against the cumulative strain, as shown in Fig. 9 for Tensar SR2, Wilding and Ward (1978). This plot clearly indicates whether the daily strain rate is reaching a minimum value or approaching zero and so determines the duration of the test. It also provides data for identifying the linear creep strain component of the rheological model and other important performance limits which will be discussed in the following sections.

From the basic strain-time plots, the short and long-term creep test data at any temperature, the instantaneous loading and unloading strain parameters for the range of applied loads are obtained as shown in Fig. 10. From the log strain rate versus strain plot, the minimum daily creep strain rate is established for the various applied loads and used to calculate the linear creep strain component of the rheological model. Knowing the instantaneous and linear creep strain components of the total strain, the non-linear creep strain at any time and load may be obtained by difference, and then character-ised using curve fitting techniques. Thus the overall strains for any load at any time and temperature can be modelled. By repeating this process for the test data at various tempera-tures then temperature corrections to the various parameters within the rheological model are obtained.

LOAD-STRAIN-TIME RELATIONSHIPS

Both the constant rate of strain test data and the creep test data, supplemented by data gene-rated from rheological modelling, may be used to identify the load required to produce any strain after a period of time at any operating tempera-ture. Due to the limitations of the period of testing in constant rate of strain tests, creep testing has been chosen as the preferred method of obtaining these data.

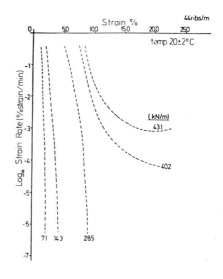

FIG. 9. LOG_{10} STRAIN RATE vs. STRAIN PLOTS FOR SR2.

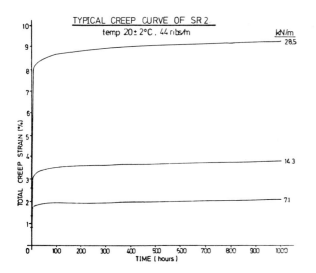

FIG. 8. TYPICAL STRAIN-TIME CURVES FOR INSTANTANEOUS LOADING FOR SR2.

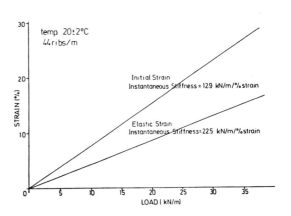

FIG. 10. RHEOLOGICAL MODEL COMPONENTS FOR SR2 (INITIAL STRAIN ϵ_I and ELASTIC STRAIN ϵ_E.

15

The data are best plotted as isochronous-strain curves for each test temperature in the manner indicated in Fig. 11 for Tensar SR2 and SS2 at 20°C. As can be seen from Fig. 11(a), the isochronous load-strain relationship for Tensar SR2 is approximately linear up to 10% strain. The log strain rate versus strain plot of creep test data for Tensar SR2 at the same tempera- ture, Fig. 9, also shows that below 10% strain, this geogrid shows no tendency to approach creep rupture for the range of applied loads. Thus for Tensar SR2 at 20°C, 10% strain represents its "Performance Limit Strain" below which rupture is not likely to occur by ductile yield. Also up to 10% strain, the isochronous stiffness of the geogrid (i.e. the ratio of load to strain at given times) may be represented by the secant slopes of the isochronous load-strain curves. These may be replotted as isochronous stiffness against time, as shown in Fig. 12 for Tensar SR2 at 20°C.

The maximum load that may be sustained for any given time by a geogrid without approaching creep rupture is simply calculated by multiplying the performance limit strain by the isochronous stiffness at the given time. In the case of the Tensar SR2 control batch, this gives for 120 year lifetime a maximum load of 29 kN/m. For any given period of time significantly less than this, the maximum load that may be carried is greater than this value.

Alternatively by assuming that creep under constant load occurs in the geogrid when it is in the soil, then for loads less than the maximum value, the strain developed at any given time is calculated by dividing the load by the isochronous stiffness at that time. Thus the isochronous stiffness data may be used to reconstruct the creep strain-time data for any load. Alternatively, should the geogrid be assumed to suffer load relaxation when in-soil (i.e. reduction in sustained load at constant strain) the rheological model may be used to generate mathematically the reduction in load in the geogrid with time. For Tensar SR2 this is shown in a normalised form in Fig. 13.

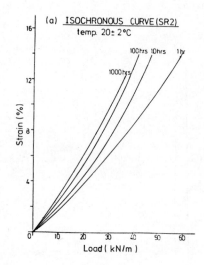

FIG. 11(a). ISOCHRONOUS LOAD-STRAIN CURVES FOR SR2.

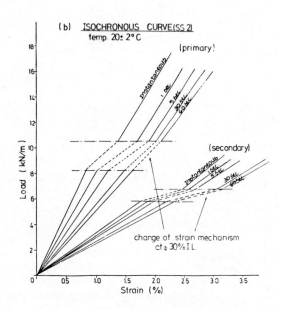

FIG. 11(b). ISOCHRONOUS LOAD-STRAIN CURVES FOR SS2 (BOTH DIRECTIONS).

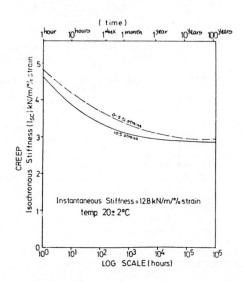

FIG. 12. ISOCHRONOUS STIFFNESS vs. TIME FOR SR2.

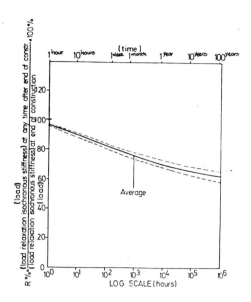

Generated Mathematically (temp. 20°C)

FIG. 13. LOAD RELAXATION NORMALISED vs. TIME
FOR SR2

USE OF DATA IN SPECIFICATIONS AND DESIGNS

The constant rate of strain test data carried
out under standardised Index conditions, provide a very useful set of data on which the
civil engineer may base his specification of
product quality. This may be based on the entire load-strain index curve, on peak values or
on specific loads to be attained for particular
strains. In each case, the characteristic
value rather than the mean value is best specified (i.e. the value which has a 95% confidence
of being achieved; CIRIA Report No. 63 (1977).

These Index data may also be used in designs
which are purely empirical or based on back
analysis, but where a more fundamental analytical approach is taken, the load-strain-time
relationships developed from rapid loading
creep testing must be used. By multiplying the
isochronous stiffness with the likely strains
to be developed in the soil system, up to the
performance limit strain, the load that may be
safely carried by a geogrid for any period of
time may be calculated and employed in the analysis of any soil-geogrid structure. Additionally the construction and post-construction
deformations may be calculated using these data
by assuming pure creep in the geogrid. Also
from mathematically generated data the possible
post-construction load-relaxation of the geogrid may be calculated. Thus the upper and
lower bound conditions of post-construction
behaviour of the geogrid can be calculated.
The method of employing these data in designs
is discussed in detail by McGown et al (1984).

CONCLUSIONS

The development work described in this paper
confirms that in common with all polymeric materials, the load-strain behaviour of Tensar geogrids when tested under constant rate of strain,
are significantly influenced by the test temperature and strain rate. The necessity of employing standardised test conditions is thus emphasised and convenient standardised Index Test
Conditions have been identified. It is suggested that these provide data which may form a
suitable basis for civil engineering specifications. Rapid loading creep tests have been
shown to be suitable for the measurement of the
load-strain-time relationship of the geogrids.
Methods of presenting data from these tests in
a form suitable for use in the design of soil-
geogrid structures have also been developed.
These allow the loads and strains in the geogrids at any time to be easily calculated.

The rapid loading creep test data allow the
identification of the performance limit strains
for the geogrids. Below these strains there is
no risk of rupture by ductile yield in the geogrids at any time during the lifetime of the
soil-geogrid structure. It must be noted,
however, that a separate analysis must be
undertaken to ensure that brittle fracture of
the polymer does not occur. This analysis
involves calculating the stresses within the
polymer and ensuring that these are maintained
at all times at levels below the critical
brittle fracture stress.

ACKNOWLEDGEMENTS

The authors wish to acknowledge that this work
was undertaken with the support of a Science
and Engineering Research Council Co-operative
Grant made in conjunction with Messrs. Netlon
Limited.

REFERENCES

CIRIA Report 63. ((1977). Rationalisation
of safety and serviceability factors in
structural codes. Jul., 55-56.

Jewell, R., Milligan, G.W.E., Sarsby, R.W. and
DuBois, D.D. (1984). The interaction
between soil and geogrids. Proc. Symp. on
Polymer Grid Reinforcement in Civil
Engineering, Mar.

Paine, N., McGown, A., Jewell, R. and DuBois, D.
(1984). The use of geogrid properties in
design. Proc. Symp. on Polymer Grid
Reinforcement in Civil Engineering, Mar.

Wilding, M.A. and Ward, I.M. (1978). Tensile
creep and recovery in ultra-high modulus
linear polyethylenes. Polymer, 19,
969-976.

Interaction between soil and geogrids

R. A. Jewell, *Binnie and Partners,* G. W. E. Milligan, *University of Oxford,* R. W. Sarsby, *Bolton Institute of Higher Education,* and D. Dubois, *Netlon Ltd*

Three mechanisms of interaction between soil and grid reinforcement have been identified to provide a framework for understanding and interpreting laboratory and field measurements. Their importance in governing the resistance to direct sliding of soil over grid reinforcement and the bond strength for different types of grid reinforcement has been examined. A qualitative model is proposed to classify the influence of soil particle size on soil grid interaction. Theoretical expressions have been developed to describe direct sliding resistance, bearing stresses, and bond strength of reinforcement in terms of well accepted soil and reinforcement parameters. The results have been compared with available laboratory and field data. Guidance is given on measuring values of grid sliding resistance and bond strength in the laboratory, and on the selection of parameters for design.

INTRODUCTION

Limit equilibrium analysis of reinforced soil focuses attention on the mechanisms or modes of behaviour which could lead to unsatisfactory performance. For example, in a reinforced slope gross outward sliding of soil over the surface of a reinforcement layer, or insufficient bond between the reinforcement and soil could lead to failure, Fig. 1. An analysis for these cases requires an understanding of the mechanism of interaction between soil and reinforcement, so that relevant parameters and their numerical values can be defined.

The case of grids reinforcing soil to provide stability in structures governed by self weight loading, such as the embankment in Fig. 1, is examined in the paper. The action of soil bearing on surfaces of the grid members plays a significant role in the interaction, but it is complex and as yet not well understood, Schlosser et al (1983).

The suitability of a grid form for soil reinforcement has been demonstrated by laboratory and field studies, Chang et al. (1977), Forsyth (1978), Peterson (1980). Over the past few years the availability of strong polymer grids with high extensional stiffness and resistance to corrosion has increased interest in grid reinforcement for soil. Their potential use in cohesive, low quality or aggressive soils is particularly attractive and has provided an incentive for research.

The results of the research have been examined to investigate the mechanics of soil grid interaction. A number of simple mechanisms have been identified which organise the experimental findings allowing simple parameters to be defined in terms of the grid geometry and conventional soil properties. These parameters can be evaluated by straightforward tests.

While it is clearly recognised that the mechanisms proposed in this paper radically simplify the complexities of real behaviour, the aim has been to provide a framework for the understanding of soil grid interaction so that it may be evaluated for design purposes.

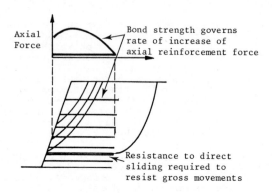

Fig.1. Two mechanisms of failure in a reinforced embankment.

GRID TYPES AND PARAMETERS

Three types of grid are shown in Fig. 2. A simple grid which has been tested in research investigations is formed by punching apertures into a plane sheet, Fig 2a. The grids formed from welded bars, Fig. 2b, and from drawn polymers, Fig. 2c, have been more widely used for research investigations and field constructions.

The parameters suggested to describe the geometry of a grid are illustrated in Fig. 3 for an element of grid length L_r and width W_r.

18

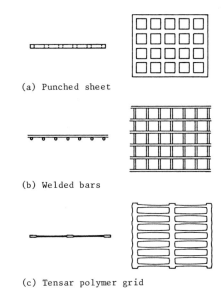

(a) Punched sheet

(b) Welded bars

(c) Tensar polymer grid

Fig.2. Three types of reinforcement grid.

The spacing between the members on which soil bearing may occur is denoted by S and the thickness of these bearing members by B. In plan, the total grid area is A_r and this comprises a fraction of solid material α_s and a fraction of apertures $(1 - \alpha_s)$. Depending on the type of grid, the full width of the members may be available for bearing, $\alpha_b = 1$ in the case of a welded grid, or only a fraction of the width $\alpha_b < 1$, as in the case of a punched sheet or a Tensar grid.

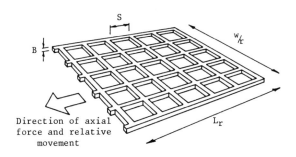

B

S

W_r

Direction of axial
force and relative
movement

L_r

$A_r = W_r L_r$ Grid surface area

α_s Fraction of grid surface area that is solid

α_b Fraction of grid width W_r available for bearing

Fig.3. Definition of terms for a grid.

MECHANISMS OF INTERACTION

Two main modes of behaviour which need to be considered for design of a reinforced embankment are illustrated in Fig.1. Reinforcement layers must have sufficient length to prevent gross outward movement of soil blocks sliding across the surface of a reinforcement layer in the soil. Sufficient reinforcement length must also be provided to ensure that the required axial force may be developed in the reinforcements through bond to maintain satisfactory equilibrium. These two cases, direct sliding and bond strength, are examined in separate sections below.

Three main mechanisms of soil reinforcement interaction can be identified. These are,

● soil shearing on plane surfaces of the reinforcement which are parallel to the direction of relative movement of the soil

● soil bearing on (bearing) surfaces of the reinforcement which are substantially normal to the direction of relative movement of the soil

● soil shearing over soil through the apertures in a reinforcement grid.

For clarity, the action of these three mechanisms and their relative importance for direct sliding resistance and bond strength are examined separately below.

DIRECT SLIDING RESISTANCE

The loading condition for direct sliding is a gross outward shearing force tending to cause a block of soil overlying a horizon of reinforcement to shear over the reinforcement and the underlying soil. The three mechanisms of interaction may each resist direct sliding as shown in Fig. 4.

Shear between soil and plane surfaces

The resistance provided by shear between soil and the plane surface areas of a grid, Fig. 4a, is similar to the skin friction that is measured conventionally for soils and construction materials. The most comprehensive series of measurements for skin friction were carried out by Potyondy (1961). He used a direct shear test with a block of the construction material mounted flush in the lower portion of the box and the soil in the upper half. The resistance to sliding is defined by an angle of skin friction δ. The test is now widely adopted as a standard method for measuring friction between soil and plane reinforcement surfaces, Schlosser & Juran (1979).

The ratio of skin friction to soil friction for sands and silts acting on smooth metallic or concrete surfaces is typically in the range $\delta/\phi = 0.5$ to 0.8, Potyondy (1961).

Soil bearing on reinforcement

The effect on overall direct sliding resistance of soil bearing on reinforcement surfaces is more difficult to assess, Fig. 4b.

To generate bearing stresses there must be relative movement between the soil and the reinforcement bearing surface. This implies that soil within the grid apertures would have to displace relative to the grid, Fig. 4b. From symmetry, the soil in the upper and lower halves of the grid apertures would need to move in opposite directions, shearing with respect to one another, so that bearing stresses could be balanced across a grid bearing member.

By choosing to concentrate shear displacement across the top of the grid, the soil contained in the grid apertures would not displace relative to the bearing members and resistance from bearing stresses could be avoided. This is illustrated in terms of rupture surfaces in Fig. 5. This suggests that soil bearing is unlikely to have a significant effect on direct sliding resistance for a grid in granular soils.

Influence of soil particle size

The relative size of the soil particles to the grid apertures can influence direct sliding resistance in four main ways, and these are shown schematically in Fig. 5 for the case of a Tensar grid.

Soil made up of small particles in the fine sand or silt sizes has greater kinematic freedom to rupture in zones of varying orientations, so that an overall displacement can be made up from several components, Arthur et al (1977). In a fine soil the size of the rupture zone, which is about 25 particle diameters, and the shear displacement required to take the soil beyond peak resistance are both reduced, Roscoe (1970). These points are illustrated by the careful measurements in the simple shear apparatus of Bhudu (1979).

Fine soil may thus take maximum advantage of the relatively smooth surface areas of a metal or polymer grid. In the case of a Tensar grid this would involve sliding across the surface of the thinner, load carrying rib members as well as across the top surface of the bearing members, Fig. 5a.

a. Shear between soil and plane reinforcement surfaces

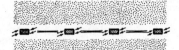

b. Soil bearing on grid reinforcement bearing surfaces

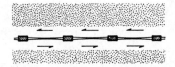

c. Soil shearing over soil through the reinforcement grid apertures

Fig.4. Three mechanisms resisting direct sliding.

Soil shearing over soil

In order for outward direct sliding to occur the overlying soil must shear with respect to the underlying soil and the grid. The resistance to shear in the aperture spaces of the grid should equal the resistance to direct shear of unreinforced soil ϕ_{ds}, which can be measured in a direct shear test.

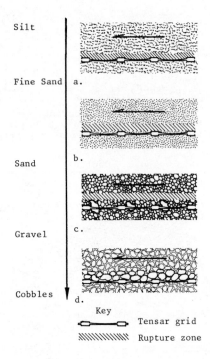

Fig.5. Qualitative effect of increasing soil particle size on direct sliding. View in cross-section across direction of sliding.

For a coarser sand, the potential advantage in terms of reduced direct sliding resistance of sliding on the smooth rib surface areas may be insufficient to compensate for the extra resistance of a rupture zone at varying orientations. The direct sliding mechanism would, in this case, change to a rupture zone formed mostly in the soil taking advantage only of the smooth top surface of the bearing members, Fig. 5b.

If the soil contains particles of a similar size to the grid apertures, then these may become lodged against the grid bearing members and extend into the soil on either side of the grid, Fig. 5c. If a number of particles were lodged in this way the soil would no longer be able to take advantage of the smooth top surface of the bearing members, and the rupture zone would be forced away from the grid fully into the soil. In this case the shear resistance to direct sliding would equal the full shear resistance of the soil.

Finally, grid reinforcement could be placed in soils with particles too large to penetrate the grid apertures, Fig. 5d. The resistance to direct sliding in this case could be very low indeed. In the extreme, shear resistance might be provided only by soil particles in contact with the plane grid surface areas. The irregularity in the surface profile of the grid in this case would be insignificant to the scale of the soil, so that little benefit from that source could be expected. The shear resistance to direct sliding could therefore be reduced to the resistance of the soil bearing on a smooth sheet of the parent reinforcement material, usually a metal or a polymer.

The influence of soil particle size on direct sliding resistance is shown qualitively in Fig. 6. Direct sliding resistance is expressed as a factored reduction on the soil direct shear resistance, $f_{ds}\tan\phi_{ds}$, as explained below.

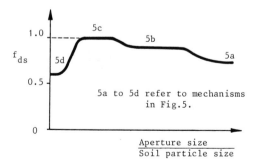

Fig.6. Schematic illustrations of the influence of particle size on the resistance to direct sliding, f_{ds} $\tan\phi_{ds}$.

Theoretical expressions for direct sliding resistance

The basic equation for direct sliding resistance can be expressed in terms of the two contributions from shear between soil and the plane surface areas of the grid, and between the soil shearing over itself in the grid apertures. Direct sliding resistance can be described by a general expression,

$$f_{ds}\tan\phi_{ds} = \alpha_{ds}\tan\delta + (1-\alpha_{ds})\tan\phi_{ds} \qquad (1)$$

where f_{ds} coefficient of resistance to direct sliding

ϕ_{ds} angle of friction for soil in direct shear

δ angle of skin friction for soil on plane reinforcement surfaces

α_{ds} fraction of grid surface area that resists direct shear with soil

The parameter α_{ds} has been introduced so that one general expression can be used to describe the four different values of direct sliding resistance envisaged, Fig.6. Equation (1) can be manipulated to give an expression for f_{ds},

$$f_{ds} = 1 - \alpha_{ds}(1 - \frac{\tan\delta}{\tan\phi}) \qquad (2)$$

The effect of the soil particle size shown in Figs. 5 & 6 can now be seen from equation (2). A reduction in the value α_{ds} results in an increase in direct sliding resistance f_{ds}, during the change from case 5a through 5b to 5c, Fig. 5. Indeed when the rupture zone is forced away from the grid, case 5c, $\alpha_{ds} = 0$ and $f_{ds} = 1.00$. In the extreme case of large particles resting directly on the grid plane surfaces rather than penetrating the apertures, case 5d in Fig. 5, $\alpha_{ds} = 1.00$ resulting in a reduction of direct sliding resistance to

$$f_{ds} = \frac{\tan\delta}{\tan\phi_{ds}}$$

For design of cases with reinforcement grids providing stability in embankments or retaining walls where the soil fill particles penetrate the grid apertures, a suitable value of maximum direct sliding resistance would be given by equations 1 or 2 adopting a value $\alpha_{ds} = \alpha_s$, where α_s is the fraction of solid surface area in a grid, Fig. 3. This corresponds to case 5a in Figs. 5 and 6.

Comparison with experimental results

A comprehensive series of tests have been carried out investigating the influence of particle size on direct sliding resistance. Two types of polymer grid Tensar SR1 and Tensar SR2 were used, Netlon Limited (1984). Seven different grading curves gave relative soil particle sizes ranging from silts to gravels.

The soil types are indicated on Fig. 7. Five of the soils were crushed limestone. Pulverised fuel ash provided a fine grained soil and crushed granite a uniformly coarse soil with an average particle size 8 mm.

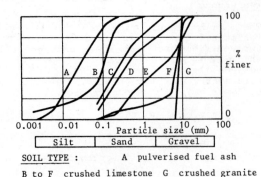

SOIL TYPE :   A pulverised fuel ash

B to F crushed limestone   G crushed granite

Fig.7. Soils tested during investigations of direct sliding resistance.

Direct shear tests were carried out in a large shear box. The lower portion of the box was first filled with compacted soil almost to the top, the grid placed so as to be flush with mid-plane of the shear box, then the upper half of the box was filled and a vertical load applied before shearing.

The results of tests on the grid Tensar SR2 with an aperture width 17.3 mm are summarised in Fig. 8. The coefficient of direct sliding resistance f_{ds} for peak conditions (the ratio of peak resistance for sliding over the grid and soil to the peak direct shear resistance of the soil alone) is shown plotted against the ratio of aperture width to the average or D_{50} particle size.

At low ratios of aperture width to soil particle size, tests were carried out on soil type G and the grid Tensar SR1. The test results were obtained with one grid type by clipping out alternate rib members and then pairs of rib members, and the results are summarised in Fig. 9.

The trend of the results shown dotted in Figs. 8 and 9 confirm the expected variation of direct sliding resistance with soil particle size.

Predictions using equation (2) for the direct sliding resistance of Tensar SR2 in crushed limestone can be compared with the experimental measurements. The ratio of peak skin friction angle to peak soil friction angle for crushed limestone (grading C in Fig. 7) on a high density polyproplene sheet was measured to be

$$\frac{\tan \delta}{\tan \phi}_{ds} = 0.63$$

The ratio of the area of solid grid surfaces to the total area of the grid for Tensar SR2 is $\alpha_s = 0.46$, Netlon Limited (1984). For the case of soil shearing over the top surface of bearing members the ratio of area reduces to $\alpha_{ds} = 0.12$.

Using equation (2), the predicted resistance to direct sliding in fine soil is,

$$f_{ds} = 1 - 0.46 \ (1 - 0.63) = 0.83 \qquad (3)$$

In coarser soil,

$$f_{ds} = 1 - 0.12 \ (1 - 0.63) = 0.96 \qquad (4)$$

When soil particles larger than the bearing member thickness can penetrate the grid apertures and lodge against the bearing member, the rupture zone is forced away from the grid so that,

$$f_{ds} = 1 - 0.00 \ (1 - 0.63) = 1.00 \qquad (5)$$

Finally for large soil particles resting on the grid surfaces and not penetrating the apertures,

$$f_{ds} = 1 - 1.00 \ (1 - 0.63) = 0.63 \qquad (6)$$

The values given by equation (2) are shown on Figs. 8 and 9 to bound the experimental data well.

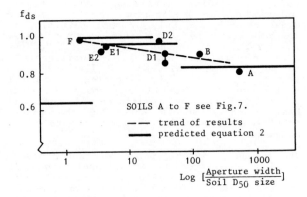

Fig.8. Measured peak direct sliding resistance f_{ds} for granular soils over Tensar SR2 grid reinforcement.

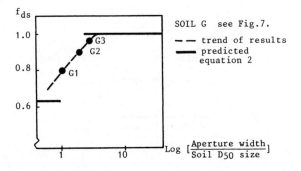

Fig.9. Measured peak direct sliding resistance f_{ds} for granular soil over Tensar SR1 grid reinforcement.

On the basis of the data available, it is recommended that equation (2) with a value $\alpha_{ds} = \alpha_s$ would give a suitably conservative value of direct sliding resistance for design when,

$$\frac{\text{Minimum Aperture Dimension}}{\text{Average Soil Particle Size}} \geq 3 \qquad (7)$$

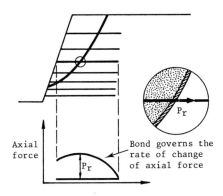

Fig.10. Bond strength governs the reinforcement length to mobilise axial reinforcement force.

BOND STRENGTH

The rate of change of axial force in reinforcement embedded in soil is limited by bond strength. For reinforcement in an embankment or retaining wall the changes in axial force close to the free end of the reinforcement embedded in the soil are important for determining the necessary lengths to enable the required axial forces to be generated. This is illustrated in Fig. 10.

Axial reinforcement force is caused by the development of linear soil strains in the direction of the reinforcement. A stiff reinforcement will resist such strains which, in the case of tensile strain in reinforced soil, results in axial tensile force in the reinforcement.

Since there is no relative displacement of soil on either side of the reinforcement, only two of the three simple mechanisms of interaction apply, namely

● soil shearing on plane surface areas of the reinforcement, and

● soil bearing on reinforcement surfaces (bearing surfaces) substantially normal to the direction of relative movement between the soil and the reinforcement.

These two mechanisms are illustrated in Fig. 11 for the case of bond, and are considered separately below.

Shear between soil and plane surfaces

The comments that were made for the case of direct sliding resistance should apply equally here. The shear stress that can be developed would depend on the angle of skin friction and the normal effective stress between the soil and the reinforcement surface.

When a rough bar or strip is pulled out from dense granular soil, volume expansion accompanies shear in the rupture zone that develops in the soil next to the reinforcement surface. This volume change can locally increase the normal effective stress and the resistance provided by shear between soil and the reinforcement surface. The phenomenon has similarities to the changes in stress that occur during a cavity expansion test in soil.

Whether the local increase in normal effective stress observed in pullout tests occurs when reinforcement acts under self-weight soil loading (rather than pullout loading), and whether, if it does, the increase in stress should be relied upon for design calculations has been and continues to be debated, Schlosser & Juran (1979), Schlosser et al (1983).

For grid reinforcement it may be prudent from the outset to estimate the normal effective stresses acting on the grid surface from self-weight and external loadings only.

The shear stress component of bond for a reinforcement grid of length L_r of width w_r placed horizontally in soil under self-weight loading would provide a maximum force,

$$(P_r)_{\text{shearing}} = 2 A_r \alpha_s \cdot \gamma z \cdot \tan\delta \qquad (8)$$

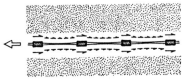

a. Shear between soil and plane surfaces

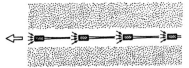

b. Soil bearing on reinforcement surfaces

Fig.11. The two mechanisms for bond between a grid and soil.

Soil bearing on grid surfaces

The bearing stress of soil on grid members is a problem similar in kind to the base pressure on deep foundations in soil. Kerisel (1961) concluded from test data that the soil density, foundation depth and size of bearing area have an important influence on bearing pressures, and that these are not a function only of soil friction angle. Vesic (1963) reported test results for which the ultimate bearing pressure for deeply embedded footings appeared to depend only on the relative density of the sand. Vesic observed a punching failure mode for footings in medium dense and loose sand models which resulted in significantly lower bearing pressures.

The bearing stresses on vertical surfaces loaded horizontally have been studied for the case of anchor plates. Rowe & Davis (1982) reported extensive numerical investigations based on the finite element method giving results which compared well with available data from anchor tests.

Selection of boundary conditions

The bearing members of a grid, can be thought to be like a line of anchors at a spacing S, Fig. 12a. The grid thickness B is likely to be small in comparison to the depth of soil z, and for most cases the bearing members can be considered deeply embedded, z/B is large. The overall normal effective stress at the level of the grid can be estimated from the depth and density of overlying soil. For the case of dry sand $\sigma_z' = \gamma z = \sigma_n'$

For shear on plane reinforcement surfaces, the shear stress acting on the reinforcement is usually expressed as a function of the normal effective stress acting between the soil and the reinforcement. The bearing stress acting on a grid may also be conveniently expressed as a function of this same normal effective stress. The problem is then reduced to the determination of the relationship between the effective bearing stress on the grid σ_b' and the overall normal effective stress in the surrounding soil σ_n', Fig 12b.

Theoretical estimates for bearing stresses

Rowe (1982) presented his results for anchors in a chart form to allow hand calculation of anchor capacity. Rowe's results can be in the form below for deeply embedded anchors,

$$\frac{\sigma_b'}{\sigma_n'} = F\gamma' \qquad (9)$$

For grids the thickness B is typically small compared to the depth of embedment z. The value F' takes account of the influence of soil dilatancy, anchor roughness and initial stress state in the soil.

Rowe's charts cover a range of soil friction angles 5° to 45°, with soil dilation angle ranging from 0 to Ø in each case. Anchor embedment z/B is in the range 1 to 8. Rowe suggests that the effect of anchor roughness and initial stress state for deeply embedded anchors can be neglected with a loss of accuracy less than 10%.

Two curves have been calculated from Rowe's (1982) charts, Fig. 13. These give a relationship between the soil bearing stress ratio σ_b'/σ_n' and the friction angle Ø for deeply embedded bearing surfaces. The curves are for the case of no dilation $\upsilon = 0$, and for soil dilating during shear according to the stress dilatancy flow rule assuming a critical state friction angle $\emptyset_{cv} = 32.5°$, Rowe (1962).

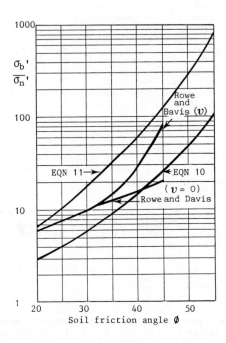

Fig.13. Theoretical relationships between bearing stress and soil friction angle.

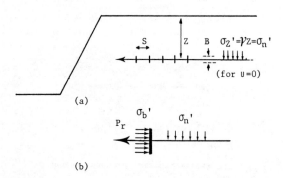

Fig.12. Definitions for bearing stresses on a reinforcement grid.

Low bearing pressures on deep footings have been associated with a punching failure mode in the soil. The stress characteristic net and boundary stress conditions shown in Fig. 14 are suggested to provide a lower estimate of the bearing stresses on reinforcement grid members. In this case $\sigma_n' = \gamma z$, and the relationship,

$$\frac{\sigma_b'}{\sigma_n'} = e^{(90°+\emptyset)\tan\emptyset} . \tan\left(45°+\frac{\emptyset}{2}\right) \quad (10)$$

Equation 10 is shown plotted in Fig. 13 and gives lower values for σ_b/σ_n' to those derived by Rowe & Davis (1982).

When a grid is acting in soil like a perfectly rough sheet, the overall orientation of the principal axis of compressive stress in the soil adjacent to the reinforcement would be significantly inclined to the vertical. In compact granular soils the overall horizontal effective stress in the soil adjacent to the reinforcement may often approximately equal the overall vertical effective stress.

An upper estimate of grid bearing stresses can be found by taking the conventional stress characteristic field for a footing (Prandtl (1921) & Reissner (1923)) rotated to the horizontal and a horizontal boundary stress in the soil $\sigma_h' = \sigma_v' = \sigma_n'$. The relationship between the grid bearing stress and the overall normal effective stress in the soil would then be given by the classical expression,

$$\frac{\sigma_b'}{\sigma_n'} = e^{\pi\tan\emptyset} . \tan^2\left(45°+\frac{\emptyset}{2}\right) \quad (11)$$

Equation 11 is shown plotted in Fig. 13 and gives higher values for σ_b'/σ_n' to those derived by Rowe & Davis (1982).

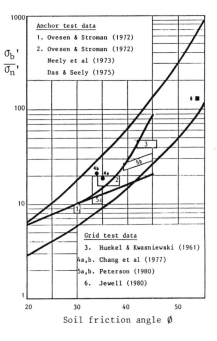

Fig.15. Comparison of test results with predicted values of bearing stress.

Comparison with experimental bearing stresses

The anchor test results summarised by Rowe & Davis (1982) have been extrapolated to the case of deep embedment and are shown plotted in Fig. 15

Four additional sets of data from tests on grid reinforcement have been plotted, Fig. 15.

The results of Hueckel & Kwasniewski (1961) are from pullout tests on short lengths of grid embedded in a body of sand and distant from rigid boundaries. Chang et al (1977) and Peterson (1980) conducted pullout tests on grids, both on a relatively large scale. Two soils were used in both investigations.

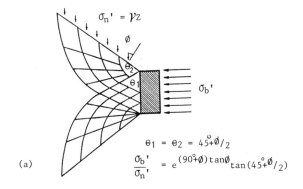

(a)

$$\theta_1 = \theta_2 = 45°+\emptyset/2$$

$$\frac{\sigma_b'}{\sigma_n'} = e^{(90°+\emptyset)\tan\emptyset}\tan(45°+\emptyset/2)$$

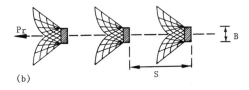

(b)

Fig.14.   Boundary conditions for bearing stresses on grid reinforcement during punching shear. Drawn for the case $\emptyset = 35°$.

Large direct shear tests with a reinforcement grid inclined across the central plane to reinforce the soil were conducted by Jewell (1980). The results of a series of tests at low stress level on dense sand have been used to derive the data point shown.

Although the spread and variability of the test results is large, the agreement with the theoretical predictions is encouraging. All the data are bounded by the lower and upper predictions given by equations 10 and 11 respectively, which is a satisfactory finding.

Theoretical expressions for bond strength

It is convenient to express the bond strength for grid reinforcement in terms of the total surface are a of the grid A_r. The bond strength can then be given as a bond coefficient multiple f_b of the angle of friction for the soil, thus

$$f_b \tan\emptyset = \alpha_s \tan\delta + f_{bearing} \tan\emptyset \qquad (12)$$

The first term represents the contribution from soil shearing stresses and the second term the soil bearing stresses. Dividing equation (12) by $\tan\emptyset$ gives a general expression for the bond coefficient,

$$f_b = \alpha_s \frac{\tan\delta}{\tan\emptyset} + f_{bearing} \qquad (13)$$

with the limiting condition $f_b < 1.00$.

If a grid were to act like a fully rough sheet through bearing stresses alone, so that $f_{bearing} = 1.00$, then the ratio of bearing member thickness B to spacing S would be given by an expression

$$2S\sigma_n' \tan\emptyset = \alpha_b B\sigma_b' \qquad (14)$$

This is illustrated in Fig. 16. Rearranging,

$$(S/B)_\emptyset = \frac{\sigma_b'}{\sigma_n'} - \frac{\alpha b}{2\tan\emptyset} \qquad (15)$$

where $(S/B)_\emptyset$ is the spacing required just to give a fully rough case from bearing stresses only.

For a grid with a greater bearing member spacing, the bond strength derived from bearing stresses would be proportionally lower,

$$f_{bearing} = \frac{(S/B)_\emptyset}{S/B} \qquad (16)$$

Manipulating equations (15) and (16) gives the expression for the coefficient of resistance due to bearing,

$$f_{bearing} = \frac{\sigma_b'}{\sigma_n'} \; \frac{B}{S} \; \frac{\alpha_b}{2\tan\emptyset} \qquad (17)$$

The coefficient of bond strength for a reinforcement grid in soil can now be written in a general form, using equations (13) and (17), as,

$$f_b = \alpha_s \frac{\tan\delta}{\tan\phi} + \frac{\sigma_b'}{\sigma_n'} \frac{B}{S} \frac{\alpha_b}{2\tan\phi} \qquad (18)$$

All the parameters describing the bond strength of grid reinforcement in equation 18 can be simply determined except, perhaps, the value σ_b'/σ_n'. The data plotted in Fig. 15 indicate a value of σ_b'/σ_n' intermediate to the values given by equations (10) and (11).

The ratio σ_b'/σ_n' may be measured directly in a pullout test. Pullout tests would be best carried out with a relatively short length of grid reinforcement embedded in soil remote from rigid boundaries. The grid bearing members should ideally be spaced widely to ensure that the full bearing stress develops on each one if a value σ_b'/σ_n' is required. Otherwise the pullout test result can be used directly to estimate the bond coefficient for the grid.

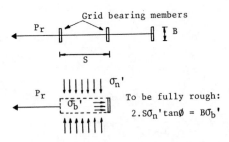

Fig.16. Basic definitions for a grid and the defining equation for the fully rough case.

Further calculations of bearing stresses

Numerical values for the ratio of bearing member spacing S to thickness B to give a fully rough grid from bearing alone may be estimated theoretically by combining equations (10) and (11) with equation (15), to give upper and lower estimates. These are shown plotted in Fig. 17, together with the logarithmic average of the two curves.

Data from grid pullout tests, and unit cell shear box tests, are shown plotted on Fig. 17 together with the value of the coefficient of reinforcement bond due to bearing. The data confirm the theoretical prediction that for values of S/B below the lower curve the grids should be fully rough, $f_{bearing} = 1.00$, and above the lower curve the roughness should decrease with increasing values S/B.

The data may also be plotted in the form suggested by equation (16), as shown in Fig. 18. The values of $f_{bearing}$ and σ_b'/σ_n' were directly back analysed from the pullout tests allowing (S/B)$\emptyset$ to be calculated from equation (15) to give the plotting point. The data are modelled well by the theoretical prediction.

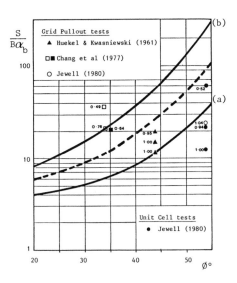

Fig.17. Lower (a) and Upper (b) predictions of S/B to give fully rough bond. Test values of f_b given next to data points.

Influence of soil particle size

Soil particle size is likely to affect bond strength in a similar way to direct sliding resistance described earlier. So, unless the average particle size is the same magnitude as the minimum aperture width, larger particles are likely to improve bond strength.

There are no test data for the influence of soil particle size on bond strength for a grid. As long as the combination of grid and soil meets the criteria specified for direct sliding.

$$\frac{\text{Minimum Aperture Width}}{\text{Average Soil Particle Size}} > 3 \qquad \text{(7 bis)}$$

then the value of maximum bond coefficient calculated from equation (18) should provide a reasonably conservative value for design.

For welded bar grids, the contribution to bond strength from shear on the plane reinforcement surfaces may typically be less than 10%. Designers may choose to ignore this contribution to bond strength when selecting a grid geometry that will act like a fully rough sheet in soil, f_b=1.00.

PORE WATER PRESSURES

The expressions for direct sliding resistance and bond strength have been derived in terms of an effective normal stress between the soil and the reinforcement grid. The magnitude of these coefficients is independent of the value of pore water pressure. The overall magnitude of stress resisting direct sliding, and bond stress giving axial reinforcement force, would however be directly reduced by pore water pressures.

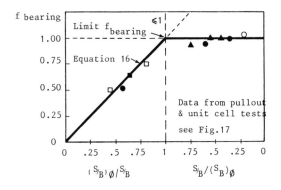

Fig.18. Relationship between the coefficient of bond due to bearing and grid geometry.

CLAY SOILS

The resistance to direct sliding and the bond strength for reinforcement grids in clay soils under drained loading conditions should be predicted from the equations given in the paper, if appropriate values of drained effective friction angle are used. Until there is data on the performance of grids in clay soils, it is recommended that direct measurements of sliding resistance and bond strength be made to select design values.

The case of clay reinforced soils subject to rapid undrained loading conditions has not been considered in the paper. Both laboratory and field measurements will be required to select design values for grids reinforcing clay in undrained loading.

DESIGN EQUATIONS

The coefficient of direct sliding resistance f_{ds} given by equation (2) applies to soil sliding over a continuous horizon of grid reinforcement.

In the practical case of reinforcement grids of width w_r spaced horizontally at s_h, the overall coefficient of direct sliding for use in a limit equilibrium analysis would be,

$$\Delta_{ds} = 1 - \frac{w_r}{s_h}(1-f_{ds}) \qquad (19)$$

27

The maximum value of axial force that may be generated by bond in a grid element of width w_r and length L_r subject to a normal effective stress σ_n', would be

$$\frac{(P_r)\max}{2.L_r w_r . \sigma_n'} = f_b . \tan\phi \qquad (20)$$

MEASUREMENT OF VALUES FOR DESIGN

The coefficient of resistance to direct sliding of a reinforcement grid in soil can be measured simply in a laboratory direct shear test. Alternatively, if the angle of skin friction between the soil and the grid reinforcement material is known a value of peak direct sliding resistance for design can be estimated from equation (2) using $\alpha_{ds} = \alpha_s$.

The coefficient of bond strength for grid reinforcement can be measured using a pullout test. To make use of equation (18), the pullout test should be carried out on a sample of the reinforcement grid with widely spaced bearing members to ensure that the full value of bearing stress is measured. The measured value σ_b'/σ_n' would be used in equation (18) with the design ratio S/B for the grid to estimate the bond coefficient.

Design at large strain

When using relatively extensible reinforcement in soil, equilibrium conditions may require relatively large soil strains to occur. Appropriate values for direct sliding resistance and bond strength may be calculated from equations (2) and (18) using large strain values of $\emptyset$, δ and σ_b'/σ_n'.

CONCLUSIONS

The complex interaction between soil and reinforcement grids has been radically simplified in order to provide a framework for the understanding and the interpretation of laboratory and field measurements.

Equations have been derived to evaluate the resistance to direct sliding of soil over a reinforcement grid embedded in soil, and bond strength for the grid. These allow for the influence of particle size to be included.

The conclusions drawn should be viewed in the context of the rapidly developing understanding of reinforced soil. As our knowledge increases, allowing a more sophisticated understanding of soil grid interaction, the simplified approach adopted in this paper may be improved.

The main conclusions drawn in the paper are as follows.

1. Three mechanisms of interaction between soil and grid reinforcement have been identified as,

    a. soil shearing over plane reinforcement surface areas,
    b. soil bearing on grid reinforcement bearing surfaces, and
    c. soil shearing over soil through grid apertures.

2. Examination of these show that mechanisms 1a and 1c govern the resistance to direct sliding of soil over grid reinforcement.

3. The relative size of the soil particles and grid apertures also affects direct sliding resistance. When coarse particles interlock with a grid, the resistance to direct sliding increases and can achieve values close to the full shear resistance of the soil alone.

4. Bond strength for grid reinforcement is governed by mechanisms 1a and 1b. The high ratio of bearing stress to normal effective stress in the soil makes a grid effective for bonding with soils, particularly with low friction angles.

5. Two equations have been presented which give values for soil bearing stresses on grid reinforcement Fig.15. These are supported by the available experimental data. However, the bearing stress on grid members is likely to vary depending on several other factors besides the soil friction angle and a range of results should be anticipated.

6. For grid reinforcement to develop a similar bond to a rough sheet in compact granular soils, the ratio of bearing member spacing to thickness is likely to be in the range,

    S/B $\sim$ 10 to 20

7. Similar to the case of direct sliding resistance, the relative size of the soil particles and the grid apertures may provide an increase in bond strength when coarse particles interlock with the grid and rest directly on the bearing members. However the bond strength is never expected to exceed the shear strength of the soil alone.

ACKNOWLEDGEMENTS

Work from several sources has been collected in this paper. Data from element tests have been reported which are from research by R A Jewell at Cambridge University sponsored by the Science and Engineering Research Council (SERC) and Binnie & Partners. This work is being continued at Oxford University under SERC sponsorship by G.W.E. Milligan and M. Dyer, whose photoelastic measurements have helped elucidate the fundamental mechanisms described. The measurements of direct sliding resistance were made by R W Sarsby and C Marshall at the Bolton Institute of Higher Education during research sponsored by Netlon Limited, and guided by D.D. Dubois who developed laboratory techniques for direct sliding measurements at Queen's University, Belfast. The theoretical studies were carried out by R A Jewell for the cooperative research project into Civil Engineering applications of polymer grids sponsored by the SERC and Netlon Limited. The authors gratefully acknowledge the sponsorship which has supported the work, and the contributions of colleagues too numerous to name individually.

REFERENCES

Arthur, J.R.F., Dunstan, T., Al-Ani, Q.A.J.L. and Assadi, A. (1977). Plastic deformation and failure in granular media. Geotechnique 27, No.1 53-74.

Bhudu, M. (1979). Simple shear deformation of sands. Ph.d Thesis University of Cambridge.

Chang, J.C., Hannon, J.B. and Forsyth, R.A. (1977). Pull resistance and interaction of earthwork reinforcement and soil. Transportation Research Record 640, Washington, D.C.

Das, B.M. and Seeley, G.R. (1975). Pullout resistance of vertical anchors. J. Geotech, Engng Div. Am.Soc.Civ. Engrs 101, GT 1.87-91.

Forsyth, R.A. (1978). Alternative earth reinforcements, Symposium on Earth Reinforcement ASCE.

Hueckel, S.M., and Kwasniewski, J., (1961). Scale model tests on the anchorage values of various elements buried in sand. Proc. 5th ICSMFE Paris.

Jewell, R.A. (1980). Some effects of reinforcement on soils. Ph.D Thesis. University of Cambridge

Kerisel, J. (1961). Deep foundations in sands. Variation of ultimate bearing capacity with soil density, depth, diameter and speed. Proc 5th ICSMFE Paris

Neely, W.J., Stewart, J.G. and Graham, J. (1973). Failure loads of vertical anchor plates in sand. J. Soil Mech. Fdns Div. Am. Soc.Civ Engrs 99 SM9 Sent 669-685

Netlon Limited (1984). Test methods and physical properties of 'Tensar' Geogrids. Technical publication.

Ovesen, N.K. and Stroman, H. (1972). Design methods for vertical anchor plates in sand. Proceedings of speciality conference on performance of earth at earth supported structures, pp. 1481-1500. New York: American Society of Civil Engineers.

Peterson, L.M. (1980). Pullout resistance of welded wire mesh embedded in soil. M.Sc. Utah State University

Potyondy, J.G. (1961). Skin friction between cohesive granular soils and construction materials. Geotechnique 11, No. 4, 339-353

Prandtl, L., (1921). "Uber die Eindringungsfestigkeit platisher Baustoffe und die Festigkeit von Schneiden." Zeitschrift fur Angewandte Mathematik und Mechanik, 1: 1, 15-20.

Reissner, H., (1924) "Zum Erddruckproblem." Proc. First Intern. Conf. Applied Mech., 295-311, Delft.

Roscoe, K.N.. (1970). The influence of strains in soil mechanics. Geotechnique 20, No.2, 129-170.

Rowe, P.W. (1962). The stress dilatancy relation for static equilibrium of an assembly of particles in contact. Proc. Roy. Soc. A, 269, 500-527.

Rowe, R.K. and Davis, E.H. (1982). The behaviour of anchor plates in sand. Geotechnique 32, No. 1, 25-41.

Schlosser, F., Jacobsen, H.M., and Juran, I. (1983). Soil reinforcement. General report. Proc. 8th Eur. Conf. Soil Mech. Fndn. Engng., Vol 3, Helsinki.

Schlosser F. and Juran, I. (1979). Design parameters for artificially improved soils. General report. Proc. 7th Eur. Conf. Soil Mech. Fndn. Engng., Vol 5, Brighton.

Vesic, A.B. (1963). Bearing capacity of deep foundations in sand. Highway Research Record No. 39. National Research Council, Washington. D.C.

NOTATION

A_r $(=W_r . L_r)$ total plan area of grid.

B thickness of bearing members

D_{50} the effective particle size which corresponds to 50% passing on the grain size diagram

$f_{bearing}$ coefficient of reinforcement bond due to bearing

f_b coefficient of reinforcement bond

f_{ds} coefficient of resistance to direct sliding

F_γ' factor for bearing on anchors (Rowe & Davis 1982)

L_r length of reinforcement available for bond

P_r axial reinforcement force

S spacing between bearing members

s_v vertical spacing between reinforcements in a reinforcement layout

s_h horizontal spacing between reinforcements in a reinforcement layout

w_r width of reinforcement

z depth of embedment

α_b fraction of grid width over which bearing surface extends

α_{ds} fraction of grid surface area that resists direct shear with soil

α_s fraction of grid surface area that is solid

$(1-\alpha_s)$ fraction of grid surface area that is aperture

γ unit weight of soil

δ angle of friction between soil and the reinforcement surface

Δ_{ds} design overall coefficient for direct sliding

ν angle of soil dilation

$\emptyset$ angle of friction of soil

$\emptyset_{ds}$ angle of friction of soil in direct shear

$\emptyset_{cv}$ angle of friction of soil at constant volume

σ_n' normal effective stress between soil and reinforcement

σ_b' effective bearing stress between soil and reinforcement

Use of geogrid properties in limit equilibrium analysis

This paper provides guidance on the choice of soil, geogrid and soil-geogrid interaction properties that should be used in limit equilibrium analysis of geogrid reinforced soil structures. It suggests that as little experience exists for the evaluation of lumped factors of safety in design, partial factors of safety should be used. The material properties to which these partial factors should be applied and the reasoning for each of them is given. Finally, the differences between overall strains, and so deformations, at working conditions and those inherently assumed in limit equilibrium designs are discussed.

A. McGown, *University of Strathclyde,*
N. Paine, *Binnie and Partners,* and
D. Dubois, *Netlon Ltd*

INTRODUCTION

As with any civil engineering construction, it is necessary to ensure that a reinforced soil structure performs satisfactorily during its lifetime, under all reasonably foreseeable circumstances. To do this, two distinct phases must be considered:

(i) The construction stage.

(ii) The post-construction stage.

Additionally, it is necessary to ensure that during each phase, the structure performs without exceeding the following two limiting conditions:

a) A limiting equilibrium condition, at which one or more of the materials within the structure ruptures.

b) A limiting serviceability condition, at which the structure deforms more than is acceptable.

To accomplish this, an appropriate means of predicting the behaviour of the structure must be identified and the operational properties of both the soil and the reinforcement material established. Additionally, soil structures incorporating relatively extensible reinforcements, such as geogrids, are strain controlled systems, McGown et al (1978), hence there must be strain compatibility between the soil and the reinforcements at all times. Thus the strength parameters for the reinforcements must be measured over the same range of strains as can occur in the soil.

Following the choice of design method and material properties, factors of safety need to be applied to allow for the inadequacies of the design method, extreme loads and variations in soil and reinforcing material properties. A single lumped factor of safety to allow for all these is the simplest approach, but at this stage in the development of soil reinforcement technology, little experience exists for

evaluation of the magnitude of such a factor. The concept of partial factors of safety can be used to concentrate attention on the various components of reinforced soil which, taken together, require a margin for safety.

In the following sections of this paper, alternative methods of design are discussed, the approach to the selection of appropriate material properties are identified for limit equilibrium designs and the parameters to which partial factors of safety should be applied are identified, together with the reasoning for the need of these factors. Also guidance is given on the estimation of overall deformations of geogrid-reinforced soil structures, in order to assess limiting serviceability conditions.

DESIGN METHODS

The available methods of analysis for reinforced soil systems can be classified as purely empirical methods, methods based on back analysis of laboratory models or field trial structures and methods based on the analysis of the internal stresses of the system. These latter methods may be categorised further into methods based on numerical techniques, such as the finite element techniques of Romstad et al (1976), Al-Hussaini and Johnson (1978) and Andrawes et al (1982), methods based on behavioural models of working conditions, such as the "Energy" method of Osman (1977) and limit equilibrium methods, such as those of Schlosser and Vidal (1969), Schlosser (1972), Lee et al (1973), Broms (1977) and Murray (1978). The finite element techniques are attractive as, in principle, they can model the stress and strain conditions throughout the reinforced system. However, these techniques are not yet commonly used in design offices, the most widely used being limit equilibrium methods. This paper therefore deals with the selection of geogrid data for use in designs employing limiting equilibrium methods of analysis.

Polymer grid reinforcement. Thomas Telford Limited, London, 1985

31

MATERIAL PROPERTIES

The material properties used in designs
employing limit equilibrium methods of analysis
for soil-geogrid systems, are the strengths of
the soil and the geogrid reinforcements
together with a parameter to describe their
surface interaction. The strength of the soil,
under various conditions of drainage, is most
commonly measured using the triaxial or shear
box apparatuses, although many other methods
are available.   McGown et al (1984) have shown
that rapid loading creep testing of geogrids
allows their load-strain-time behaviour to be
quantified and Jewel et al (1984) have shown
that the shear box is an effective method of
quantifying soil-geogrid interaction.   Just as
important as the method of testing, however, is
the presentation and choice of the data from
these tests.

In order to ensure strain compatibility between
the soil and the geogrid at all times, the
likely strain levels mobilised at the limiting
equilibrium condition must be known.   Most
reinforced soil structures are built with
compacted granular soils and their axial
strains at peak strength are associated with
lateral tensile strains of 8 per cent at most.
Although for dense sands, many investigators
have shown that the lateral tensile strains at
peak strength (stress ratio), in axial compression may often be less than 6 per cent, Jewell
(1980).   Thus if peak strength parameters are
used for the soil, the forces in the geogrid

inclusions lying in or close to the directions
of principal tensile strains, must be limited
to the forces developed at corresponding strain
levels.

Although once passed their peak strength,
compacted granular soils reduce their strength
with continuing straining, they eventually
attain a constant volume at which strength does
not vary with continued straining, Fig. 1

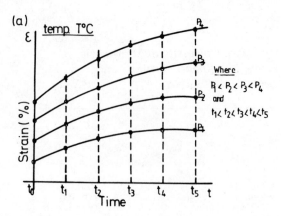

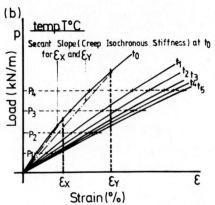

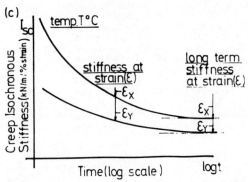

FIG. 2 THE DEVELOPMENT OF GEOGRID CREEP ISO-
CHRONOUS STIFFNESS CURVES FROM CREEP TEST
DATA: a) Creep Test Data; b) Isochronous
Load-Strain Curves and c) Creep Isochron-
ous Stiffness Curves.

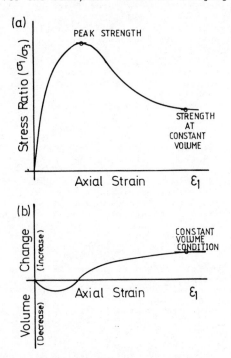

FIG. 1. THE STRESS-STRAIN BEHAVIOUR OF A
COMPACTED GRANULAR SOIL.

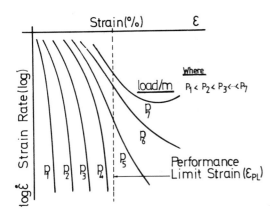

FIG. 3. PLOT OF STRAIN AGAINST LOG. STRAIN
RATE TO DETERMINE THE PERFORMANCE LIMIT
STRAIN.

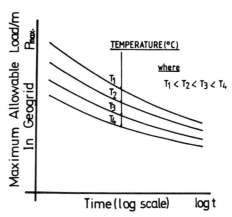

FIG. 4. PLOT OF MAXIMUM ALLOWABLE LOAD/m
AGAINST TIME FOR DIFFERENT TEMPERATURES.

Thus strains beyond those required to mobilise peak strength need not immediately induce conditions of limiting equilibrium. As the soil strains and its strength reduces from its peak value to that at constant volume, so the geogrid mobilises more force and overall the soil-geogrid system may remain in equilibrium. Only when lateral tensile strains in the soil induce strains in the geogrid which cause it to approach rupture, will a limiting equilibrium condition be reached. Thus for geogrids within compacted granular soils, the most likely limiting equilibrium condition occurs when the soil has strained beyond its strain to peak strength and has reached a state of constant volume. The geogrid may then be approaching rupture. Hence for designs employing limit equilibrium methods for soil-geogrid systems, the mobilised strength of compacted granular soils should be taken as their angle of friction at constant volume, (ϕ_{cv}).

The data from rapid loading creep tests on geogrids may be replotted as isochronous load-strain plots, as shown in Figs. 2(a) and (b).

The secant slopes of these isochronous load-strain curves for various levels of strain are termed the "Creep Isochronous Stiffnesses" and should be plotted against time as in Fig. 2(c), for various operational temperatures, with zero time in each plot taken as the time to reach half full load, McGown et al (1984). Thus for strains up to rupture of the geogrid, the operational load at any time and temperature is taken as the product of the strain in the geogrid and the appropriate isochronous stiffness.

In order to identify the maximum load that can safely be applied to the geogrid, the maximum strains that may be induced without risk of rupture must be identified for various temperatures. These maximum strain values are termed "Performance Limit Strains", McGown et al (1984), and are evaluated from plots of cumulative strain against the logarithm of strain rate at different temperatures, after Wilding

and Ward (1978), Fig. 3. Knowing these, then the maximum allowable load in the geogrid at any time and temperature is obtained by taking the product of the appropriate performance limit strain and creep isochronous stiffness, as shown in Fig. 4.

The soil-geogrid interlock data from the shear box tests may be presented as equivalent soil-geogrid friction angles (δ). It is also convenient to state these data as soil-geogrid interaction coefficients (α), which are the ratios of the equivalent soil-geogrid friction angles to the soil friction angles. In each case, the density of the soil must be indicated as the efficiency of soil-geogrid interlock is not independent of soil density, Jewell et al (1984). Also the soil-geogrid interaction coefficient is not necessarily the same for peak and constant volume soil friction angles, often approaching unity for the latter. Thus in many designs employing limit equilibrium methods for soil-geogrid systems, the soil-geogrid friction angle may be taken to be that for the soil at constant volume where the assumed shear failure surface passes along the geogrid, but in areas away from the failure surface it may be taken as an appropriate fraction of the peak soil friction angle, for example when calculating anchorage lengths.

PARTIAL FACTORS OF SAFETY
The basis of designs employing limit equilibrium methods of analysis, for soil-geogrid systems, is the establishment of the out-of-balance force that exists in the structure at limiting equilibrium conditions. To calculate this, the loads applied and the mobilised soil strength must be estimated, then the forces that should be generated in the reinforcements may be established. Partial factors of safety are employed at each part of the calculation process in order to account for factors such as variability of loads and materials, durability of materials, lack of accuracy of the design model and influence of construction methods.

33

The loads applied to reinforced soil structures may either be dead or live loadings. For dead loadings a high degree of confidence can be placed on the estimation of the largest expected values of these hence it is generally possible to apply a partial factor of safety of unity. The partial factors of safety to be applied to live loadings vary with the purpose of the structure but is generally encompassed in the choice of the worst combination of load conditions that can be envisaged.

As stated previously, it is necessary for soil-geogrid systems to ensure that strain compatibility exists between the soil and the geogrid. Thus for compacted granular soils, their limiting strength is likely to be their strength at large strains, i.e. constant volume (ϕ_{cv}) Fig. 1. As this can be measured relatively simply with a high degree of confidence for most granular soils, then it is often possible to use a partial factor of safety of unity on the soil strength. Thus for most types of limit equilibrium analysis, the angle of friction at constant volume (ϕ_{cv}) can be used directly in calculations.

The forces that can safely be generated in the geogrid reinforcement depend on the maximum load in the grid not causing rupture and on there being sufficient anchorage lengths away from the critical shear surface within the structure. Thus partial factors of safety need to be considered for both the strength of the geogrid and the soil-geogrid interaction

coefficient used in the design.

The strength of the geogrid that can be taken in the design calculations is the product of the operational strain and the appropriate isochronous stiffness of the material. The value of isochronous stiffness chosen should not only take account of the operational temperature and design life of the structure but also of the variability of the production materials. C.I.R.A. Report No. 63 (1977) suggests that in order to do this for construction materials, the strength used should be the "characteristic strength", defined as the strength which has a 95 per cent probability of being attained.

Partial factors of safety should be applied to this value of production materials characteristic strength which will usually be provided by the manufacturer. For geogrids the inadequacies of the design models, particularly the assumption that pure creep occurs from the time of achieving half full load with no account taken of the possibility of load relaxation, requires a partial factor to be applied. The possibility of variations occurring during construction in the vertical spacing and lengths of reinforcement layers must be similarly allowed for. Additionally the possibility of site damage to the geogrid during compaction and the possible subsequent deterioration of the material with time must all be allowed for by applying a partial factor of safety to the strength of the geogrid. These various partial factors may be lumped together and should then be related to the type of soil in which the geogrid is compacted, to the method of construction and to the design lifetime of the structure involved.

The possibility of applying a partial factor of safety to soil-geogrid interaction should also be considered. In many forms of reinforced soil structures, the soil-geogrid interaction does not dominate the stability calculations and the assumption that the soil-geogrid friction angle is that of the soil is thus often satisfactory.

Where the depth of the geogrid reinforcement is less than one or two metres below the surface, a considerable length of reinforcement may be necessary to anchor the force that can be safely taken in the geogrid. In such cases a partial factor of safety might be applied to soil-geogrid friction angle. However, in practice the need to mobilise the full allowable force in a shallow layer of reinforcement normally arises from the possibility of a deep shear surface developing, which intersects several other reinforcing layers. Practical geogrid reinforcing layouts provide excess capacity and any apparent shortfall in anchorage near the surface will therefore usually be more than compensated for at depth.

ESTIMATE OF OVERALL DEFORMATIONS

When partial factors of safety for uncertainties in design models, site damage, inaccuracies in construction and long term durability have been applied to the production materials

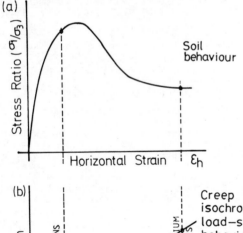

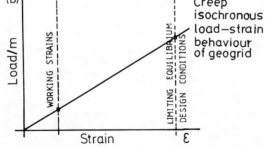

FIG. 5. THE POSSIBLE DIFFERENCE IN WORKING STRAINS AND THE STRAINS AT LIMITING EQUILIBRIUM CONDITIONS ASSUMED IN DESIGNS.

characteristic strength and this is used in conjunction with the soil angle of friction at constant volume and its related soil-geogrid friction angle, the equilibrium strains at working conditions are likely to be much less than those assumed in the design. Indeed as indicated in Fig. 5, it is likely that the working strains will be less than the horizontal strains required to bring the soil to peak strength. For well compacted granular soils, equilibrium strains in the zone of maximum shear stresses are likely to be of the order of only 3 or 4 per cent. The distribution of forces and so strains along geogrid reinforcements is as yet not known. Measured distributions of forces and strains in reinforcements have principally been associated with metal strips. These of course rely on a surface friction stress transfer mechanism rather than interlock, and so it is possible that the distributions of force and strain along the strips will be different from those along grids. It may, however, be suggested that at working conditions the average forces and strains along the grids will be a fraction of the maximum working forces and strains. The overall working deformations of geogrid reinforcements will therefore be very small and will not approach the deformations that would develop were the strains assumed in the limitint equilibrium condition to be generated.

DISCUSSION

The paper has made clear recommendations for the selection of appropriate material properties for use in designs employing limit equilibrium analysis and for the partial factors of safety that should be used in conjunction with these. These recommendations may be summarised as follows:

1. Soil Property (Compacted Granular Soil)

   Angle of Soil Friction $= \dfrac{\phi_{cv}}{\gamma_S}$

   where ϕ_{cv} is the angle of friction at constant volume (large strains).

   γ_S is the partial factor of safety for soil friction and is often taken to be unity.

2. Geogrid Property

   Safe working load at time (t) and temperature $(T^\circ C) = \dfrac{P_{PM}}{\gamma_M}$

   where P_{PM} is the production material characteristic strength at time (t) and temperature ($T^\circ C$), taken as the product of the performance limit strain and the appropriate creep isochronous stiffness, (this includes an allowance for production variability).

   γ_M is the partial factor of safety for the geogrid which takes account of deficiencies in the design model, site damage, construction inaccuracies and durability.

3. Soil-Geogrid Friction (Using Compacted Granular Soil)

   Angle of Soil-Geogrid Friction $= \dfrac{\delta}{\gamma_I}$

   where δ is usually taken to be equal to angle of soil friction.

   γ_I is the partial factor of safety for soil-geogrid friction and is often taken equal to unity.

4. Applied Loads

   Partial load factors may or may not be applied to dead loads and live loads depending on the nature of the loads and the structure.

   For certain structures it may be necessary to undertake an analysis for both short and long term if the soil or loading conditions are likely to change with time. Nevertheless for most cases, geogrid reinforced soil structures need only be designed for limiting equilibrium conditions in the long term. The strains assumed in this design are not, however, likely to occur as equilibrium working conditions are likely to be achieved at strains less than those required to generate peak strength in the soil.

ACKNOWLEDGEMENTS

The authors wish to acknowledge that this work results from the efforts of the Soil Reinforcement Design Group set up as part of the Science and Engineering Research Councils Co-operative grant made in conjunction with Messrs. Netlon Ltd.

REFERENCES

Al-Hussaini, M.M. and Johnson, L.D. (1978). Numerical analysis of a reinforced earth wall. Proc. Sym. on Earth Reinforcement. ASCE Annual Convention Pittsburgh. 98-126.

Andrawes, K.Z., McGown, A., Wilson-Fahmy, R.F. and Mashhour, M.M. (1982). The finite element method of analysis applied to soil-geotextile systems. Proc. 2nd Int. Conf. on Geotextiles, Las Vegas 3 695-700.

Broms, B.B. (1977). Polyester fabric as reinforcement in soil. Int. Conf. on the Use of Fabrics in Geotechniques. Paris 1. 129-135.

C.I.R.I.A. Report No. 63 (1977) Rationalisation of safety and serviceability factors in structural codes. Construction Industry Research and Information Association. 226 pp.

Jewell, R.A. (1980) Some Effects of Reinforcement on the Mechanical Behaviour of Soils. PhD Thesis, University of Cambridge. (unpublished).

Jewell, R.A., Milligan, G.W.E., Sarsby, R.W. and DuBois, D.D. (1984). The interaction between soil and geogrids. Proc. Symp. on Polymer Grid Reinforcement in Civil

Engineering. Mar.

Lee, K.L., Adams, B.D., and Vagneron, J.J. (1973). Reinforced earth walls. Jour. S.M. and F.E. Div. ASCE. 99. 10. 745-763.

Murray, R.T. (1978). Design of reinforced earth walls. T.R.R.L. Int. Report 18 pp.

McGown, A., Andrawes, K.Z. and Al-Husani, M.M. (1978), Effect of inclusion properties on the behaviour of sand. Geotechnique 28. 3 327-346.

McGown, A., Andrawes, K.Z., Yeo, K.C. and DuBois, D. (1984). The Load-Strain-Time Behaviour of Tensar Geogrids. Proc. Symp. on Polymer Grid Reinforcement in Civil Engineering. Mar.

Osman, M.A. (1977). An Analytical and Experimental Study of Reinforced Earth Walls. Ph.D. Thesis University of Glasgow (unpublished).

Romstad, K.M., Herrmann, L.R., and Shen, C.K. (1976). Integrated study of reinforced earth - 1: theoretical formulation. Jour. Geot. Eng. Div. A.S.C.E. 102 GTS 457-471.

Schlosser F. (1972). La terre armee, researches et realisations. Bull. Liason. Lab. Ponts. et Chauss. 62.72-92.

Schlosser F. and Vidal H. (1969) Reinforced Earth. Bull Liason. Lab. Ponts. Chauss.41.

Wilding, M.A. and Ward, I.M. (1978). Tensile creep and recovery in ultra-high modules linear polyethylenes. Polymer.19, 969-976.

Introduction to polymer grids: report on discussion

J. E. Templeman, *Netlon Ltd*

The possible increases in tensile strength of Tensar grids (given as an illustration to the keynote speech) were further described by Prof. Sir Hugh Ford as an indication that one could proceed much further up the stiffness curve provided by Prof. Ward (paper 1.1). The practical limits for these increases were subjects of research and discussion.

Prof. Ward was encouraged by these possible improvements and stressed the need for continued development on the production process and stated that he had no doubt that polymers could be made intrinsically as stiff as glass or aluminium in large section.

The induction of increased strain hardening during the manufacturing process was explained by Prof. Ward to occur as the result of the production of chemical cross-links. Radicals are produced during the stretching process, which possibly react with other chemicals present, but in any case succeed in joining molecular chains together to form a chemical network.

Orientation improves many properties, not just strength and stiffness. In particular, it reduces the permeability to gases, so it improves chemical resistance, and it makes the material more stable, both in the sense that it has a much lower thermal coefficient of expansion, and that it has less permanent shrinkage.

In answer to a further question posed specifically on any ageing or hardening effect caused by organic agents or bacteria in the soil, Prof. Ward stated that polyethylene had been well examined by the medical community, had been found inert and not subject to biological attack, to the extent it is used in internal prostheses.

Other contributors suggested that polyethylene had been used in soil for 50 years - as cable covering and water piping - with the only real problem being one of disposal.

Studies had recently been completed on a range of pulverised fuel ash materials, to check whether any chemicals present in these chemically aggressive materials would cause deterioration of polymer grids.

None were effectively found.

Dr McGown, presenting paper 1.2 emphasised that the properties of the soil as well as the polymer grid have to be considered. Polymer grids have to be designed which are compatible with the soil, not to be too brittle, and not too extensible.

If equilibrium is attained at low strains under working conditions, and we design for failure at large strains, considerable warning will be provided before the structure fails.

We are designing a composite.

Dr Bassett agreed that we are dealing with a composite and that compatibility of strain is important, but disagreed with the use of classic limit analysis mechanisms with the reinforcement assumed to be an additional "tie force".

The parameters of a soil are not altered when we reinforce it, but the composite is restrained to behave in a different manner.

If an ordinary active retaining wall were considered purely on the basis of stresses, it was agreed that the failure mechanism would be on the basis of a wedge at 45 degrees + $\phi'/2$.

On the basis of strains, ruptures are constrained to take place eventually in the direction at right angles to the zero extension planes. These directions are related to the dilation characteristic ψ .(ψ and ϕ' are often assumed equal in "upper bound" calculations).

The stiffness of the reinforcement will control the orientations of the zero extension planes.

If the case of a reinforced wall is considered, the reinforcement is conventionally placed in a horizontal direction and the mechanism described by Jewell et al (paper 1.3) can be acceptable, - sliding on the surface of the reinforcement could occur, driven by an active wedge outside it.

This is similar to a concrete block sliding on a layer of weak material left in the concrete, it does not do justice to reinforced earth.

If the soil lying parallel to the grid is strained and the reinforcement locked into the soil system then displacements in the soil will cause the grid to be pulled in two directions. It is therefore not clear whether or not shear tests or pull out tests had much relevance with what happened in the composite.

If the grid will remain unslipping with respect to the soil, then it is the compatibility of strain in the horizontal, or in the reinforcement direction, that is important.

If we deal with the compatibility of strain then the zero extension planes must be moved and the potential rupture plane at right angles to these will also be in a different orientation (Fig. A).

The behaviour of the reinforced earth wall is therefore a system in which there is a much modified failure mechanism with different slip planes operating.

In the limit when the reinforcement will not strain at all the mechanism for failure becomes very different from the classic active case as we normally understand it.

Polymer grid reinforcement. Thomas Telford Limited, London, 1985

37

DISCUSSION

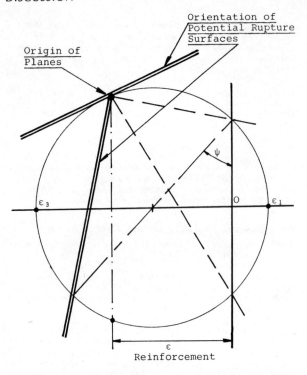

Fig. A. Mohr's circle of strain

The value of ϕ' would still be valid but the σ_h' and σ_v' values would have to agree with the principal stress directions as re-orientated to comply with strain compatibility.

In reinforced earth therefore, there is a rotation of the principal stress and strain directions, which must alter our whole concept of how this system works. As the strain in the soil and the strain in the grid are comparable with each other then the grid will only carry the amount of stress that the soil strain will generate in it, and the resulting change will be an increase in and a rotation of principal directions.

Dr Bassett emphasised that it was comparative strain and compatible strain that would define the whole ability to produce reinforced earth and therefore that reinforcement stiffness was a significant parameter.

In discussing paper 1.3, Dr Jewell expressed disagreement with Dr Bassett in:-

a) Relevance of the pull out test.

    The pull out test was performed to find the limiting value of the bond strength, so that the maximum value of the force could be calculated - assuming it to be inextensible - or in the regions near the end of the reinforcement, where forces and therefore strains are very low.

    In substituting values in limit equilibrium calculations the value of force used may be governed by limiting the magnitude of tensile strain to 2% or

3% - to comply with allowable working conditions. Pull out tests were therefore considered to have relevance.

b) Overall failure modes

    Internal surfaces drawn through reinforced blocks are surfaces upon which equilibrium can be estimated - by estimating the equilibrium looking at the magnitude of mobilised reinforcement across the surfaces, and the effect the mobilised force would have on shear resistance.

c) If there is no tensile strain in the direction of inextensible reinforcement, then the reinforced soil is not going to fail, and so this mechanism of failure would never be observed.

d) Looking at inextensible reinforcements in soil, if the reinforcement is in the direction of zero extension and the reinforcement bonded perfectly with the soil, you may hope to see a pattern of strain, giving a zero extension line along the reinforcement.

    Caution was however advised in using uniform patterns of stress and strain to describe discontinuous phenomena in the field.

    Relevant movement was required in the case of grids to generate bearing forces.

Dr Jewell advised that in the case of a Geogrid being placed at the interface between two dissimilar soils, the bond strength could be estimated as the average of the bond strengths of two soils with the Geogrid, or if a conservative value was required, then the bond strength between the Geogrid and the weaker soil should be taken.

In discussing the possible variation between the established method for calculating the effective adherence length for non-extensible reinforcement - and a method for calculating the effective adherence length for extensible reinforcement, the author suggested:-

i)   Limit equilibrium equations might show that in reinforced soil, the locus of maximum force, is a consequence only of the reinforcement providing equilibrium, on all potential surfaces through the soil and that there is no one failure mechanism to which we can ascribe the presence of that locus.

ii) A procedure could be, to estimate, using soil interaction, what the maximum force could be along each reinforcement in its position in the soil increasing from the end embedded deepest in the soil, perhaps maintaining a force in the case of a rigid face, or dropping again to zero, where a loose face was used.

     From the maximum force envelope, one could then use limit equilibrium calculations to calculate the mobilised value of force.

     There would not be a calculation of bond length. It would simply be that each reinforcement would contribute a portion of the maximum force envelope.

iii) When extensible reinforcement was used, the maximum force and maximum force envelope will be governed by serviceability considerations in the maximum allowable tensile strain.

Answering a question relating to Paper 1.4 about the strength of the crossbars of the Geogrid, Dr DuBois explained that the development of bond stress occurred as a buttress effect on the crossbar such that the acceptance of stress, occurred by transmitting positive pressures.

The critical strength of the Geogrid must therefore be its ability to accept the tensile stresses generated by interaction along the length of the reinforcement.

Creep test values of the Geogrid had been obtained from the whole material being examined in this mode.

An alternative combination of value of ϕ and partial factor of safety was suggested, instead of the use of $\phi\,c\,v$ and a partial factor of unity. The author explained that, in accepting there should be strain compatibility between the soil and the Geogrid, at the limiting equilibrium condition at high strain, it must be assumed that the two materials are comparable at the worst conditions.

The actual profile of the soil (Fig. 4 of Paper 1.4) was indicative of a well compacted material, and would not likely be worse in performance than the condition indicated by $\phi\,c\,v$.

A partial factor of safety of unity was therefore recommended for the soil in high strain condition.

PARTICIPANTS:

Professor Sir Hugh Ford FRS

Professor Ward FRS

Dr Bonaparte

Dr McGown

Dr Jones

Professor Nordal

Dr Bassett

Dr Jewell

Professor Jarrett

Dr Juran

Dr Osborn

Dr DuBois

Dr Craig

Introduction to polymer grids: written discussion contributions

Pull-out resistance T, can be evaluated using an expression in the form of equation (2).

$$T = N_c \, C_u \, \Sigma a + \beta C_u \Sigma a_s \qquad (2)$$

The first term in this expression relates to the bearing force generated by transverse grid members. This is the product of a bearing factor N_C, the undrained shear strength of the clay and the sum of the areas of the grid members Σa, normal to the direction of pull-out. The second term relates to surface adhesion βC_u, generated on the sum of the surface areas Σa_s, parallel to the direction of pull-out. Equation (2) is a general expression which can be developed for a planar punched grid or a welded rod grid. For example, if the sum of the normal and parallel components of the members' lengths are Σl_n and Σl_p respectively, then for an orthogonal grid of rod of diameter D

$$T = N_c C_u D \Sigma l_n + \beta C_u \pi D \Sigma l_p \qquad (3)$$

The adhesion factor α has been previously defined as the ratio of apparent surface shear stress to undrained shear strength. Thus for reinforcement with an embedded plan area A

$$\alpha = T/2(AC_u) \qquad (4)$$

Combining equations (2) and (4) leads to a redefinition of the adhesion factor

$$\alpha = (Nc\Sigma a + \beta \Sigma a_s)/(2A) \qquad (5)$$

For the specific case of the rod grid under consideration

$$\alpha = (N_c D \Sigma l_n + \beta \pi D \Sigma l_p)/(2A) \qquad (6)$$

The above equation was evaluated for the Weldmesh grid using values of 7.5 and 0.5 for N_C and β respectively. This gave rise to a calculated adhesion factor of 0.38, which is very close to the mean measured value of 0.39 (Ingold, 1983). This close agreement is fortuitous. Since the values of the bearing capacity factor and skin adhesion factor are reasonable, it is postulated that the general equations are of the correct form. One omission in equation (5) is the upper limit for the adhesion factor. If consideration is limited to saturated clays, where shear strength is independent of the ambient stress level, then the upper limit of the adhesion factor must be unity, i.e. there will be a critical shear plane discontinuity on both sides of the reinforcement

Professor T. S. Ingold, John Laing Construction Ltd and Queen's University, Belfast

In Fig. 1 of paper 1.3, Jewell et al. define two failure mechanisms in granular soil, namely direct sliding of soil over reinforcement and pull-out of the reinforcement. These same mechanisms would apply in the undrained loading of saturated clay where both sliding and pull-out resistance can be expressed in terms of the undrained shear strength C_u. Direct sliding resistance can be expressed in terms of an average adhesion αC_u, where α is an adhesion factor that can never exceed unity. For a planar or thin grid this average adhesion will comprise two components. Firstly, in the clay filled apertures of the grid there will be direct clay to clay contact for which the local adhesion factor is unity. Secondly, there will be adhesion mobilised between the clay and the material of the grid for which the local adhesion factor is β. Taking the open area ratio of the grid to be ρ the average adhesion factor can be expressed by equation (1).

$$\alpha = \varrho + \beta (1 - \varrho) \qquad (1)$$

To illustrate the results that can be obtained using equation (1), two grids are considered - Netlon 1168 and Weldmesh 5119. Table 1 gives grid data together with theoretical values of α from equation (1) and measured values obtained from direct shear box tests (Ingold, 1980).

GRID	APERTURE SIZE (mm)	CELL SIZE (mm)	β (Measured)	α (Theoretical)	α (Measured)
Netlon 1168	5 x 5	6 x 6	0.6	0.88	0.89
Weldmesh 5119	24 x 11	25 x 12	0.5	0.94	1.07

Table 1. Grid data

40

Polymer grid reinforcement. Thomas Telford Limited, London, 1985

Fig. 1. Special test piece

in the plane of the reinforcement. This thesis was tested using an orthogonal 'grid' whose transverse members were 50 mm deep and 200 mm wide. These transverse members were very stiff, being made from a 6 mm thick steel plate. The two longitudinal members, which were also very stiff, were 12.5 mm diameter steel rods on which the transverse members were attached at 92 mm centres. The general appearance of the test piece including the special pull-out bracket can be seen in Fig. 1. Application of equation (2) indicated an adhesion factor of 2.27. Pull-out tests were conducted together with associated triaxial testing of the clay involved in the tests and resulted in a measured adhesion factor of 0.98.

In the case of an orthogonal grid formed from circular rods of diameter D, a simple analysis can be carried out to assess the relationship between longitudinal and transverse rod spacings, the embedded plan area and the potential adhesion factor. Fig. 2 shows a diagrammatic representation of a single cell of a planar grid with transverse and longitudinal member spacings shown as multiples n_x and n_y, respectively, of the rod diameter.

The hatched area shows the one transverse member and the two halves of the longitudinal members that would react against the pull-out force ΔT. The lagging transverse rod is not included in the grid cell since this forms the leading transverse rod in the adjacent cell. Based on

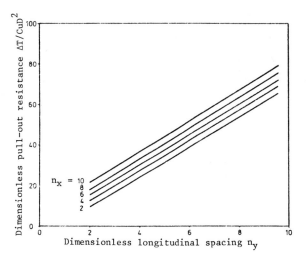

Fig. 3. Variations of $\Delta T/C_u D^2$ with n_x and n_y

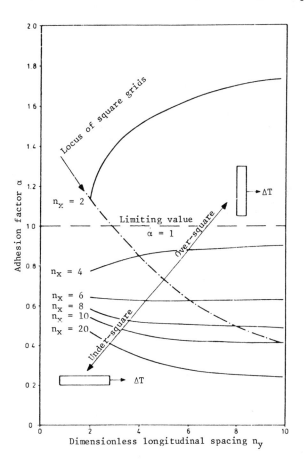

Fig. 4. Variation of adhesion factor with n_x and n_y

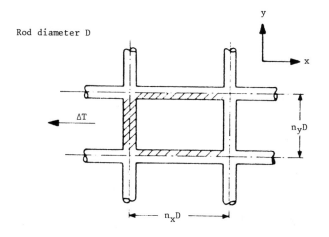

Fig. 2. Geometry of unit grid cell

41

DISCUSSION

this geometry it is possible, using the same argument in deriving equation (3), to obtain an expression for ΔT.

$$\Delta T = N_c C_u D^2 (n_y - 1) + \beta C_u \pi D^2 (n_x - 1) \qquad (7)$$

Obviously ΔT would vary with the square of the rod diameter. However, when equation (7) is rendered dimensionless as shown in equation (8), it becomes apparent that ΔT varies linearly with n_x and n_y.

$$\Delta T / C_u D^2 = N_c (n_y - 1) + \beta \pi (n_x - 1) \qquad (8)$$

This variation, which is plotted in Fig. 3 for the previously employed values of $N_c = 7.5$ and $\beta = 0.5$, shows that n_y has a much more significant effect than n_x, since a larger value of n_y is associated with a longer length of the more efficient transverse member per unit cell. A less expected result emerges if the expression for ΔT is substituted into equation (4) together with the plan area of the grid cell $n_x n_y D^2$ to give an expression for the adhesion factor.

$$\alpha = [N_c (n_y - 1) + \beta \pi (n_x - 1)] / (2 n_x n_y) \qquad (9)$$

The variation of adhesion factor with n_y is plotted in Fig. 4 with a family of curves for n_x. This shows that for low values of n_x, i.e. transverse rods at close centres, the adhesion factor increases with n_y as might be expected. However, since the plan area $n_x n_y D^2$ is also increasing, α levels off. The trend of α increasing with n_y becomes less marked as n_x increases, until at $n_x = 6$ the value of α is almost independent of n_y. For values of $n_x > 6$, the adhesion factor decreases with increasing n_y. This is a function of two things. Firstly, as n_x increases the less efficient longitudinal members, which only generate surface adhesion, are becoming more prominent.

Secondly, as n_y increases the plan area increases and consequently reduces α. The interplay between these factors is quite intriguing. Although a grid with low n_x and high n_y might be thought efficient, its

potential pull-out resistance is curtailed by the limiting value of the adhesion factor of unity. Thus, over-square grids ($n_y > n_x$) may not be as efficient as was first thought because of this upper bound limit. It also appears that square grids and under-square grids of high n_y are not efficient. From this theoretical analysis it would appear that an over-square grid with $n_x = 4$ and $n_y > 6$ is most efficient, giving $\alpha = 0.9$. Since α tends to level out for $n_y > 6$, it follows that large n_y, i.e. large transverse rod spacing, might be very economic.

References

Ingold, T. S. (1980). Reinforced clay. PhD thesis, Surrey University.

Ingold, T. S. (1983). A laboratory investigation of grid reinforcements in clay. ASTM Geotechnical Testing Journal, Vol. 6, No. 3, 112-119.

Mr. A. R. Dawson, University of Nottingham
The paper by Jewell et al. (1.3) discussed two types of potential failure mechanism: direct sliding across the top of a grid and pull-out of the grid from between soil. Methods by which load might be transferred between grid and soil were suggested for each mechanism, and these were to some extent confirmed by the slides of stress fields obtained in research work at Oxford with crushed glass.

However, the original work of Jewell's doctoral thesis showed by means of X-ray techniques that a trapezoidal shear zone developed around a small grid inclusion placed at large angles to the plane of shear. Thus a third potential failure mechanism appears to exist at the intersection of grid and shear plane. It may be argued that in a real soil-reinforcement situation the soil-grid interaction local to the shear plane will form an insignificant part of the influence of the whole grid inclusion. However, this remains to be demonstrated.

It is conceivable that this third mechanism is the predominant means by which the soil is strengthened when small randomly-oriented mesh elements are included in the soil as described by Mercer et al. (Paper 8.1).

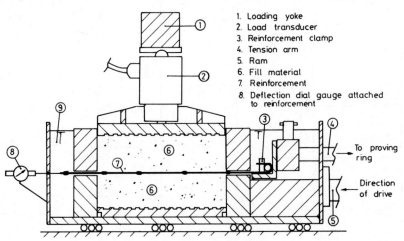

1. Loading yoke
2. Load transducer
3. Reinforcement clamp
4. Tension arm
5. Ram
6. Fill material
7. Reinforcement
8. Deflection dial gauge attached to reinforcement

Fig. 5. Modified shear box apparatus

Dr R. W. Sarsby and Mr C. B. Marshall, Bolton
Institute of Higher Education

PULL-OUT TESTS ON TENSAR REINFORCEMENT

Introduction

Reinforced Soil is a composite material formed
by the addition of relatively inextensible
reinforcement to a soil mass so that the
inherent low tensile strength of the soil is
enhanced in a manner analogous to the use of
steel reinforcement in concrete. The amount of
'bond' or friction which is developed between
the soil and the reinforcement is thus a very
important design parameter.

The interaction between reinforcement and fill
is usually determined experimentally using
modified direct shear box tests (Schlosser &
Juran, 1979) or pull-out tests (Schlosser &
Elias, 1978). A method for testing geogrids in
direct shear has been developed by Sarsby &
Marshall (1983) and the results are presented in
paper 1.3 by Jewell et al. Some preliminary
pull-out tests have been conducted at Bolton
Institute of Higher Education on Tensar geogrid
SR2 contained within a medium sand and the
results are presented herein.

Apparatus

The apparatus used was a modified direct shear
box with a plan area of 300 mm by 300 mm. Both
halves of the box were clamped together with
spacer pieces between them so that a gap was
left in the front and rear faces for the
reinforcement to pass through (see Fig. 5). The
reinforcement was anchored, via tension arms, to
a proving ring and pull-out was achieved by
pushing the shear box outwards relative to the
reinforcement. To monitor deformation of the
reinforcement thin wires were attached to
various ribs of the grid. These wires passed
out through the rear of the box and were
themselves attached to deflection dial gauges.

Normal stress was applied hydraulically to the
sample and this was monitored using a load
transducer.

Sample preparation and testing

The sand fill was placed and compacted in four
equal layers using a vibrating hammer and
tamping plate. Each layer was formed by
compacting a known weight of soil to a specific
thickness to obtain a sample dry density
corresponding to 92% of the maximum achievable
value. The second layer of the sand was
compacted flush with the surface of the bottom
half of the shear box and the reinforcement was
fed through the gaps in the box and laid flat on
the sand. The steel wires were attached to the
ribs, after which they were passed out of the
box and pulled straight. Two more layers of
fill were subsequently placed and compacted.

After the shear box had been mounted in the
carriage the reinforcement was attached to the
proving ring. The steel wires were tensioned
and fastened to the appropriate deflection dial
gauges. The requisite normal stress was then
applied and the sample was saturated with water.

Samples were sheared (under drained conditions)
by driving the shear box outwards at a nominal
rate of displacement (relative to the leading
edge of the reinforcement) of 0.6 mm/min.
Shearing was continued until the limit of travel
of the apparatus was reached.

Test data

Details of the tests performed are contained in
Table 2.

The stress-strain behaviour of the reinforced
samples is presented in Fig. 6 in terms of the
equivalent shear stress τ_{mean} acting on each
face of the reinforcement and the displacement
of the leading edge of the reinforcement, i.e.
that part closest to the proving ring, relative
to the fill Δ_{mean}. The stress-strain curve for
a typical plain sample (test code U3) is
included for comparison.

The development of shearing resistance with
displacement is very slow and peak resistance is
only achieved after large movements. For the

Test Code	Test Type	Normal Stress (kPa)	γ_{dry} (kN/m³)	$\left(\dfrac{\gamma_{dry}}{\gamma_{dry\ max}}\right)$ (%)
R1	Reinforced	3	15.7	90
R2	"	15	15.8	91
R3	"	25	15.8	91
R4	"	56	16.1	93
R5	"	75	16.1	93
R6	"	100	15.8	91
U1	Plain	29	15.8	91
U2	"	43	16.0	92
U3	"	80	15.9	91
U4	"	102	16.4	94
U5	"	117	16.0	92

Table 2. Test data

DISCUSSION

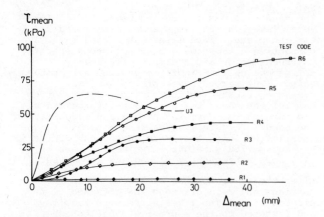

Fig. 6.   Stress-strain behaviour

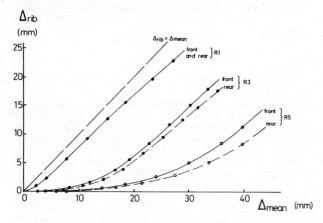

Fig. 7.   Reinforcement displacement

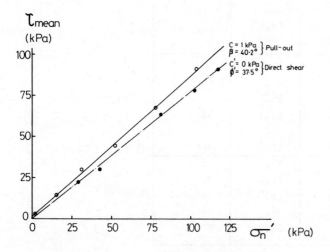

Fig. 8.   Effective shear strength parameters

same normal stress the displacement to failure in the pull-out tests was approximately 3.5 times that for soil-on-soil sliding.

At high normal stresses the peak resistance to pull-out was developed only slightly before the limit of travel of the apparatus was reached. However, even at low normal stress levels, there was no marked peak in the stress-strain curves and no significant drop in resistance at large displacements. This is in contrast to the behaviour reported by Alimi et al. (1977) and Jewell (1979) for relatively stiff reinforcements. However, during pull-out of the Tensar grid there was appreciable deformation of the reinforcement, as indicated in Fig. 7. This graph shows the horizontal movement of ribs of the geogrid relative to the movement of the leading edge of the reinforcement for different normal stress levels. The front and rear ribs were approximately 100 mm and 200 mm distant from the leading edge respectively, so that extension of the geogrid can be gauged from the separation of the respective front and rear curves.

In Fig. 8 the peak value of τ_{mean} has been plotted against effective normal stress σ_n. For the pull-out tests the slope of the resultant line is an equivalent mean shearing angle β, such that $\tan \beta = f_b \tan \phi'$, where f_b is defined in equation (12) of paper 1.3 by Jewell et al. For these pull-out tests the value of f_b is 1.10 as opposed to the theoretical maximum value of unity. This discrepancy is believed to be due to boundary effects in the apparatus, whereby the bearing pressure on some reinforcement ribs is enhanced due to their proximity to the rigid faces of the shear box.

Published data for pull-out tests have indicated values of f_b significantly greater than unity, but it is not certain whether these high values are due to the mechanism of soil-reinforcement interaction or to boundary effects in the testing systems.

It has been proposed in paper 1.3 that the coefficient bond strength f_b can be expressed as

$$ f_b = \alpha_S \frac{\tan \delta}{\tan \phi'} + \frac{\sigma_b'}{\sigma_n'} \frac{B}{S} \frac{\alpha_B}{2 \tan \phi'} $$

For the sand and reinforcement (Tensar SR2) used, $\phi' = 37.5$ and $\tan \delta = 0.63$. Thus (σ_b'/σ_n') is in the range 15 to 20 approximately and this leads to a predicted f_b in the range 0.82 to 0.93 (as opposed to the measured value of 1.10). Back-analysis of the reinforcement extensions shown in Fig. 7 generates values of 10 to 18 for (σ_b'/σ_n') at peak shearing resistance.

Acknowledgements
The authors wish to acknowledge the patience and dedication of Mr B. Porter and Mr E. Carroll, without whom this work would not have been undertaken.

References
Alimi, I., Bacot, J., Lareal, P., Long, N. T. & Schlosser, F. (1977). Study of adherence between soil and reinforcements. Proc. 9th Int. Conf. Soil Mech., Tokyo.

Fig. 9. Soil wedge on pulled out anchor

Fig. 11. Tension curl-up on top-split behind the anchor

Fig. 10. Pile-up in front of anchor

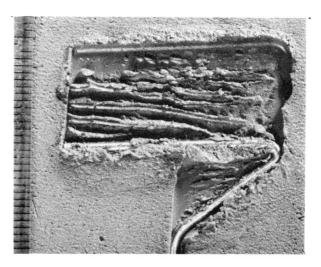

Fig. 12. Tension curl-up on bottom split behind the anchor

Jewell, R. A. (1979). Discussion. Design parameters for artificially improved soils. Proc. 7th Eur. Conf. Soil Mech., Brighton.

Sarsby, R. W. & Marshall, C. B. (1983). A method for determining the interactive behaviour of polymer grids and granular soils. Report No. BCS/Gl/2A, Bolton Institute of Higher Education, Bolton.

Schlosser, F. & Elias, V. (1978). Friction in reinforced earth. Symp. on Earth Reinf. ASCE, Pittsburgh.

Schlosser, F. & Juran, I. (1979). General report. Design parameters for artificially improved soils. Proc. 7th Eur. Conf. Soil Mech., Brighton.

Dr R. B. Singh, University of Glasgow

The symposium has highlighted the uses, design methodology, and construction technology pertinent to polymer grids in civil engineering. However, uses of polymer grids are very rare in developing countries and there would appear to be a necessity of cheaper alternatives, or at least of using them in what may be called a 'combined system'. Towards this end the contributor has proposed a system of using TRRL z-type patented anchors (Murray & Irwin, 1981) alternating with embankment situations. The combined system would utilise the advantages of the two systems, viz. reduce the rigidity of a fully anchored earth system and increase the rigidity of a fully reinforced system.

DISCUSSION

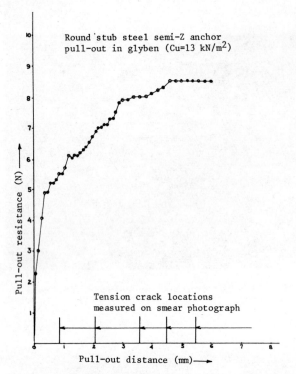

Fig. 13. Pull-out resistance of model anchor versus pull-out distance

More importantly, such a combined system would reduce costs by cutting down on the import cost factor of geogrids. In highly cohesive soils, a suitable filter membrane could possibly be attached to the geogrid layer so that one face of the geogrid would still be available for the soil reinforcement while the other face would promote the dissipation of pore pressures in the plane of the membrane.

With a view to having a better understanding of the failure and resistance mobilisation mechanisms operating in the pull-out of anchors in reinforced soils, the contributor has run small model pull-out tests in a modified direct shear box (inside dimensions 86 mm x 80 mm x 50mm) using an artificially prepared purely cohesive soil called Glyben (Davie & Sutherland, 1977). A special technique of 'split-model' and 'smear traces' which is possible with Glyben has been used. Studies of the soil wedges and pile-ups in front of the pulling out anchors and of tension curl-ups behind the anchor on the smear traces are continuing. It is hoped that the study will eventually help in more exact design formulations of cohesive soils reinforced with anchors. Figs 9-13 present typical information.

References
Murray, R. T. & Irwin, M. J. (1981). A preliminary study of TRRL anchored earth. TRRL supplementary report 674.
Davie, J. R. & Sutherland, H. B. (1977). Uplift resistance of cohesive soils. J. Geotech. Div. Am. Soc. Civ. Engrs 103, GT 9, 935-952

Reinforcement techniques in repairing slope failures

R. T. Murray, *Transport and Road Research Laboratory*

A recent survey of slope stability in selected lengths of motorway has indicated that shallow failures have already occurred on significant lengths of both cutting and embankment slopes constructed in over-consolidated clays. Moreover the survey also revealed that substantial further lengths are potentially at risk. The usual method of repairing slope failures in such conditions has involved replacement of the foundered region of soil with free-draining granular material, but where granular soils are not locally available, this approach can incur large haulage costs.

To minimise costs the Department of Transport has recommended re-using the existing material whenever possible, perhaps in conjunction with lime, if necessary, when the soil is too wet to be compacted.

An alternative method of repairing slip failures is described in which the foundered soil is re-used in conjunction with layers of geotextile or geogrid reinforcement. A number of design charts are presented to assist in assessing the reinforcement requirements for a particular situation. The paper also gives details of the application of the techniques to the repair of a cutting in London clay.

INTRODUCTION

Because of the economic constraints on both materials and land usage, embankment and cutting slopes constructed in cohesive soils are generally designed with relatively low factors of safety. In the longer term, therefore, failures may develop as a result of changes in the pore pressure conditions or in the site drainage characteristics.

Some preliminary results of a survey of 300 km of selected lengths of motorway are presented in Table 1 (Transport and Road Research Laboratory, 1983)

Although the results highlight the more severe examples of the stability survey, they nevertheless demonstrate that significant problems can be expected with overconsolidated clays.

The usual method of repairing such slope failures has involved the replacement of the foundered material by a free-draining and easily compacted granular soil of high frictional strength. Where a source of suitable granular soil is not locally available, however, considerable haulage costs can be incurred and to minimise costs the Department of Transport has proposed re-using the existing material whenever possible (Department of Transport, 1983). One possible approach involves re-using the foundered soil together with reinforcement layers of geogrid or geotextile to ensure the integrity of the reinstatement.

In this paper the technique of repairing slopes employing reinforcement layers is briefly described and some details are provided of the application of the method to

Geological Stratum	Age of Earthworks (year)	Slope Gradient (Vertical : horizontal)	Total length of slope surveyed (m)	Percentage of length failed
Cutting				
Boulder clay	22	1:2½ – 1:2	8010	4.0
Reading clay	10	1:3½ – 1:3	7750	6.8
Gault clay	22	1:3½	1240	19.6
Oxford clay	22	1:2	1970	16.1
Embankment				
London clay	10	1:2½ – 1:2	22820	12.6
Reading clay	14	1:2½ – 1:2	2460	9.2
" "	10	1:2½ – 1:1½	23380	15.5
Gault clay	22	1:3 – 1:2	4420	10.7
Kimmeridge clay	10	1:2	5420	13.8
Oxford clay	22	1:2½ – 1:2	7620	9.5
" "	10	1:2	3540	17.3
Kelloways clay	10	1:2 – 1:1½	4340	5.3

TABLE 1 More severe examples from TRRL Slope Stability Survey on Selected Motorways

Polymer grid reinforcement. Thomas Telford Limited, London, 1985

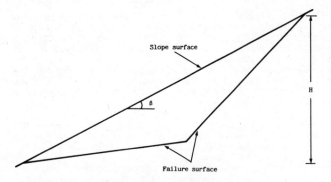

Fig 1 REPRESENTATION OF FAILURE SURFACE AS BILINEAL SLIP PLANE

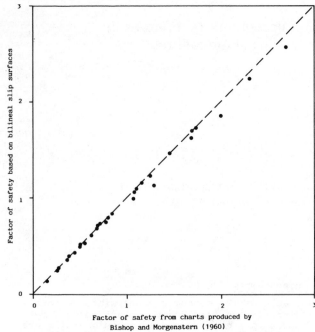

Fig 2 COMPARISON OF FACTORS OF SAFETY BY TWO DIFFERENT METHODS OF ANALYSING UNREINFORCED SLOPES

the repair of a cutting in London clay.   A set of design charts are also provided to enable a rapid assessment to be made of the reinforcement requirements for a range of typical slope situations.

Method of Analysis

Details of the method of analyses are provided in earlier publications (Murray et al 1982; Murray, 1982) but for completeness a brief outline is given below.

The failure surface is represented by a bilineal slip plane as shown in Fig. 1, in preference to the more conventional circular or non-circular slip surfaces; much of the complexity of analysis is then removed and moreover the volume of computation is greatly reduced.   It is possible to provide a reasonable representation of most types of failure surface by the use of bilineal slip planes and the accuracy of the results obtained from the analysis agree favourably with those obtained by other methods.   This is demonstrated by the comparison provided in Fig. 2, which relates to stability assessments based on the above method together with the results obtained from stability coefficients produced by Bishop and Morgenstern (1960).   As can be seen from the figure the results are in very close agreement.

The results shown in Fig. 2. relate to the situation where the slope is unreinforced, and the inclusion of a tensile component into the analysis produces a significant increase in the complexity of the assessment of the stability of the reinforced soil system.   It is normal practice to consider two aspects of internal stability, namely adherence and tensile resistance.   However, because geotextiles or geogrids are more strain susceptible than metallic reinforcing elements, additional reinforcement layers are generally incorporated to limit these strains.   The adherence or pull-out resistance of the tensile elements is determined by the coefficient of friction between the soil and reinforcement. It is frequently the case that the friction coefficient at the soil geotextile interface is a high proportion of the soil friction co-

efficient, particularly where geogrids are used.   The possibility of an adherence failure occurring in such circumstances thus appears very unlikely.   For purposes of the stability assessments therefore, the influence of the reinforcement layers is treated purely as a mobilised or permissible tension component and no consideration has been given to pull-out or adherence type failures.

The result of a series of analyses for 3 different values of pore water pressure ratio (r_u), have been used to compile the charts shown in Figs. 3 - 5; a unit weight for the soil of 20 kN/m^3, is assumed.   These charts can be employed to obtain a rapid assessment of the reinforcement requirements for different slopes.   It is assumed in using the charts, that the vertical spacing, S_v, of the reinforcement layers is constant.

In selecting a value of permissible tension to use in a particular situation, due consideration must be given to both the short-term and long-term strain and creep characteristics of the geotextile.   On this basis, therefore, the value chosen will generally be much smaller than the known ultimate tensile strength of the material and will frequently correspond to that required to produce no more than about 5 per cent **strain** in the long term.   Some consideration of the procedures for determining the potential deformations of a reinforced slope is given in an earlier publication (Murray, 1982).

The following example has been included to demonstrate how the charts are used:

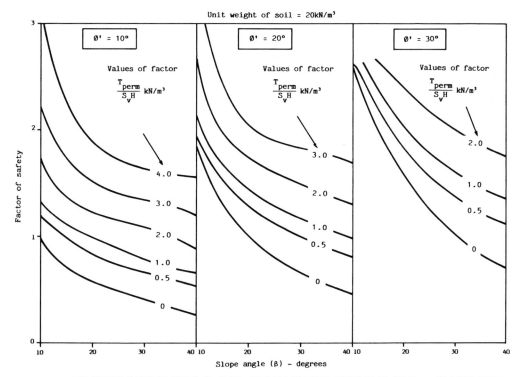

Unit weight of soil = 20kN/m³

Fig 3   RELATION BETWEEN FACTOR OF SAFETY AND SLOPE ANGLE FOR DIFFERENT VALUES OF $\emptyset'$ AND T_{perm} AT A PORE PRESSURE RATIO OF 0.0

(a) Select a value of permissible tension in each reinforcement layer, in the appropriate units of kN/m width. The value selected is intended to avoid problems of both short and long term serviceability - say a value of 30 kN/m width has been chosen.

(b) Estimate the height (H) in metres from the lowest to the highest point on the anticipated slip surface. An initial estimate could involve only the height of slope at risk, however it is possible that the slip surface will be more deep-seated and actually dip below the lowest point on the slope face. Assume a value of 12m has been selected in this case.

(c) Select a suitable value of S_V - say 1.0m.

(d) For the known values of pore pressure ratio (r_u) and effective angle of internal friction $(\emptyset')$, select the appropriate chart in Figs. 3-5; say values of 0.25 and 20° for r_u and $\emptyset'$ respectively apply in this case.

(e) Evaluate the factor $T_{perm}/(H \times S_V)$: for the selected values this is $30/12 \times 1 = 2.5$kN/m³. Obtain a factor of safety for the given slope angle β from the chart (Fig 4). For purposes of the example a value of β equal to 30° will be assumed, and on this basis the factor of safety from Fig 4 is about 1.4. If this factor of safety is not considered adequate then either T perm must be increased or Sv reduced. A reduction in the vertical spacing, S_V, of the reinforcement layers to 0.8m would increase the value of T perm/H^V x S_V to 3.1, and the corresponding factor of safety is a little more than 1.6. It is interesting to note that the same chart gives a factor of safety of about 0.4 when the slope is unreinforced.

Practical Considerations in Slope Repair

Although a substantial proportion of the total cost of repairing a slip failure by the conventional method of replacing the failed region of soil by granular material arises from the haulage distances involved, it is also necessary to add to these costs those resulting from the impedance to the free flow of traffic that occurs while the construction work is in progress; these are particularly significant where this involves the removal and importation of a large quantity of material along the section of road affected by the failure. The difficulties are likely to be compounded where the repair applies to minor roads or to other roads with limited access.

The repair of a slope by re-use of the foundered soil could thus alleviate much of the hindrance to the free flow of traffic by reducing the amount of construction traffic utilising the affected road. Of course, even with the soil reinforcement method of repair, it will be necessary to remove some of the foundered soil at the start of the reinstate-

49

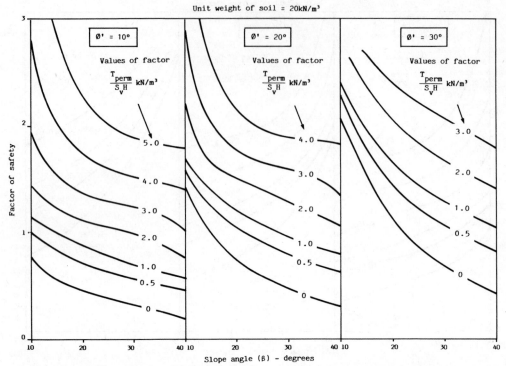

Fig 4   RELATION BETWEEN FACTOR OF SAFETY AND SLOPE ANGLE FOR DIFFERENT VALUES OF Ø AND T$_{perm}$ AT A PORE PRESSURE
RATIO OF 0.25

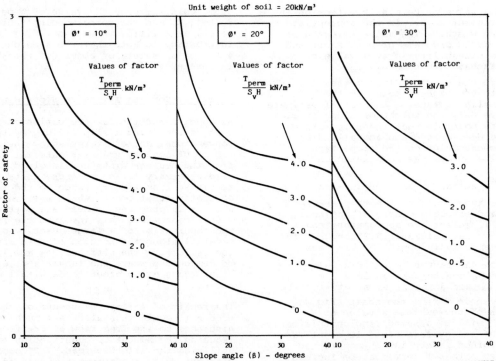

Fig 5   RELATION BETWEEN FACTOR OF SAFETY AND SLOPE ANGLE FOR DIFFERENT VALUES OF Ø' AND T$_{perm}$ AT A PORE PRESSURE
RATIO OF 0.5

ment operation. The procedure envisaged in this method involves excavating a strip of soil up the failed slope. The width of the excavated zone would be dictated by the requirements to operate the construction plant efficiently in addition to the need to prevent further slips developing as a result of excavating too large a region; it is self-evident that a narrow excavation will assume a greater degree of stability than is produced by a wide excavation. The excavated zone would then be reinstated together with sheets of reinforcement at the required vertical spacing. The soil used for the reinstatement would be obtained from a similar excavation at the other side of the failure region. This procedure would then be repeated until the whole of the foundered soil had been treated as a series of strips.

The depth of the treated zone would need to extend beyond the slip surface in order to permit a sufficient length of reinforcement to be installed to avoid adherence failures occurring. It would also be necessary to check that the stability of the region beyond the treated zone is adequate, otherwise this region also would need to be reinforced to prevent deeper seated failures developing. It is normal practice in placing each re-inforcement layer to continue the geotextile up the surface of the slope and for a distance of one or two metres onto the next soil level where a reinforcement layer is to be placed.

Because the foundered soil is generally in a very soft condition, it will usually prove difficult, or even impossible, to operate with conventional construction plant. A technique which has proved very successful in these circumstances is to add quicklime during the placement operations. The quicklime is normally only mixed into the upper part of the compacted layers in order to induce a rapid improvement in the working surface so that the construction plant can operate more effective-ly. Both the strength and permeability characteristics of the lime treated soil will be increased resulting in a further enhancement of slope stability. However, because the underlying material in each compacted layer remains virtually unchanged, the geotextile reinforcement layers are nonetheless required to ensure the integrity of the reinstated material.

Once having prepared a surface on which a reinforcement layer has to be placed, the geotextile will be positioned over the surface and pegged down, if necessary, to prevent it hampering the construction work. To prevent the geotextile from being damaged during the filling operation, it will normally be necessary to end-tip the fill and work from one edge by spreading in front of the plant.

To prevent the exposed surfaces of geotextile on the slope face from being damaged by ultra-violet radiation, it is advisable to spread a layer of top soil or other suitable material over the surface of the slope soon after installation.

Application of the method to the repair of a failed cutting

The method was first applied to the repair of a cutting slope in London clay. As this work has been previously reported (Murray et al, 1982) only a brief outline will be given:

The cutting, which extended to a depth of 20m, formed part of the M4 motorway some ten miles west of Reading. Over a period of several years a series of slips had occurred at the particular section and this culminated in a major slope failure, over the top 10m of the cutting, in 1979, with further movement in 1980.

The work of reinstatement, which took place in late 1980, was carried out by Berkshire County Council by direct labour contract and involved the excavation and replacement of some 7000 m^3 of soil. It was possible to excavate all of this material in a single operation after the owner of the adjacent land had given permission for temporary stockpiling of the excavated material. Clearly greater product-ivity can be achieved by dealing with larger regions in a single operation rather than as a series of strips but, as discussed previously, the risk of initiating further slips is increased. However, as the work was carried out during a relatively short period and in dry weather, no serious problems were encountered in this instance.

The excavation was taken down to some 30cm below the observed slip plane. Drainage trenches of 1.2m depth and at 5m centres were installed at this level to reduce the possib-ility of further slips developing below the treated zone. The first layer of reinforce-ment was pegged down over the base of the excavation and sufficient length was taken beyond the face of the slope to allow it to be folded back with an adequate lap onto the next level of reinforcement. The reinforcement used was a polymer mesh (Netlon CE 131) which was supplied in rolls 30m long and 2m wide.

The fill was placed and compacted to layers of about 25cm thickness. Prior to compaction, about 2 per cent by weight of quicklime was mixed into the upper part of the layer. The reinforcement was spaced at vertical intervals of 0.5m in the lower levels which was increased to 1m in the upper levels. Plate 1 shows the reinforcement spread over the excavated slope. To avoid damage by construction plant, a protective layer of top soil was spread over the lower portion of the slope when this slope had been completed. The work continued in the same fashion until the reinstatement was completed and thereafter the remainder of the slope was top-soiled.

Site observations have indicated that no movement has taken place since completion of the work in 1980.

An estimate of the relative costs of the repair using the soil reinforcement method indicated that these were just over half those estimated

Plate 1   REPAIR OF CUTTING SLOPE ON M4 MOTORWAY

for the conventional approach employing granular soil (£70,000 as against £120,000). Thus very considerable savings were achieved by the former method and in view of the fact that Table 1 shows that significant lengths of motorway cuttings and embankments are at risk the method may well find wide applications for the repair of slopes.

Conclusion

(1)   A recent survey into problems of slope stability in motorway cuttings and embankments has indicated that significant lengths of slope constructed in over consolidated clays have failed and, by implication, further lengths are liable to failure in the longer term.

(2)   The repair of slope failures is traditionally carried out by replacing the foundered soil with a free-draining granular material.   However, where a source of such material is not locally available, considerable expenditure can be incurred by the haulage costs involved in the transportation of suitable fills to the site over relatively long distances.

(3)   An alternative procedure has been described for the repair of failed slopes which involves re-use of the foundered soil in conjunction with reinforcement layers of geotextiles or geogrids.   It may also prove necessary to use a relatively small proportion

of quicklime in carrying out the reinstatement to enable the construction plant to operate effectively when the soils are very soft.   To assist in determining the reinforcement requirements for a particular problem, a number of design charts have been prepared which permit a rapid assessment to be made of slope stability for a range of typical situations.

(4)   The application of the technique to the repair of a failed cutting in London clay has been described.   The use of a polymer mesh reinforcement (Netlon CE 131) in conjunction with 2 per cent by weight of quicklime enabled the soft, foundered, soil to be re-used and since the reinstatement in 1980, site observations have shown no indication of further movement.   The cost of the repair by this method amounted to just over half that estimated for the conventional approach: this reflects the fact that haulage distances of some 30 miles were anticipated for importation of the granular fill.

(5)   In view of the lengths of motorway, slopes in over-consolidated clays potentially at risk from instability, the application of the method could result in significant savings nationally.

Acknowledgements

This paper is published by permission of the Director, Transport and Road Research

Laboratory. The author is particularly grateful to Mr D M Farrar who was responsible for preparing the charts given herein.

References

Bishop, A W and N Morgenstern (1960) Stability coefficients for earth slopes. Geotechnique, 10, 129 - 150.

Department of Transport (1983). Maintenance of Highway Earthworks. Roads and Local Transport Directorate, Departmental Advice Note HA 26/83, London. (Department of Transport).

Murray R T, Wrightman J and A Burt (1982) Use of fabric reinforcement for reinstating unstable slopes. Department of Environment. Department of Transport, TRRl Report SR 751 Crowthorne.

Murray RT (1982) Fabric reinforcement of embankments and cuttings. Proc. 2nd. Int. Conf. on Geotextiles, Vol III, Lag Vegas, 707-713

Transport and Road Research Laboratory (1983). Studies of slope stability problems in highway earthworks, Leaflet LF943, February.

La Honda slope repair with geogrid reinforcement

As a result of a series of storms almost unprecedented in their intensity and duration in January 1982, the toe of the highway embankment on Route 84 near La Honda was eroded by the action of a contiguous stream causing a slipout. *Site geometry required restoration of the embankment with oversteepened (greater than 1:1) slopes that were strengthened by utilizing Tensar geogrid reinforcement. Embankment design parameters and calculations are presented and design features to mitigate drainage problems and to prevent future erosion at the embankment toe are described. The results of laboratory pullout tests are summarized. Construction, which has been suspended for the winter will be completed the summer of 1984.

R. A. Forsyth and D. A. Bieber, *California Department of Transportation*

INTRODUCTION

As a result of a series of storms almost unprecedented in their intensity and duration beginning in January 1982 and continuing through the past winter, the California highway system sustained severe damage. Consequently, the Department faced significant repair and restoration costs. Extensive damage occurred south of San Francisco on Route 84 near La Honda, California caused by the action of a contiguous stream that eroded the highway embankment causing a slipout. The site geometry and right-of-way constraints required restoration of the original embankment to a slope which was somewhat steeper than that which would assure long-term stability for the soils in the area. The site cross-section diagram (Figure 1) illustrates the critical section where the embankment must be reconstructed. After an initial investigation, the Materials and Hydraulics Engineers recommended the use of earth reinforcement to develop embankment stability. After consideration of several systems, Tensar geogrid was selected. Subsequently, a cooperative research agreement between Caltrans and Netlon was negotiated.

SITE PARAMETERS AND BACKGROUND

The slipout site is 70 m in length requiring slopes varying from 1.5:1 to approximately 1:1. The streambed is 14 m below the freeway grade and 11 m laterally from the hinge point on freeway grade at the critical cross-section. Access to the area is limited. The water table fluctuates up to 3 m and corresponds with changes in the stream elevation. Acquisition of additional right-of-way was not possible.

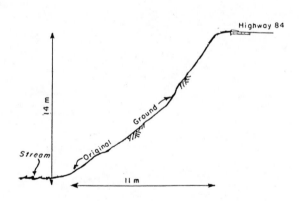

Fig. 1   Cross-section Diagram

The initial storm damage repair report called for rock slope protection 4.5 m high on a 1.5:1 slope placed at the embankment toe. Maintenance forces cleared the streambed of log jams and debris and placed 376,500 kg of rock slope protection as an interim repair. Permanent repair required rebuilding the upper embankment to slope ratios as steep as 0.9:1. Earth reinforcement would be necessary for reconstruction. The Caltrans' Transportation Laboratory initiated design of the reinforced embankment utilizing slope stability analysis and information generated from previous large scale laboratory pullout tests conducted on Tensar ER-2 (renamed Tensar SR-2).

PULLOUT TESTS

Large scale laboratory pullout tests were performed in 1980 on Tensar ER-2[1]. The test apparatus consisted of a rigid steel box 46 cm deep, 92 cm wide and 137 cm long in which soil is compacted half way, the geogrid material placed on this layer, and the remainder of the box filled with soil and compacted. A hydraulic ram located above the test box simulates overburden loads up to an equivalent of 15 m of earthfill. A horizontally positioned hydraulic ram attached to the geogrid provides the pullout force. Displacement is adjusted to maintain a controlled strain rate of approximately 2%/min. The pullout tests were performed on Tensar ER-2 in the direction the fabric is drawn from the roll.

Utilizing decomposed granite from an unspecified site as fill (ϕ equal to 35°) and imposing an overburden load equal to 34.5 kPa, the geogrid was pulled to failure. The Tensar ER-2 failed in tension outside the soil block (Plates 1 and 2) at a load of 44,000 newtons/meter indicating design would be limited by the geogrid material's maximum tensile strength. Load versus deformation curves were developed from the pullout tests including a comparison between Tensar ER-2 and bar mesh reinforcement (Graph 1). Bar mesh reinforcement of soil has been used by Caltrans to strengthen wall supported embankments[2]. The bar mesh (constructed from 0.95 cm diameter reinforcing bar welded to form 10 cm by 20 cm spacings) has sufficient steel to preclude tensile rupture and force a slippage failure within the soil block. Thus, the bar mat fails in pullout producing a cone shape failure near the soil face. From the graph, the ultimate strength of the Tensar can be obtained and used in design calculations.

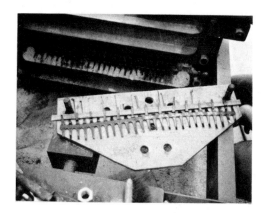

Plate 1   Tensile Break, Pull Section

Plate 2   Tensile Break, Face Plate

DESIGN

Computer slope stability analysis was used to determine the safety factor of the reconstructed embankment without reinforcement. From triaxial compression testing of samples of the native soil, properties at the La Honda site were determined to be:

$$\text{Cohesion} = 2.5 \text{ kPa}$$
$$\phi = 32°$$

These parameters, along with site dimensions, were input into SOILX, a circular slope stability program utilizing the modified Bishop technique. From the computer analysis, a minimum safety factor of 0.78 was calculated for the unreinforced embankment. Because an overall safety factor of 1.2 or greater was desired, additional resisting moment due to the soil strength increases from the reinforcement had to be quantified and new safety factors generated.

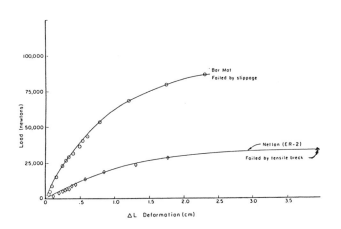

Graph 1:   Load Versus Deformation Characteristics of Bar Mesh & Tensar ER-2

55

Utilizing the computer generated overturning moment, and friction and cohesion resisting moments, the required reinforcement to increase the resisting moment later was determined. The following equations illustrate the calculations used for estimating safety factor (S.F.) increases as a result of the added reinforcement.

$$S.F. = \frac{Resisting + Tensar\ Moment}{Driving\ Moment} \qquad \dots\dots 1)$$

$$1.2 = \frac{2.01 \times 10^6\ newton\text{-}meters + \bar{a} \cdot \Sigma T_n}{2.57 \times 10^6\ newton\text{-}meters} \qquad \dots\dots 2)$$

$$\bar{a} \cdot \Sigma T_n = 1.2 \cdot (2.57 \times 10^6) - 2.01 \times 10^6 = 1.07 \times 10^6$$
newton-meters ......3)

where: 2.01×10^6 (nt-m) = Resisting moment @ S.F. = .78

2.57×10^6 (nt-m) = Driving moment @ S.F. = .78

$\bar{a} \cdot \Sigma T_n$ = Total Tensar moment

The coordinates of the failure arc and the location of the centroid of the reinforcement are used to determine the distance ($\bar{a}$) to the centroid of reinforcement. In this case, $\bar{a} = 14.3$ meters (Figure 2).

$$Tensar\ moment = \bar{a} \cdot \Sigma T_n = 1.07 \times 10^6\ newton\ meters \qquad \dots\dots 4)$$

$$T_n = \frac{1.07 \times 10^6\ newton\ meters}{14.3\ m} = 74720\ newtons \qquad \dots\dots 5)$$

The allowable working strength of the Tensar was limited to 6.670 newton/meters (15% of the ultimate strength). Knowing the working tensile strength contributed per meter, the number of reinforcing layers was determined.

$$\frac{74,720\ newtons}{6,670\ newtons/layer} = 11.2\ layers \qquad \dots\dots 6)$$

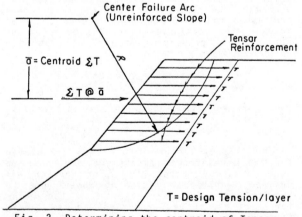

Fig. 2   Determining the centroid of Tensar Moment

Based on these calculations, vertical spacing of the Tensar was standardized at 0.6 meter. Circular failure arcs were plotted to assure that the embedment depth of the reinforcement was sufficient to preclude pullout of the Tensar. Figure 3 illustrates the geometrics of the critical cross-section. The embankment is approximately 14 m high. The lower 4.5 m has a slope ratio of 1.5:1 and is covered with 1 meter of rock slope protection to prevent water scour. The slope ratio in the upper reinforced portion of the embankment is steeper than 1:1. Permeable material lined with filter fabric is placed at the interface of the original ground and the reconstructed embankment. The permeable blanket drains into a horizontal outlet pipe located at the embankment base. The Tensar geogrid extends from the permeable material to the slope face.

Each layer of reinforcement will be folded back a minimum of 1.5 m and anchored in place. The face of the embankment is lined with compacted straw. The total amount of Tensar SR-2 required to complete the slope is 6000 square meters.

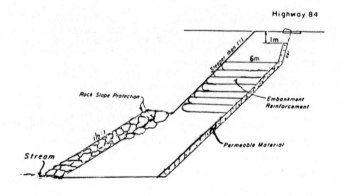

Fig. 3   Final Design Geometrics

INSTRUMENTATION

In order to monitor the stability of the reinforced embankment and the performance of Tensar SR-2, instrumentation was incorporated in the fill.

The instrumentation placed in the embankment includes:

1. One Inclinometer installed through the reinforced embankment structure.

2. Extensometers at three levels to monitor lateral movement and internal strains within the reinforced embankment.

3. Survey reference points at the hinge line and toe of the embankment to monitor vertical and horizontal surface deformations.

ALTERNATIVE REINFORCEMENT

Before reaching an agreement with Netlon, tire sidewall reinforcement was considered at the La Honda site. Tire sidewall reinforcement consists of recycled tire sidewalls hooked together with steel rods(3). The system's material costs are low, but construction is labor intensive. The embedment depth necessary to ensure slope stability was equivalent to that necessary to achieve slope stability with Tensar SR-2. Due to variations in limiting lengths of reinforcement, a total of 4800 m^2 of tire sidewall reinforcement was considered sufficient to stabilize the embankment to the desired safety factor.

COST COMPARISON

Comparing the cost of Tensar SR-2 to the cost of the required amount of tire reinforcement revealed a substantial cost savings for the Tensar project. Bar mat cost comparison is included, though actual design was not initiated.

Tensar Grids
    6000 m^2 @ $4.50/$m^2$ = $27,000
        U.S. Currency 1982

Tire Reinforcement
    4800 m^2 @ $13.45/$m^2$ = $65,000
        U.S. Currency 1982

Bar Mat Reinforcement
    2060 m^2 @ $67.30/$m^2$ = $138,600
        U.S. Currency 1982

By using Tensar geogrid reinforcement, Caltrans was able to realize the greatest cost savings.

CONSTRUCTION

Due to inclement weather and restrictions in the bid process, the construction at La Honda has been shut down until the spring of 1984. When construction resumes and the project is complete, Caltrans will publish research results of the slope repair at La Honda using Tensar geogrids.

CONCLUSIONS

As a result of pullout tests and design cost analysis, Caltrans has found that geogrid fabrics can be an economical earth reinforcement system. The long-term benefits realized by using geogrids in corrosive environments are considerable. Though not as strong in tension as other types of reinforcement, the La Honda project should demonstrate that satisfactory results can be obtained with geogrid reinforcement at a cost savings.

ACKNOWLEDGEMENTS

The authors would like to thank the following people for their service and technical assistance in completing this paper: Ken Jackura, Darla Bailey, Morris Tatum and Bernie Hartman.

REFERENCES

1. Hannon, Joseph B. and Pete Mundy (1980). Pullout Tests on Tensar ER-2, Transportation Laboratory.

2. Forsyth, Raymond A. and Joseph B. Hannon (1984). "Performance of an Earthwork Reinforcement System Constructed with Low Quality Backfill," prepared for presentation at the 63rd Meeting of the Transportation Research Board, Washington, D.C., January.

3. Yee, W. S. (1979) "Fill Stabilization Using Non-Biodegradable Waste Products - Phase II," Final Report No. FHWA/CA/TL-79/22, Caltans Laboratory, Sacramento, California, November.

4. Chang, Jerry, Raymond Forsyth and Joseph Hannon (1977). "Pullout Resistance and Interaction of Earthwork Reinforcement and Soil," TRB Report 640.

5. Ingold, T. S. (1983). "Laboratory Pull-Out Testing of Grid Reinforcements in Sand," Geotechnical Testing Journal, GTJODJ, Vol. 6, No. 3, September, pp. 101-111.

Stabilization of Canadian Pacific Railway slip at Waterdown, Ontario, using Tensar grid

J. R. Busbridge, *Golder Associates*

This paper presents a case history of the use of Tensar geogrid to construct a railway embankment on the side of The Niagara Escarpment, at Waterdown, Ontario. The railway was on a sidehill embankment built on an old landslip in the clayey silt till overlying the dolomitic limestone caprock of the escarpment. The track had a history of movement extending back over many years and it required frequent maintenance. Sudden significant movement of the embankment in December, 1982 caused rail traffic to be suspended. The old failure surface was excavated and the embankment reinstated at a slope of 45 degrees using granular fill reinforced with Tensar SR2 geogrid.

INTRODUCTION

The Town of Waterdown is located about 60 km south west of Toronto and 6 km north west of Lake Ontario (Figure 1). South of the town a C P Rail track follows a route along the west side of a deep ravine and at "Mileage 4.0" it is located on a sidehill embankment constructed about 6 m from the top of the ravine. About 12 m below the elevation of the tracks, rock outcrops on the side of the ravine. From about 6 m below the top of the rock outcrop, a scree slope extends down to a creek over a height of about 35 m at an average gradient of 2 horizontal to 1 vertical (Figure 2).

The track in the area of "Mileage 4.0" had a movement history extending back over many years. The movement was reported to occur generally after periods of heavy precipitation such as during spring melt or after a heavy rainfall. Regular inspection and maintenance were necessary and a reduced speed limit of 8 km per hour was imposed by C P Rail.

In December, 1980, Golder Associates was retained by C P Rail to investigate the reasons for the movement and provide recommendations for stabilizing the track. The investigation carried out included boreholes put down to bedrock in the immediate vicinity of the track and a shallow borehole put down by portable drilling equipment at a location downslope of the track. This information was supplemented with a slope survey and mapping of the bedrock outcrop. In addition, the movements were monitored by means of an inclinometer placed at the shoulder of the embankment.

SITE DESCRIPTION AND INVESTIGATION

Waterdown is located close to the ridge of the Niagara Escarpment. The escarpment is a massive topographic feature following the edge of the "Michigan Basin" which was formed during the Silurian Period (Figure 1). Sediments deposited in the basin were eventually compressed into rocks of varying hardness ranging from soft shales and sandstones to the more durable dolomite which generally forms the caprock.

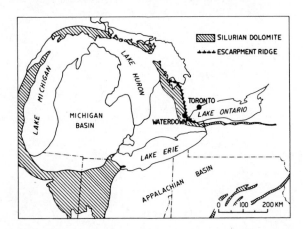

Fig. 1. Site location

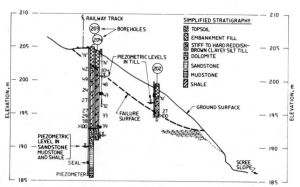

Fig. 2. Slope prior to stabilization

Polymer grid reinforcement. Thomas Telford Limited, London, 1985

Various erosive agents have subsequently removed the softer shales underlying the dolomite. As the underlying material is eroded the dolomite becomes undermined and blocks break off creating the near vertical face of the present day escarpment.

The ravine along which the C P Rail track is aligned has been cut through the escarpment by the erosive action of the creek. The scree slope below the bedrock outcrop is formed from the debris arising from the cycle of escarpment formation. Above the bedrock, periodic glacier advances deposited ablation till on top of the bedrock. As the bedrock is being eroded the overlying till is being continuously degraded by a process of gradual instability.

The subsurface investigation indicated that the rail track was supported on loose fill overlying a very stiff to hard silt till. The till was in turn underlain by a 3 m thick layer of dolomite which forms the caprock of a series of sandstones, mudstones and shales (Figure 2). Monitoring established that the lateral movement beneath the embankment shoulder was confined to the fill and the main movement was taking place at the till/fill interface. Over a 3 month period from March to June, 1981, a total downslope movement of 35 mm was recorded.

Piezometers sealed into the till and the underlying bedrock indicated that while the water pressures in the sandstones, mudstones and shales were very low (piezometric level at elevation 190 m, Figure 2), the water pressure in the till was high with the piezometric level close to the top of the stratum. It appeared that the dolomite caprock was preventing underdrainage of the till by the more permeable underlying rock layers.

Periodic monitoring continued over a period of two years. It was concluded that the railway embankment had been placed on an old failure in the till. High groundwater pressure in the till had probably been a major cause of the initial failure. Since then, surficial run-off from snow melt or rain would ingress to the base of the fill and activate movement along the old failure surface, which in turn would reflect in movement of the track.

Effective stress analyses of the slope were carried out using laboratory measured shear strength parameters for the till and fill. Having established the marginal stability of the slope in the computer model, the analysis was extended to investigate the relative merits of alternative remedial schemes. Details of these analyses are given by Busbridge, Chan and Sward (1984).

With private property immediately above the rail track and the near vertical escarpment rock outcrop located 12 m below, there was no room for regrading the slope. It was apparent that the problem of the high groundwater pressures which had precipitated the original failure would have to be solved in

any stabilization scheme. The best opportunity for reducing this pressure seemed to be to utilize the sandstone and shale bedrock as an underdrain by drilling a series of vertical drainage holes down through the till and dolomite caprock into the underlying more permeable layers. These holes would intercept the recharge through the till and lower the phreatic surface. In addition to the drainage scheme, it was proposed to rebuild the railway embankment using earth reinforcement to maintain a side slope of 45 degrees.

On December 25, 1982 an exceptional thaw resulted in a 200 to 250 mm vertical movement of the track and a crack developed along the shoulder of the embankment. At that point, stability of the track could not be assured and rail traffic was suspended. It was decided to implement the track rebuilding immediately, using earth reinforcement.

STABILIZATION WORKS

Excavation of the failed slope section was carried out from December 29 to 30, 1982 using two excavators. Benches, 1.2 to 3 m wide, with steps of about 2 m were made in the till along the natural ground profile below the old failure surface (Figure 3). The excavation was 37 m long. Immediately under the tracks the railway fill was found to rest on undisturbed till. Downslope of the track a discrete shear surface within the till was identified and this was found to extend down to the surface of the bedrock, which was 4.5 m higher than originally inferred from the bedrock outcrop mapping. The extent of the old shear surface and the higher elevation of the bedrock called for a revision to be made to the design. The excavation was taken down to the bedrock and all pre-sheared surfaces were removed (Figures 3 and 4). A 2 m high rockfill toe founded on bedrock was built to form the toe of the new embankment. Compacted granular material was placed as a filter between the till and the rockfill. Construction of the rockfill toe and granular filter was completed on December 31, 1982.

The slope was then built up at a gradient of 1.5 horizontal to 1 vertical using the same granular material. A typical grain size distribution curve of the granular material is given in Figure 5. The exposed surfaces of the natural till were cleaned of loose and softened materials prior to filling. Horizontal surfaces of the till were trimmed to a downslope fall of not less than 5 per cent to assist natural drainage into the granular fill. Compaction of the granular material to a density of not less than 95 per cent of the Standard Proctor density was carried out using vibratory smooth drum rollers. The average dry density of the compacted fill as determined by an in situ nuclear density testing device was 2.0 t/cu. m. Two layers of reinforcing elements were placed at 1.2 m height intervals within the lower rebuilt slope. The reinforcement used was Tensar SR2

Fig. 3.   Benched slope after excavation of
          old failure surface

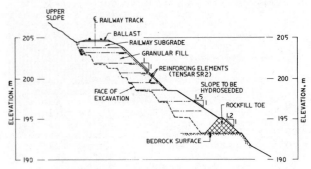

Fig. 4.   Stabilized slope

soil reinforcing grid supplied by Tensar
Incorporated. Each strip was 3.7 m long and
was placed against the face of the excavation.
This portion of the fill slope, which actually
forms the base for the steeper earth reinforced
slope, was completed on January 2, 1983.

The upper earth reinforced slope is 5.5 m
high, and was built at a gradient of 45
degrees. Six layers of reinforcing grid
(Tensar SR2) were laid at height intervals of
1.2 m. The length of the reinforcing layers
varied from 3.5 to 5.5 m as shown in Figure 3.
Each strip of 1 m wide reinforcing grid was
cut to the design reinforcing length, plus an
additional length for wrapping around the
slope surface and overlapping with the next
layer. The strips were laid with a transverse overlap of 50 to 100 mm and granular
fill was placed with some overbuilding beyond
the design gradient to enable adequate compaction near the finished slope surface
(Figure 6). When the elevation for the next
layer of reinforcing grid was reached, the
alignment of the slope was set out and the
slope surface was trimmed to the required
gradient. The trimmed slope surface was
covered by reinforcing grid from the previous (lower) layer, and the grid taken into
the embankment for a minimum length of 1 m.
The grids were held in place by 0.2 m long
steel pins driven into the fill (Figure 7).
This process was repeated for each layer of
reinforcing grid until the top of the
embankment was reached, on January 8, 1983.
Train traffic resumed on January 9, 1983,
which was 12 days after starting the works.
A total of 62 rolls (each 1 m wide and 30 m
long) of reinforcing grid was used, while
the actual length of Tensar SR2 laid in the
slope was about 150 linear metres. Wastage
due to cutting the reinforcement lengths from
the 30 m long rolls was about 6 per cent.

DESIGN OF THE EARTH REINFORCED SLOPE

As in any earth reinforced structure, the first
design consideration was to ensure overall
stability of the reinforced mass. This is
particularly important in stabilization of a
natural landslip where critical conditions
originally existed in the natural soils. Where
groundwater pressures are present, it is
necessary to ensure good drainage within and
behind the reinforced zone. In addition, the
low winter temperatures in Ontario mean that
a non-frost susceptible layer of soil should
be placed in the outer 1.2 m of fill to avoid
stresses in the reinforcement caused by frost
heave. In order to meet both these criteria,
free draining granular material was specified
for the fill.

The design, which had been prepared prior to
the emergency which precipitated the construction, had included vertical drains to
lower the piezometric level in the till.
Excavation of the till down to bedrock at
the bottom of the slope and replacing it with
granular fill resulted in greater drawdown
of the piezometric level than had been
considered in the original design. Analyses
of the revised section indicated that the
vertical drains could be dispensed with.

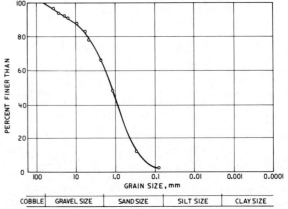

Fig. 5.   Typical grain size distribution curve
          of fill

Fig. 6. Compaction of fill on top of reinforcing layer

Fig. 7. Wrapping reinforcement grid up face and anchoring at elevation of next layer

Slope stability analyses were carried out using a computer program based on the method developed by Sarma (1973). A back analysis of the original slope profile assuming limiting equilibrium along the observed failure surface indicated an average cohesion along the failure surface of 3 kPa if a friction angle of 31 degrees is assumed (Figure 8). This is only 4 kPa lower than the cohesive intercept measured together with the same friction angle in a laboratory drained triaxial test on an "undisturbed" sample of till.

Since the upper section of the failure surface was between railway fill and undisturbed till, the figure calculated by the back analysis does not accurately reflect the residual shear strength of the till mobilized in the bottom section of the failure surface.

The shear strength of the granular fill used in construction of the earth reinforced slope was established by three direct shear tests. These gave friction angles varying from 32 to 38 degrees with zero cohesive intercept.

For design, a friction angle of 35 degrees and a unit weight of 21 kN/cu.m were assumed for the fill.

Using these parameters for the fill and a friction angle of 31 degrees, together with a cohesion of 7 kPa for the undisturbed till, the factor of safety along the critical failure arc extending behind the earth reinforced structure was calculated to be 1.3 (Figure 9).

An important design consideration was that any remedial measures would not adversely affect the stability of the overall slope including the property above the rail track. Because of the improved drainage of the slope provided by the granular fills, the stability of the area immediately above the track was increased. An analysis of deep seated stability involving the upper slope indicated that the factor of safety for this mode of failure was 1.2 which is the same value calculated for the slope prior to stabilization (Figures 8 and 9).

Internal stability of the earth reinforced slope was established by designing the reinforcement to resist the maximum horizontal earth pressure within the reinforced mass. The general procedure adopted was the "tied-back" method which has been described by Murray (1980) for vertical walls. The design force was obtained by investigating

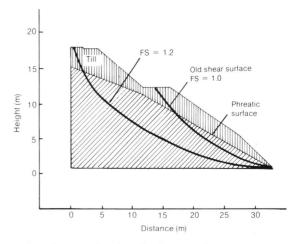

Fig. 8. Analysis of slope prior to stabilization

a series of trial failure surfaces exiting at the toe and at various points on the slope. For critical surfaces, the tensile capacity of the reinforcement intercepting the failure surface was checked together with the pull-out resistance of reinforcement extending behind the critical surface.

The computer analysis adopted has the facility of computing the horizontal force required to maintain stability of any failure surface investigated. Circular arcs and two-part wedge failure surfaces were investigated together with Coulomb planar surfaces. It was found that for the 45 degree slope of the embankment, the planar surface underestimated the restraining force necessary to maintain equilibrium by about 50 per cent. For calculating the maximum forces in the reinforcement, circular failure arcs were adopted.

The following design parameters were adopted:

Minimum Factor of Safety for overall stability	= 1.3
Allowable Tensile Capacity in Tensar SR2	= 16 kN/m
Minimum Factor of Safety against pull-out	= 2
Coefficient of friction between soil and geogrid	= 0.8 tan ϕ'

COMMENTS

The Waterdown stabilization, to the best of the author's knowledge, was the first use of geogrid in Canada for slope stabilization. In the months prior to the works described, Golder Associates had been working closely with the Ontario Ministry of Transportation and Communications in designing the earth reinforced slopes for the Highway 410 embank-

ments, which are described in Mr. Devata's paper to this conference (Devata, 1984). The method of face support adopted at Waterdown, which involved turning the horizontal reinforcement up the face and lapping with the next layer, proved to be labour intensive and time consuming. Using the experience gained at Waterdown, and after consultation with Tensar Incorporated engineers, this detail was revised for the Highway 410 embankment. The detail described by Mr. Devata of providing short horizontal layers of reinforcement at compaction lift intervals between the main reinforcement layers proved to be easier to construct.

Difficulty was experienced at Waterdown in achieving adequate compaction near the edge of the 45 degree slope. Although this is helped to some degree by providing intermediate horizontal strips, in cases where it is necessary to specify clean granular material at the face of slopes steeper than 45 degrees, facing panels or temporary formwork is required to provide the necessary reaction for compaction.

For vertical walls, the use of Coulomb's earth pressure theory for determining the maximum force in the reinforcement is adequate for design. However, as the wall/slope becomes flatter, Coulomb's assumption of planar failure surfaces leads to significant underestimation of the forces involved and circular arcs or two-part wedges must be investigated. For the case described in this paper, the method of stability analysis developed by Sarma (1973), which is readily available in computer programs, proved to be a convenient and reliable design method.

As the slope flattens from the vertical, the forces in the reinforcing elements reduce significantly and the use of polymer geogrids as earth reinforcement provides an attractive and economical solution to the problem of constructing steep slopes. In the case described, the geogrid reinforcement allowed a railway embankment to be constructed in the limited space and at the same time provided effective stabilization of an old landslip.

ACKNOWLEDGEMENTS

The author is grateful to C P Rail; particularly to Mr. R. Smith and Mr. R. F. Sward for permission to publish this paper and for their readiness to accept a novel technique for the stabilization work described. Thanks are due to Mr. P. Chan who carried out the analyses and supervised the construction and to Mr. V. Milligan for his insight and advice.

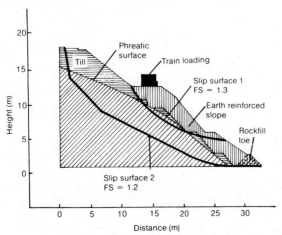

Fig. 9. Overall stability of stabilized slope

REFERENCES

Busbridge, J.R., Chan, C.N.P., and Sward, R.F. (1984). "Stabilization of an Old Landslip on the Niagara Escarpment." Paper to be presented to the International Symposium on Landslides, Toronto.

Devata, M. (1984). "Geogrid Reinforced Earth Embankments with Steep Side Slopes." Symposium on Polymar Grid Reinforcement in Civil Engineering, London.

Murray, R.T. (1980). "Fabric Reinforced Earth Walls: Development of Design Equations." Ground Engineering, October.

Sarma, S.K. (1973). "Stability Analysis of Embankments and Slopes." Geotechnique 23, No. 3.

Repair of landslides in the San Fransisco Bay area

R. Bonaparte, *The Tensar Corporation,*
and E. Margason, *Edw Margason
Geotechniques*

California has experienced two consecutive winters of exceptionally heavy rainfall. In the hills surrounding San Francisco Bay, the rainfall has resulted in numerous landslides. Along transportation routes and in populated areas, these slides have required prompt repair. This paper describes the use of polymer geogrid reinforcement in the repair of a large and potentially disasterous slide along Hillcrest Road in the City of San Pablo. The slide repair had two components: (i) an emergency winter repair consisting of driven steel piling to temporarily stabilize endangered homes at the headscarp of the slide; and, (ii) a permanent summer repair. The permanent repair consisted of excavation of a deep prism of soil below the road right-of way. The road excavation was refilled with compacted on-site soils reinforced with horizontal layers of geogrid. The steep face of the reinforced soil prism was formed using the "geogrid wrap-around" technique. The soil-geogrid composite, keyed into unweathered claystone beneath the deepest active slide plane, is designed to provide passive resistance against further movement in the headscarp area of the slide.

INTRODUCTION

Geogrids are high strength, orientated polymer grid structures used to reinforce soils. This paper describes the use of geogrids in a large landslide stabilization and road repair project in San Pablo, California. The repair extended over a road length of about 150m (500 ft), with excavation depths in the road right-of-way of 15m (50 ft). Geogrids were used because they were found to be more cost effective than other considered repair alternatives including gabion retaining walls, concrete crib walls, and conventional concrete retaining walls. The paper begins by presenting background on the geologic conditions and history of the slide area, and follows with a site description and results from site soils investigations. The landslide repair is then discussed. An emergency repair to temporarily stabilize endangered homes is described first. A subsequent permanent repair is presented next. The primary components of the permanent repair are a geogrid reinforced soil prism anchored into hard claystone and an extensive subsurface drain system. The design and construction of the reinforced soil prism is discussed in detail. Conclusions are drawn regarding the suitability of geogrids for the stabilization of landslides.

BACKGROUND

The City of San Pablo, California is situated near the east shore of San Francisco Bay, about 15 miles northeast of San Francisco. A hillside area within the city is located along the San Pablo Ridge. This ridge, which lies at the northwest end of the Diablo Range (part of the Pacific Coastal Ranges), is comprised of Orinda Claystone with interbedded stringers of sandstone and conglomerate, overlain in areas by surficial soils derived from the parent rock. Regional tectonic activity has caused intense shearing and folding of these rock strata, resulting in variable geologic and ground-water flow conditions. Surface faulting reflects the nearby presence of the Hayward Fault.

Hillcrest Road is situated in a north facing hillside area along the San Pablo Ridge and has long been identified as having zones with actively unstable or potentially unstable slopes. Through the years, numerous geologic and engineering studies have identified many of these zones. In spite of the recognized potential for slope instability, portions of the hillside area have been developed for residential and commercial purposes.

SITE DESCRIPTION AND HISTORY

Fig. 1 shows a site plan of Hillcrest Road and its surroundings. In the area of interest, the road runs in an east-west direction approximately 400m (1300 ft) south of San Pablo Dam Road. From the "Dam" Road, the ground slopes upward, at first gently, then more steeply, to a series of homes that line the south (upslope) side of Hillcrest Road. With the exception of a church and an apartment complex adjacent to the Dam Road, the downslope area is undeveloped.

Hillcrest Road had been disrupted by landslides in the past. In the 1950's, a slide damaged two homes and destroyed the road at the location shown in Fig. 1. A second slide disrupted the road in 1969. Smaller slides, slumps, tension cracks, and creep movements have been observed periodically, and geologic hazard studies of the hillside neighborhood (Woodward-Clyde Consultants, 1978 and 1983a,) ranked this area as having a high potential for slope instability. In mid-February, 1983, a major slide scarp appeared along the south shoulder of Hillcrest Road just west of the 1950's slide. This large headscarp, associated with the heavy winter rains of 1981/1982 and 1982/1983, endangered the houses along the south side of the road. In order to save the homes, the City of San Pablo commissioned an immediate emergency headscarp stabilzation program, followed by a permanent repair of the road and headscarp area the following summer.

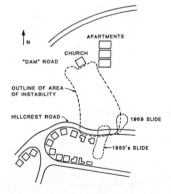

Fig. 1   Hillcrest Road and surrounding area.

Polymer grid reinforcement. Thomas Telford Limited, London, 1985

SITE AND SOIL INVESTIGATION

Five exploratory soil borings were made during February and March 1983 (Fig. 3). The borings identified a road fill and an upper soil zone of stiff to very stiff, light brown sandy and/or silty clay, underlain by a thinner zone of weathered, sheared, siltstone and claystone. Bedrock consisting of hard gray Orinda Claystone lay below the weathered siltstone and claystone. Slope indicators were installed in each finished boring.

Analysis of the slope indicator data suggested several active slide planes beneath Hillcrest Road, (Fig. 2). An upper and most active slide plane was found about 3m (10 ft) below road grade. A second active slide plane was found at a depth of about 6m (20 ft) to 9m (30 ft) below road grade. Movements along both of these slide planes were of the order of 0.03m (0.1 ft) per day. Deep-seated movements were observed in the weathered zone, but these were much slower than in the upper more active zones. As Fig. 2 indicates, the slide planes dipped to the north. The shallow active slide plane was situated approximately at the base of the old road fill, while the lower active slide plane was controlled by the boundary of the light brown clay and the weathered, sheared, siltstone and claystone. The deep, slower movements appeared to be taking place on top of unweathered Orinda Claystone.

The site investigation revealed that the Hillcrest Road slide was actually the head scarp of a much larger slide covering a hillside area of about 50,000m² (12 acres). The toe of the slide was evidenced in the parking lots of the church and apartment complex shown in Fig. 1. At both locations, ground heave resulted in structure distress and/or disruption of parking facilities.

REPAIR OF HILLCREST ROAD SLIDE

The Hillcrest Road slide repair had two components: (i) an emergency winter repair to temporarily stabilize the endangered homes; and, (ii) a subsequent permanent road repair. The emergency stabilization plan called for the driving of a series of steel H-piles along the upslope curbline of Hillcrest Road. The permanent repair plan consisted of excavation of the slide mass in the road right-of-way north of the pile line, followed by construction of a drained prism of reinforced or retained engineered fill anchored into the hard claystone. While the conceptual design of the permanent repair was integrated with the emergency stabilization plan from the start, final design of this permanent repair was carried out after the emergency program had been completed.

Emergency Stabilization of Hillcrest Road Slide

Emergency stabilization consisted of driving 63 steel H-piles along the south curbline of Hillcrest Road, adjacent to the residential properties. The piles were driven through the active slide planes into hard claystone. The layout of the 12WF53 steel H-piles is shown in Fig. 3, and the slide scarp, pile-driving rig, and the endangered homes are shown in Fig. 4. The steel piles had two functions: (i) to act as temporary cantilever supports to prevent further uphill headscarp movements (and therefore movements of the homes) until the permanent repairs; and, (ii) to provide a shoring system so that the slide material in the road right-of-way could be safely excavated during the subsequent permanent repair. The piles were spaced at 1.2m (4.0 ft) centers and, as a general rule, were driven to depths of about 10m (33 ft) to 14m (46 ft), or at least twice the depth of the lower active slide plane beneath the pile line. The piles stabilized the upslope headscarp area. By winter's end, continued movement downslope of the pile line had caused a 1.0m (3 ft) high scarp along the downslope edge of the pile line.

Permanent Repair of Hillcrest Road

In the assessment of permanent repair alternatives, two requirements were: (i) the repair structure had to be built within the road right-of-way; and, (ii) the repair had to be completed quickly, before the 1983/1984 rainy season. The road was in the headscarp portion of the slide so repair plans based on active lateral earth pressures were considered. It was concluded that these pressures could be resisted by a reinforced or retained soil prism (soil buttress), keyed into hard claystone, and extending up to original road grade. The prism would be designed to stand unsupported, should future slope movements occur downslope of the road. Four alternative soil prism designs were considered: (i) geogrid reinforcement; (ii) gabion retaining wall; (iii) concrete crib retaining wall, and (iv) conventional concrete retaining wall. The decision to use the geogrid reinforcement was based on its lower cost.

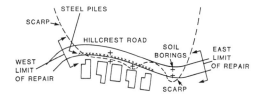

Fig. 3   Hillcrest Road showing the slide scarp and the line of emergency stabilization steel H-piles.

Fig. 4   Hillcrest Road (looking west), showing a portion of the slide scarp.

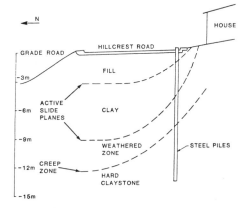

Fig. 2   Slide planes beneath Hillcrest Road (identified from soil borings and slope indicator readings).

DESIGN OF PERMANENT REPAIR

A typcial design cross-section of the slide repair is shown in Fig. 5. Plans called for construction to proceed with excavation of the area encompassing the reinforced soil prism as well as an area north of the reinforced soil prism extending at a 45° angle from the base of the prism, as shown in Fig. 5. The area between the anchored prism and the 45° slope was to be filled using random on-site soil compacted to 85% maximum dry density determined by ASTM Test Method D-1557. It was recognized that if the hillside area downslope of Hillcrest Road continued to move, the unanchored zone of random fill would also move, and the face of the reinforced soil prism would be unsupported and exposed. These factors were considered in the design.

An important component of the permanent slide repair was the subsurface drain system. The functions of the drain system are: (i) to keep the prism fill drained; and, (ii) to keep water pressures along potential slide planes to a minimum. As shown in Fig. 5, the drain system beneath the reinforced soil prism consisted of a minimum 0.3m (1.0 ft) thickness of crushed gravel. Along the face of the piles, the drain system comprised a 6mm (0.25 in) thick plastic net (Netlon DN1) wrapped in a needlepunched, nonwoven geotextile (Trevira Spunbonded). The function of the geotextile is to act as a filter, allowing the flow of water into the plastic net drainage layer while preventing the movement of soil particles. Additional drain system details are given subsequently.

Design of Geogrid-Reinforced Soil Prism

The internal stability of the reinforced soil prism was analyzed using active Rankin lateral earth pressures. Geogrid spacings were based on the geogrid Tensar SR2 in the lower portion of the soil prism, and the lower-tensile strength geogrid Tensar SS2 in the upper portion. The Tensar SS2 was used in the upper portion of the soil prism for two reasons; (i) the larger roll size would allow quicker placement of the geogrid; and, (ii) the quantity of reinforcement needed in the upper portion of the prism was small. The design was based on an allowable working stress in the geogrid equal to 40% of the strength measured in constant rate-of-strain tensile tests.

The reinforced soil prism was constructed using the on-site silty and sandy clay soils compacted to 92% of maximum dry density (ASTM Test D-1557). From experience on previous projects in the area, an effective cohesion of 24 kN/m^2 (500 lbs/ft^2) and an effective friction angle of 20° was used for the compacted fill. The factor of safety against sliding and overturning of the reinforced soil prism was evaluated using the sliding wedge of soil shown in Fig. 5. Because of eventual pile corrosion, support provided by the steel piles was neglected in the sliding and overturning analyses. A large strain effective friction angle of the presheared soil along the slide plane of 10° was assumed, based on large-strain direct shear tests.

The design called for a key trench to be excavated through the lower active slide plane into hard claystone. To speed construction, sections of the key trench deeper than 12m (40 ft.) were to be backfilled with imported graded aggregate. Above 12m (40 ft.), the reinforced soil prism was to be constructed with compacted on-site fill and horizontal layers of geogrid. The vertical spacings of the geogrid layers were as follows: from a depth of 8.5m (28 ft.) to 12m (40 ft.), Tensar SR2 at 0.4m (1.3 ft.) lifts; from a depth of 4.9m (16 ft.) to 8.5m (28 ft.) Tensar SR2 at 0.6 (2.0 ft.) lifts; between the ground surface and a depth of 4.9m (16 ft.), Tensar SS2 at 0.6m (2.0 ft.) lifts.

Due to limited right-of-way, the geogrid-reinforced soil prism was designed with an average batter of 4:1, vertical to horizontal. To simplify construction, the battered face was stepped, (0.3m (1 ft) set-back at 1.2m (4 ft) vertical steps). Alignment of the key trench was controlled by the north curbline of Hillcrest Road and the 4:1 prism batter.

Geogrid Wrap-Around Facing Method

The geogrid wrap-around construction technique was used to form the face of the reinforced soil prism. This method had been used on previous geogrid projects, with scaffolding or formwork required to temporarily support the wrap-around face. On this project, the random fill provided the temporary support and no special scaffolding or formwork was needed.

The following procedure was developed to construct the wrap-around face for each 1.2m (4 ft) prism step: (i) compact the random fill one full step above the prism fill; (ii) using a small dozer, trim a vertical face in the random fill along the north boundary of the reinforced prism; (iii) place the geogrid reinforcement across the width of the prism with an extra length of geogrid draped up and over the vertical face of the random fill; (iv) place and compact the required thickness of prism fill; (v) pull the extra length of geogrid back over the prism fill to form the wrap-around and tail of the reinforcement layer; (vi) secure the geogrid tail to the prism fill using 0.15m (0.5 ft) steel pins; (vii) repeat iii through vi for the next one or two layers of geogrid to complete the prism step; (viii) repeat i through vii to form subsequent prism steps.

CONSTRUCTION OF PERMANENT REPAIR

Excavation of Slide Material

Repair of the Hillcrest Road slide commenced in September 1983. Excavation of the roadway was carried out to a depth of 3m (10 ft), with the excavated soil being stockpiled in an open downslope area. A line of drilled-in anchored tiebacks was then installed along the pile alignment to provide construction-phase pile support. Each tieback was anchored (grouted) in unweathered claystone, connected to an adjacent pile and preloaded. After tieback installation, excavation continued according to the plan outlined in Fig. 5. Below 6m (20 ft), a series of soil benches were left standing against the pile wall to provide resistance

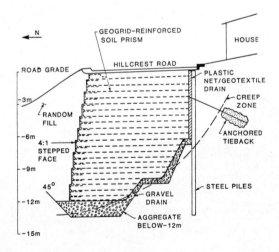

Fig. 5   Typical design cross-section of the geogrid-reinforced slide repair.

against kicking out of the toe. The maximum excavation depth varied from approximately 6m (20 ft) at the west end of the 150m (500 ft) long excavation, to 16m (52 ft) near the east end. These depths were required to get the key trench anchored into hard claystone.

The trench was constructed in two sections corresponding to the east and west sides of the site. Fig. 6 shows the excavation (looking west) at a time just after the east key trench had been excavated and filled, and prior to excavation of the west key trench. The random fill, raised one step above the geogrid-reinforced prism fill can be seen in Fig. 6, as can the geogrid wrap-around tail. The small dozer in the right foreground was used to cut the vertical face for the random fill step.

Construction of Geogrid-Reinforced Soil Key Trench

The east key trench was constructed first, in one long section. On the east side, the maximum excavation depth was about 16m (52 ft.). Since the east trench was greater than 12m (40 ft) deep along its entire length, it was filled with graded aggregate. No geogrid reinforcement was used in the east trench.

As the long west key trench was excavated, several failures developed in the side slopes of the trench in front of the pile line. Blocks of soil began to move from the adjacent soil benches into the trench. These failures were associated with the loss of support of the slide material on the two sides of the trench. On the south side of the trench the failures were taking place on the pre-existing slide planes. Movement occurred both north and south of the trench, but the south movement was of greater concern because it reduced the support at the toe of the steel piles. The scarp that developed along the south side of the key trench can be seen in Fig. 7. As soon as the instability was indentified, the west key trench was refilled.

The construction procedure for the west key trench was modified to reduce the risk of trench collapse and pile toe failure. The modified procedure included the following steps, designed to keep short sections of trench open for short periods of time: (i) excavate a 3m (10 ft.) section of key trench using a backhoe (Fig. 8); (ii) place preassembled segments of 3-roll wide geogrid along the bottom and face of the trench ; (iii) quickly dump 2m (6 ft.) of gravel on top of the geogrid ; and (iv) wrap the geogrid back over the gravel. This modified construction procedure, while slowing progress, allowed the key trench to be completed without further stability problems.

Fig. 7 Failure of south slope of key trench and soil bench. Movement is occuring on pre-existing slide plane.

Construction of Geogrid-Reinforced Soil Prism

Construction of the main portion of the reinforced soil prism was carried out as soon as the west key trench was completed. Figs. 9 and 10 shows the methods and equipment used to construct the prism. Each step of the reinforced soil prism was started by cutting a vertical face into the random fill and placing the geogrid (Tensar SR2 in Fig. 9 and Tensar SS2 in Fig. 10). The geogrid extended from the gravel drain (on the left side of both Figs. 9 and 10) and up and over the vertical face of the random fill. The prism fill was placed using a $6m^3$ ($8 yd^3$) scraper (Fig. 9). To speed construction, the scraper was allowed to drive directly on the geogrids. The acceptability of direct contact between geogrid and scraper was established through site trials. The prism fill was placed in 0.2m (0.67 ft) lifts and compacted using a sheepsfoot roller (Fig. 10). At the soil prism face, compaction was carried out using a hand-operated vibratory compactor. After compaction, the geogrid was pulled back over the prism fill to form the wrap-around and tail (Fig. 6), as described previously. The minimum geogrid tail length was 1.2m (4 ft). Where required, the geogrids were cut with a hand-held power saw. To maintain geogrid continuity during construction, adjacent rolls of Tensar SR2 were connected with metal rings, and adjacent rolls of Tensar SS2 were overlapped a minimum of 0.15m (0.5 ft).

Fig. 6 Construction of lower portion of geogrid-reinforced soil prism looking west (note the step between the geogrid -reinforced soil prism to the left, and the random fill to the right).

Fig. 8 Modified construction procedure for west key trench (to minimize potential for soil failure into trench).

Fig. 9   Construction of geogrid-reinforced soil prism with Tensar SR2 (note the geogrid covered step used to form the wrap-around face).

Fig. 10   Construction of reinforced-soil prism with Tensar SS2 (this view is just before the Tensar SS2 is pulled back to from the geogrid tail).

Fig. 11 shows the plastic net/geotextile drain curtain draped from the steel channel section. The 1.6m (5.2 ft) wide drain sections were prefabricated on-site, connected to a steel channel at the top of the pile line, and then dropped down the face of the piles. The bottom of the drain curtain was buried in the gravel drain. The gravel drain contained 0.15m (0.5 ft) perforated plastic subdrain pipes placed at the back edge of each soil bench. The pipes were designed to carry water to a concrete manhole (Fig. 9). The water flows from the manhole, through an outlet pipe, to a nearby stream. Flushing risers (foreground, Fig. 10) were connected to the subdrain pipes at each end of the slide repair to allow periodic flushing.

Earthwork construction was completed in early November, 1983 with the raising of the fill to road grade and the cutting-off of the tops of the steel piles. A slope indicator was installed through the prism fill and into hard claystone at the location of the 1950's slide. Heavy early winter rain precluded replacement of the Hillcrest Road pavement before spring 1984.

Fig. 11   Geotextile wrapped plastic net drain system.

CONCLUSIONS

The Hillcrest Road project represents the first use of geogrids to repair a landslide in California. Significant aspects of geogrid use on this project include: (i) of the reinforcing and retaining systems evaluated, the cost of the geogrid-reinforced system was lower than the estimated costs of the other systems; (ii) construction with the geogrids was rapid, allowing the project earthwork to be completed prior to the winter rainy season; (ii) geogrids allowed the use of the on-site clay fill materials; (iv) as shown by the slope problems in the west key trench, geogrid design and construction was adaptable to the specific site conditions; and, (v) geogrids provide a means of constructing cost-effective, steep-faced reinforced fills.

Based on the successful completion of the Hillcrest Road project and a second smaller project as well, geogrids are now being considered as a reinforced soil repair alternative for a number of other slides in California.

ACKNOWLEDGEMENTS

Both authors began work on the Hillcrest Road slide while members of the Geotechnical Engineering Group of Woodward-Clyde Consultants. Preliminary design of the geogrid-rienforced soil prism was carried out with the assistance of J. Dixon and J. Paul of Netlon, Ltd., England. The authors are indebted to J. Dixon for valuable comments, and S. Hartmaier and N. Stuart for assistance in the preparation of this paper. The authors thank Gary Leach and Gil Zermino from the City of San Pablo for providing information used in the paper.

REFERENCES

Woodward-Clyde Consultants, "Phase II Geologic Assessment, Hillside Neighborhood, San Pablo, California," unpublished report for San Pablo Redevelopment Agency, City of San Pablo (1978).

Woodward-Clyde Consultants, "Phase III Geologic Assessment, Hillside Neighborhood, San Pablo," unpublished report for San Pablo Redevelopment Agency, City of San Pablo (August, 1983a).

Woodward-Clyde Consultants, "Geotechnical Study and Recommendations for Repair Options, Hillcrest Road Slide," unpublished report for San Pablo Redevelopment Agency, City of San Pablo, (May, 1983b).

Reinstatement of slopes: report on discussion

R. T. Murray, *Transport and Road Research Laboratory*

With reference to paper 2.1, it was noted that in his presentation Dr Murray referred to the use of quicklime when reinstating the slope with Netlon reinforcement mesh. As quicklime or hydrated lime is very corrosive and yet commonly used in building sites and road sites, Dr Murray was asked whether any research has been carried out into this topic and whether any adverse effects were noted where the mesh and quicklime were in contact.

It was confirmed that no research is being carried out at TRRL into the corrosivity of quicklime. As described in the paper, the quicklime was employed to improve the properties of the London Clay such that the construction plant could operate effectively, particularly after wet weather. No adverse effects were noted to the exposed mesh in direct contact with the quicklime and as the inclinometer measurements through the reinstated slope showed no movements subsequent to completion, it is reasonable to suppose that the mesh at deeper levels was unaffected.

Mr A Burt, Project Engineer for the scheme, stated that there was no evidence of the quicklime having had any ill effects on the mesh. In any case the high moisture content of the clay produced a rapid and complete hydration of the lime such that this would have been unlikely.

Dr Murray confirmed that in his paper bilineal slip surfaces are sliding surfaces formed by two planes, corresponding to a two-part wedge. A figure is included in the paper showing the arrangement.

Dr Bonaparte was asked to comment on the reference in paper 2.4 to the scraper being permitted to operate directly on the Geogrid. In other papers the authors have placed considerable emphasis on trying to prevent this occurring.

In reply Dr Bonaparte agreed that he was aware of these recommendations and at the outset had specified that plant should not be permitted to work directly on the Geogrid. However, time was very short to complete the project before the rainy season, with the result that construction had to be expedited as much as possible. Field trials were therefore carried out to ascertain whether the requirement could be relaxed.

The trials demonstrated that the relatively small (7 cub yard) scraper with large rubber tyres which was to be employed did not damage the Geogrid and therefore work proceeded accordingly. He did make the recommendation that without the field trials such an approach is not acceptable.

It was not obvious to another contributor where the excavation from the east trench was stockpiled. As there appeared to be quite a large volume involved, this could have proved to be a destabilising factor if located in the wrong place.

Dr Bonaparte stated that there were a number of off-site areas to stockpile. Permission was obtained to stockpile in these areas, although it was not permitted to erect the permanent structure on the owner's property. Some of the stockpile areas were on the 12 acres of land that was actively sliding and the contractor stockpiled some of the fill here initially. Some tension cracks developed at the top of the fill and action was taken quickly to move those stockpiles.

Dr Bonaparte, commenting on the relationship between estimated and actual costs, confirmed that the estimated cost of the Geogrid repair alternative was 60 per cent of the next closest alternative which involved the use of plastic coated steel gabions. It transpired that the final costs were fairly close to the estimated figures. However, the size of the job, and therefore the amount of soil and Geogrid involved, had changed dramatically during the course of the reinstatement works.

Participants: Dr Murray

Dr Bonaparte

Mr Burt

Mr Strudley

Mr Rankilor

Mr Paine

Polymer grid reinforcement. Thomas Telford Limited, London, 1985

Design methods for steep reinforced embankments

R. A. Jewell, N. Paine and R. I. Woods,
Binnie and Partners

Design calculations for steep reinforced slopes are described. Charts for preliminary design are presented for the case of steep slopes over a competent foundation reinforced by horizontal layers of polymer grids. Design evaluations require a mix of calculations and performance criteria to ensure satisfactory behaviour. Descriptions of the calculations and criteria used to derive the design charts are given. A simple procedure to determine reinforcement spacings is introduced, based on a spacing constant Q. The relationship between the stiffness of reinforcement and the soil strength deformation behaviour influence the selection of design values. A design value of soil strength at large strain is recommended for polymer grid reinforced soil, although it is anticipated that equilibrium may be established sooner at smaller strain magnitudes.

INTRODUCTION

The availability of strong polymer reinforcement materials which are resistant to corrosion provides attractive possibilities for the use of poor and aggressive soils to build steep embankments and retaining structures.

Current knowledge and experience from reinforced earth walls, and the findings of research into the behaviour of reinforced soil with reinforcement materials other than metal, and soils other than sand, can be combined to provide design calculations for reinforced embankments. This paper presents calculations for steep reinforced embankments.

Using the calculations described in the paper, a number of design charts for steep reinforced slopes founded on a strong level foundation are presented in order to facilitate preliminary design studies.

EFFECT OF REINFORCEMENT IN SOIL

The purpose of reinforcement used to strengthen concrete, cement composites or polymers is well understood. The reinforcement is positioned to alter advantageously the stresses in the matrix material to be strengthened. Observations at model and field scale on reinforced soil walls, and on unit cell tests in the laboratory, Fig. 1, have shown that reinforcement acts to alter the pattern of stresses in soil to enable greater applied loading to be supported (Schlosser et al 1983).

The pattern of stress (and strain) in reinforced soil is complex and non-uniform, even in laboratory unit cell tests. Simplifying idealisations are required for calculation purposes. A direct idealisation that satisfies force equilibrium examines potential failure surfaces through the reinforced soil and the resultant reinforcement force acting across these, as shown in Fig 1 for laboratory tests.

The normal component of the tensile reinforcement force mobilises additional frictional resistance in the soil and the tangential component acts directly to resist applied shear loading.

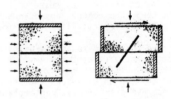

(a) Axial Compression Test
(b) Direct Shear Test

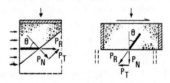

(c) Effect of reinforcement force, P_R

P_N increases normal effective stress

P_T reduces applied shear stress

Fig.1 Unit cell tests on reinforced soil (a & b) and idealisations for the effect of reinforcement (c)

Both compressive and tensile strains occur when soils are subject to shear loading, and reinforcement acts advantageously when placed in directions in which tensile strains occur. Tensile reinforcement forces are mobilised by the tensile soil strains. The amount of tensile force which may be generated in low modulus or extensible reinforcement may be governed by allowable tensile strains in the

Polymer grid reinforcement. Thomas Telford Limited, London, 1985

soil. Care is needed in design to ensure that the assumed reinforcement forces and soil shear resistance are compatible with the expected strains in the reinforced soil. This point is discussed further in the context of selecting design values for soil and reinforcement properties.

STEEP REINFORCED SLOPES

To build a slope steeper than the naturally stable angle requires additional stabilising forces. These can be provided by reinforcements placed horizontally in the slope. The reinforcements carry axial forces which increase the shear resistance of the soil and improve stability.

The case of a steep embankment built over a competent foundation is examined in this paper. The aim of design is to determine the number and disposition of reinforcements to provide overall equilibrium in the slope and to avoid any local overstressing.

The strength characteristics of reinforced soil can be highly anisotropic, and potential failure surfaces passing between reinforcement layers are as important as those intersecting reinforcement layers, Fig, 2a. The length of the reinforcement in a steep embankment must provide sufficient bond to mobilise the maximum permissible force where the reinforcement is intersected by critical surfaces in the soil. The length must also be sufficient to prevent the occurence of bodily outward sliding of the reinforced zone over an underlying layer of reinforcements, Fig. 2a.

The design problem can be expressed as the search for the spacing and length of reinforcements in the slope that provides equilibrium with

(i) a distribution of required axial force along each reinforcement that nowhere exceeds the maximum available force which may be governed either by the factored reinforcement or bond with the soil, Fig.2 b & c and

(ii) a reinforced soil zone of sufficient dimensions to adequately support the unreinforced interior of the slope.

CHOICE OF ANALYSIS

Most studies and experience with reinforced fills to date have been on compacted granular soils reinforced uniformly with metal strips and built vertically to form retaining walls and abutments, Schlosser et al (1979,1983). A mix of analyses are used for design calculations, which seems appropriate given the complex and indeterminate nature of reinforced soil.

Limit equilibrium analysis is used to calculate a gross reinforcement quantity and distribution to provide equilibrium. A plane potential failure surface is adopted in the British (1979) and French (1979) National Transportation Department codes for vertical walls.

The value of reinforcement force can be locally estimated by assuming that the product of the horizontal stress in the soil and the vertical area of soil served by the reinforcement is balanced by the reinforcement axial force, Fig. 3. The horizontal stress depends on the vertical stress and the local value of the coefficient of earth pressure.

The combination of limit equilibrium analysis and local checks on individual reinforcement spacings appears to offer a rational basis for design which can be adopted for reinforced embankments.

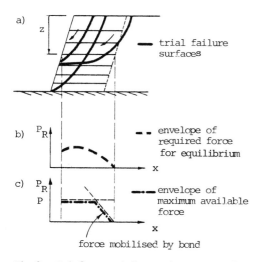

a) z trial failure surfaces

b) P_R — envelope of required force for equilibrium

c) P_R P — envelope of maximum available force

force mobilised by bond

Fig 2. Reinforcement forces in a steep slope

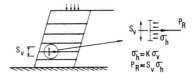

$\sigma_h = K \sigma_v$
$P_R = S_v \sigma_h$

Fig.3 Reinforcement forces calculated from local stresses

ANALYSIS FOR REINFORCED SLOPES

A limit equilibrium analysis for reinforced slopes has been developed and is performed by the computer program WAGGLE (1982). Four important aspects of the analysis are summarised below.

Two-part wedge surfaces are examined. This shape is suitable to check both mechanisms shown in Fig. 2 a. Selecting a grid of trial wedge nodes within and behind the reinforced zone, the full range of potential failure mechanisms may be investigated. Current lack of knowledge about interslice forces in reinforced soil make a simple mechanism attractive.

At each point that a trial wedge surface crosses reinforcement in the soil, the mobilised value of reinforcement force is calculated and resolved into components normal and tangential to the surface (as shown in Fig. 1) and included in the equilibrium equations.

A conventional definition of safety factor can be adopted. In this case the soil shear strength available to provide equilibrium equals the design peak strength for the soil divided by the overall safety factor. Similarly the reinforcement force in each layer intersected that is available to provide equilibrium equals the maximum design value divided by the overall safety factor. (The maximum design value of reinforcement force is governed by either strength or bond parameters depending on where along its length the reinforcement is intersected by the wedge mechanism, see Fig. 2c). Iterations are required with the conventional definition of safety factor in order to find the combination of wedge angles giving the lowest value.

The safety margin may equally well be expressed in terms of partial factors. Typically the loadings (which include surcharge, soil unit weight and pore water pressures) would be increased by partial factors, while the resistances (which include soil shear strength, reinforcement strength and bond) would be reduced by partial factors. An additional residual factor may be used to allow for the consequences of failure, uncertainties in the analytical method, etc. In this case the factored loadings and resistances would be used directly in the WAGGLE computation, and a reinforcement layout sought which provides the desired residual partial factor.

Pore water pressures are included in the overall equilibrium equations, taking account of the observations by Turnbull & Hvorslev (1967), and Whitman & Bailey (1967) on the resolution of pore water pressures. Pore water pressures also reduce the normal effective stress and hence the bond shear stress between the reinforcement and the soil.

The design of a reinforcement layout to give a desired overall safety factor for a slope proceeds by trial and error, like the design of an unreinforced slope. The total number of reinforcement layers and their spacing can be arranged to give more or less equal safety factors on mechanisms passing through the reinforced zone, and on mechanisms passing mostly through the unreinforced interior of the slope and passing out to the face between reinforcement layers, Fig 2a. Such an arrangement usually keeps to a minimum the total quantity of reinforcement required.

Comparison with published results

Results from the WAGGLE program have been compared with published solutions for unreinforced slopes. For example, the spread of results found by comparing the minimum safety factor computed by the WAGGLE program and values given by Bishop and Morgenstern (1960) for a range of slopes, soil shear strengths and pore water pressures are summarised in Fig. 4. The deviation of the calculated safety factor from the published value is shown plotted against the coefficient of interslice roughness, which lies in the range zero to unity.

Taking the coefficient of interslice roughness equal to zero for WAGGLE gives results of the order 2% to 6% on the safe side of published solutions. The coefficient was set equal to zero for the computation of design charts described below.

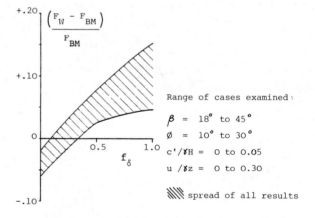

Range of cases examined:

$\beta = 18°$ to $45°$

$\phi = 10°$ to $30°$

$c'/\gamma H = 0$ to 0.05

$u/\gamma z = 0$ to 0.30

▨ spread of all results

Fig 4. Comparison of safety factor calculated by WAGGLE (F_W) to Bishop & Morgenstern (1960) published results (F_{BM}).

DESIGN CHARTS

Preliminary design investigations are greatly facilitated if solutions for standard cases can be derived from design charts, tables or equations of closed form. To this end, a chart design procedure has been developed for the preliminary design of steep slopes reinforced by polymer grids. The derivation of the design charts is described below, and the basic assumptions are listed.

An interpretation of equilibrium compatability between polymer reinforcement and soil in steep reinforced embankments is given in a final section, to provide a basis for selecting design values of parameters and safety margins.

Cases examined

The charts have been devised for the following cases, Fig. 5

- Embankment slopes built over a competent level foundation that will not be overstressed by the constructed slope.

- Uniform slope with a horizontal crest.

- Uniform surcharge w_s along the slope crest.

- Slope angles β in the range 30° to 80°.

- Soils with effective stress strength parameters in the range $c' = 0$ and $\phi' = 20°$ to 40°.

- Pore water pressures in the slope expressed by the coefficient r_u in the range 0 to 0.5 (See Bishop & Morgenstern (1960) and Fig. 5).

- Polymer reinforcement grids with constant length placed adjacently in horizontal layers. The design strength for the reinforcement grid allows for construction effects, environmental conditions in the soil and time effects on reinforcement mechanical behaviour during the design life of the structure.

Main steps for chart design

There are three main steps in the chart design procedure.

Firstly, the maximum horizontal force T required to hold the slope in equilibrium when the soil and pore water pressures are at their design values is determined. If each reinforcement layer can support a maximum force P per unit width, then the minimum number of reinforcement layers N required for equilibrium is given by the ratio T/P.

Secondly, the minimum length L for the reinforcement layers is determined so that the reinforced zone is not overstressed by pressures from the unreinforced interior of the slope, and to ensure adequate bond lengths.

Thirdly, as practical reinforcement layouts are likely to be divided into zones containing layers at an equal vertical spacing, a calculation is required to derive a practical spacing arrangement which will not lead to local overstressing in any reinforcement layer.

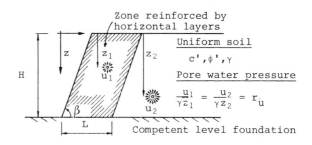

Fig.5. Definitions for the slope cases examined

GROSS HORIZONTAL FORCE REQUIRED FOR EQUILIBRIUM

The maximum horizontal force required to hold a slope in equilibrium has been estimated using the two-part wedge program WAGGLE. For a given slope, a search is made both for the worst wedge point location and the worst combination of wedge angles which give the greatest required force T, Fig. 6. This magnitude of force just provides equilibrium on the worst or critical two-part wedge when the design value of soil shear strength is fully mobilised.

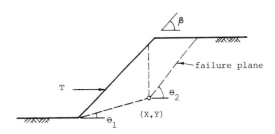

Fig 6. Definitions for two-part wedge mechanisms

The results of analyses are shown in Fig.7. The force T is plotted in a non-dimensional form against slope angle,

$$K = \frac{T}{\frac{1}{2}\gamma H^2} \qquad (1)$$

where K is the coefficient of earth pressure.

Pore water pressures and effective soil cohesion both affect the magnitude of the gross horizontal force for equilibrium. A chart of the type shown in Fig. 7, applies for fixed values of the non-dimensional parameters for cohesion $c'/\gamma H$ and pore water pressure $u/\gamma z$. The results shown in this case are for $\frac{c'}{\gamma H} = 0$ and $\frac{u}{\gamma z} = 0$.

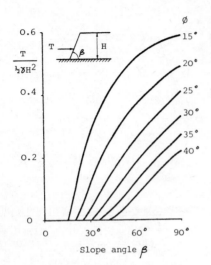

Fig 7. Gross horizontal force required
        for equilibrium in a steep
        unreinforced slope.

A check on the results can be made by comparing the calculated value of the coefficient of earth pressure K for slopes with the limit analysis results published by Chen (1975). For the case of inclined retaining walls, the results cover the cases shown in Fig. 7 for slope angles in the range 50° to 90°. The values of K calculated by WAGGLE are typically 5% greater than the values given by Chen for smooth walls, except for the vertical case where the coefficients are equal. This finding bears out at steep slope angles the finding for flatter slopes that the selection of an interslice roughness coefficient equal to zero for WAGGLE gives results that are slightly conservative.

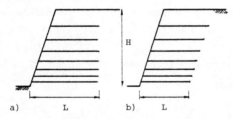

Fig 8. Reinforcement truncation -
        a) at a vertical line
        b) parallel to slope face

REINFORCEMENT LENGTH

Two simple arrangements for reinforcement result from either truncating reinforcement layers to a common vertical line in the slope, or by keeping the length constant throughout the slope, Fig. 8. Both methods give the same result for the vertical case.

Comparisons were made between these two layouts for a number of slope cases using WAGGLE, with

the finding that parallel truncation generally gave a more efficient use of reinforcement, a smaller quantity to provide the desired safety factor against all potential failure mechanisms. The use of reinforcement layers all of a constant length Fig.8b has been selected for the chart designs.

Three main criteria were identified which govern the selection of the minimum allowable reinforcement length. The three undesirable consequences envisaged if there is insufficient reinforcement length are,

1. Reinforcement layers near the top of the slope have insufficient length to support the required design forces which are shed to lower layers resulting in overstressing, Fig. 9a.

2. The reinforcement length is insufficient to prevent outward sliding along the interface between the soil and a reinforcement layer, Fig. 9 b.

3. The reinforced zone, acting as a rigid block, is not wide enough to resist the outward thrust of the unreinforced soil in the slope interior without developing tensile vertical effective stresses anywhere along its base, Fig. 9 c.

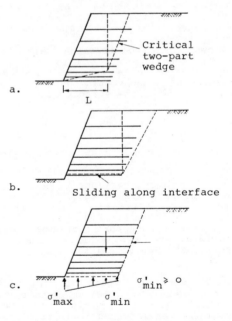

Fig.9. Criteria to determine the minimum
       allowable reinforcement length

Overstressing of reinforcement layers

The maximum tensile force which may be generated by frictional bond over a length of reinforcement is conventionally estimated from the product of the surface area of the reinforcement over which bond develops, the average normal effective stress acting on this area and an angle of friction that can be generated between the reinforcement and the soil. For grid reinforcement it is convenient to treat the whole plane surface area to be available for bond, selecting an appropriate angle of friction to represent overall bond shear stress. At best, the reinforcement grid may be as effective as a rough sheet in which case the bond angle of friction would equal that of the soil, (Jewell et al, 1984). For the design chart calculations the reinforcement bond angle of friction was assumed conservatively to equal half the design friction angle for the soil.

For a range of slope cases the profiles of required reinforcement force in the reinforcement layers were calculated with a variety of assumed reinforcement lengths. This showed that when the reinforcement was shortened so that the critical two-part wedge mechanism could pass outside the reinforced zone near the top of the slope, the reinforcement layers lower in the slope were required to carry proportionally greater forces, leading to overstressing. The criteria adopted to control overstressing was that the reinforcement length should always be sufficient to contain the critical two-part wedge, which is the mechanism in the slope requiring the largest gross horizontal force to maintain equilibrium.

The position of the node and the back wedge angle θ_2, Fig.6, define the location of the critical mechanisms found previously when calculating the coefficient of earth pressure values. The minimum ratio of reinforcement length to slope height to contain the worst mechanism is shown plotted against slope angle in Fig.10 for the case of no cohesion and no pore water pressures.

Outward sliding

Reinforcement grids provide good resistance to outward direct sliding by soil along the interface between the grid and the overlying or underlying soil (Jewell et al, 1984). Resistance to direct sliding includes the contribution of soil shear resistance mobilised by the shear of soil over soil through the area of the reinforcement grid apertures. The resistance to direct sliding over grid reinforcement may often be close to the full shear resistance of the soil. For design chart calculations the value of frictional resistance to direct sliding over grid reinforcement was assumed equal to eighty percent of the design friction angle for the soil.

For a layout of reinforcement with layers of constant length, the worst case for outward sliding occurs over the lowest reinforcement layer, Fig. 9 b. The required reinforcement length to provide equilibrium on the most critical direct sliding mechanism was calculated using WAGGLE, and results for the case of no cohesion and no pore water pressures are shown in Fig. 11.

No tension on the base

A simple analysis assuming the reinforced zone to act as a rigid block has been carried out. The forces acting on the rigid block are self weight and the outward thrust from the unreinforced slope interior, Fig 9c. The magnitude of outward thrust, the point of action on the reinforced block and the roughness on the back of the reinforced block are all unknown. For design chart calculations

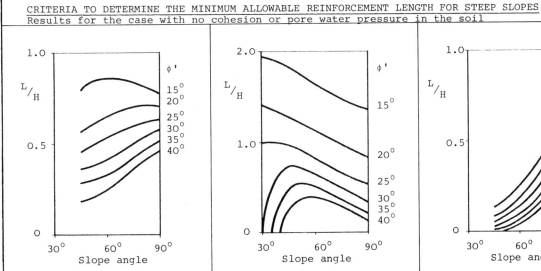

CRITERIA TO DETERMINE THE MINIMUM ALLOWABLE REINFORCEMENT LENGTH FOR STEEP SLOPES
Results for the case with no cohesion or pore water pressure in the soil

Fig. 10. Minimum length to contain the critical two-part wedge mechanism

Fig.11. Minimum length to prevent outward sliding

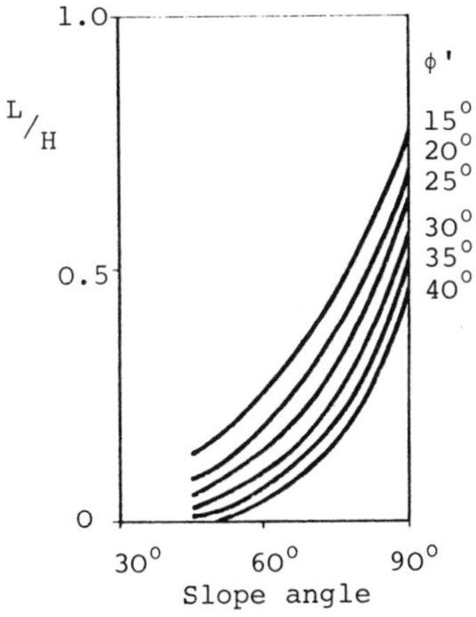

Fig.12. Minimum length to prevent tensile effective stresses on the reinforced zone base

the outward thrust was assumed to equal T and assumed to act horizontally on the reinforced block at one third of the slope height above foundation level.

The forces on the reinforced block were resolved to give vertical, horizontal and momement components acting at the centre of the base. Each of these forces was assumed to be balanced by a uniform distribution of stress between the reinforced block base and the foundation surface. The minimum ratio of reinforcement length to slope height to avoid effective tensile stresses occuring on the base of the reinforced block is shown in Fig. 12, for the case of no cohesion and no pore water pressure.

Composite charts for reinforcement length

Results for the three criteria were combined on one chart to give the required reinforcement length, expressed as the ratio L/H, as a function of slope angle β, for a range of design soil friction angle values. Charts are given in the appendix for the cases of zero soil cohesion $c'/\gamma H = 0$ and pore water pressures $u/\gamma z = 0$, 0.25 and 0.50.

PRACTICAL REINFORCEMENT SPACING

An aim of design should be to distribute reinforcement layers within the slope so that no zone of soil or layer of reinforcement is unduly stressed in comparison to the remainder. If reinforcement were to be locally overstressed, mobilising axial tension is excess of the design value, then the risk of rupture in the reinforcement which could lead to progressive collapse of the reinforced slope would be greatly increased.

The concept of local stress equilibrium in reinforced soil slopes is illustrated in Fig. 3. Locally, a continuous reinforcement layer at a vertical spacing S_v could be required to hold in equilibrium the horizontal soil stresses σ_h, in which the case the local mobilised value of reinforcement force P_R would be,

$$P_R = S_v \sigma_h = S_v K \sigma_v \qquad (2)$$

For steep slopes where the vertical stress approximately equals the overburden pressure, the reinforcement layers would carry approximately equal forces P throughout the slope if the vertical spacing was varied as the inverse of depth. For $v = z$, and $P_R =$ constant value P, from (2),

$$S_v = \frac{P}{K\gamma z} \qquad (3)$$

A separate deduction based on the results of limit equilibrium analysis leads to a similar conclusion.

The gross horizontal force that the reinforcement must provide to maintain equilibrium in the slope can be described by equation (1). For a slope, the increase in required horizontal force for equilibrium can be found by differentiating equation (1) with respect to depth z to give,

$$\frac{dT}{dz} = K\gamma z \qquad (4)$$

If the increased requirement for force is provided by reinforcement layers each able to support a constant force P, placed at a variable vertical spacing S_v, then to satisfy equilibrium the reinforcement spacing would have to be varied, from (4), as

$$\frac{P}{S_v} = K\gamma z \qquad (5)$$

which reduces to equation (3).

Finally, results from WAGGLE analysis of reinforced slopes with uniformly spaced reinforcement layers typically give higher than average values of safety factor on mechanisms in the upper parts of the slope, and lower than average values on more deep seated mechanisms passing through the reinforced zone lower in the slope.

With the current lack of knowledge about the effect of reinforcement spacing on reinforcement forces and overstressing, it would be prudent to space reinforcement layers on the assumption that each layer may locally have to support the horizontal stresses in the soil as shown in Fig. 3, and described by equation (2). The maximum expected local force in the reinforcement would have the value,

$$P_R = S_v K\gamma z \qquad (6)$$

Thus the ideal spacing arrangement to give a balanced distribution of reinforcement in the slope is when for each layer the value of maximum force calculated from equation (6) equals the design value of reinforcement force P. This is shown diagramatically in Fig. 13.

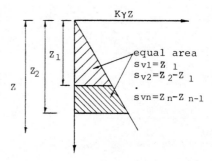

Fig.13 Change of vertical spacing with depth to equally load reinforcements.

The calculation of spacing arrangements for the reinforcement is simplified by defining a spacing constant Q for the slope, in terms of the minimum spacing v to be used,

$$Q = \frac{P}{K \gamma v} \qquad (7)$$

Using Q, the relationship between the maximum allowable spacing and depth z below the crest of the slope can be reduced to a line of gradient unity in the non-dimensional form shown in Fig. 14.

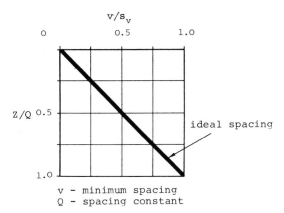

Fig.14 Variation of spacing s_v with depth to load reinforcement layers equally.

In practice, the available spacings may be selected as multiples of the compaction layer thickness. Setting this thickness equal to v, the maximum depth to which reinforcement spaced at v, 2v, 3v...nv may be used can be calculated from the equation,

$$Z_{nv} = \frac{Q \, v}{nv} = \frac{Q}{n} \qquad (8)$$

To use reinforcement efficiently, the largest allowable spacing should be used at any depth, to give the step wise distribution shown in Fig. 15.

A tabular procedure can be used to calculate the numerical values of reinforcement layer depths below the slope crest. Start at the bottom of the slope (depth H) and place the first reinforcement layer at foundation level. The number of layers in the first zone of equal spacing can be calculated by dividing the thickness of the zone by the required spacing of the reinforcement. The result is unlikely to be a whole number, and it is sufficient to round down to the nearest integer number of layers, and add the remaining thickness to the overlying zone, repeating the process to the top of the slope.

Allowance for vertical surcharge

Vertical surcharge w_s on the slope crest is treated as an additional thickness of fill, and an increased slope height H' used for design chart calculations,

$$H' = H + \frac{w_s}{\gamma} \qquad (9)$$

When calculating the reinforcement spacing by the tabular procedure described above, the depth of fill representing the surcharge should be subtracted from the thickness of the zone of equal spacing at the top of the slope.

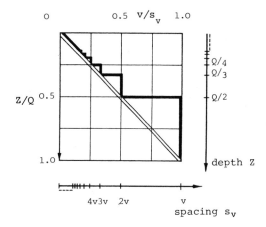

Fig.15 Zones for reinforcement layers spaced equally at v,2v,3v.....nv.

SOIL AND POLYMER GRID PROPERTIES FOR DESIGN

A vertical stress in soil can give rise to an inclined shear stress and a horizontal tensile strain. If the shear stress is insufficient to maintain equilibrium in a soil slope, the introduction of horizontal reinforcement may prevent collapse. If there is no slippage between the soil and the reinforcement, the soil will strain until it produces the force needed in the reinforcement to attain equilibrium.

The relationship between soil shear stress and horizontal strain can be derived from triaxial or plane strain tests. Limit equilibrium analysis can give the horizontal out of balance stress in the soil. Thus the force needed in the reinforcement can be expressed as a function of strain in the zone of maximum shear stress.

The slope of the load/extension line (the stiffness) of the reinforcement and the degree of compaction of the sand, together with the amount of reinforcement, will determine whether equilibrium is achieved before the sand has strained to give its peak strength or whether equilibrium will depend upon the sand strength at large strain. When choosing values of soil and reinforcement properties and factors of safety for use in design, consideration should be given to the relationship between the stiffness of the reinforcement and the soil strength characteristics.

Taking the example of the polymer grid Tensar SR2 at 20°c, the work of McGown et al (1984) indicates that at creep loads lower than 29 kN/m the extension will be proportionally less than 10% within 120 years. Extensions in the range of 4 to 7% might be expected depending on the design force P. However, the loading on the reinforcement is unlikely to be constant throughout its working life.

The reinforcement will be loaded during construction, first by the compaction effort and subsequently by the height of fill and any surcharge. Equilibrium will be established by the soil tending to shear and thus extending the reinforcement until the balancing force is provided. The initial response of Tensar SR2 reinforcement can be estimated from the isochronous creep curve for a time corresponding to the end of construction McGown et al (1984). The loaded grid will extend further during the life of the structure allowing an increase in horizontal strain in the soil. Generally this will increase the shearing resistance of the soil and reduce the balancing force needed in the geogrid. At worst the soil will reach its angle of shearing resistance with tensile strains of the order 10% and the grid would subsequently experience pure creep.

At this stage in the development of reinforced soil design, prudence dictates that for polymer reinforcement grids, the soil strength at large strain should be used to check equilibrium.

The design force for the polymer reinforcement should take account of construction effects, environmental conditions in the soil and the effects of time on mechanical properties during the design life of the structure.

An additional overall safety factor of a conventional magnitude 1.3 to 1.5 would provide for the uncertainties of construction alignments, unexpected external or pore water pressure loadings and inadequacies in the current understanding of reinforced soil and its analysis.

SELECTION OF DESIGN VALUES

Design values for parameters should be used directly with the chart design procedure. Guidance on the selection of design values is given below.

Geometry

The slope height and slope angle can usually be well defined and the expected values used for design.

Loads

The maximum expected value of loads which can occur simultaneously should be taken. The soil unit weight, pore water pressures and surcharge loading may all attain their maximum values at the same time.

Soil strength

A large strain value of soil strength would be appropriate for design with polymer grid reinforcement, and the critical state friction angle ϕ'_c may be used. For long lived clay fills it would be appropriate to assume zero effective cohesion.

Polymer reinforcement strength

The strength appropriate to service conditions in the ground at the end of the design life of the structure would be suitable for design. The value should apply to the most severe conditions for the reinforcement that are expected.

Overall safety factor

A conventional overall safety factor of the order 1.3 to 1.5 would be appropriate. For simplicity with the chart design procedure, the overall safety factor can be applied to the design value of reinforcement strength to give a factored design force.

CONCLUSIONS

Limit equilibrium analysis and local stress calculations may be used together for the design of reinforced soil slopes. A number of criteria can be established to ensure the satisfactory performance of reinforced soil. A set of design charts for reinforced slopes have been developed together with a procedure for their use, and these are summarised in the appendix. The shear strength of the soil at large strain, and the strength of the polymer grids in the soil environment at the end of the design life are suggested as design values to be used with a conventional safety factor.

ACKNOWLEDGEMENTS

The two first named authors have benefited from membership of both the Soil Reinforcement Design Group and the main Steering Committee coordinating the Science and Engineering Research Council's cooperative research project with Netlon Limited into the Civil Engineering applications of polymer reinforcement grids. The work on design charts was carried out as a part of this programme. Grateful acknowledgement is given to the members of these two committees for many stimulating discussions.

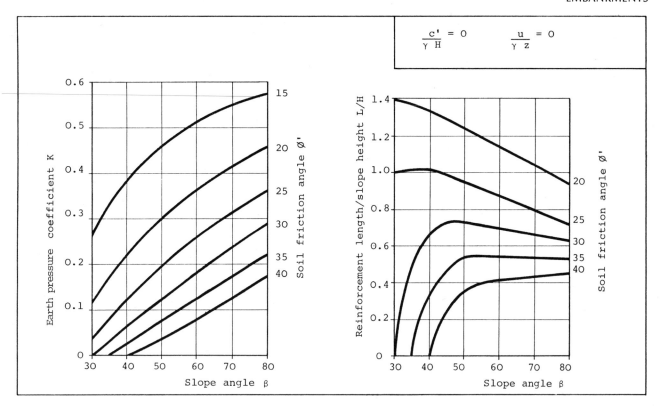

$$\frac{c'}{\gamma H} = 0 \qquad \frac{u}{\gamma z} = 0$$

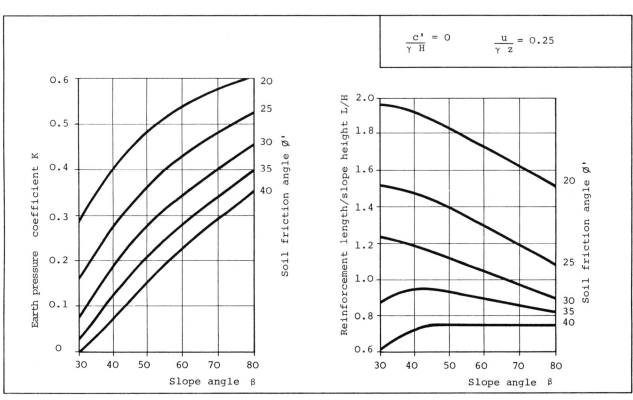

$$\frac{c'}{\gamma H} = 0 \qquad \frac{u}{\gamma z} = 0.25$$

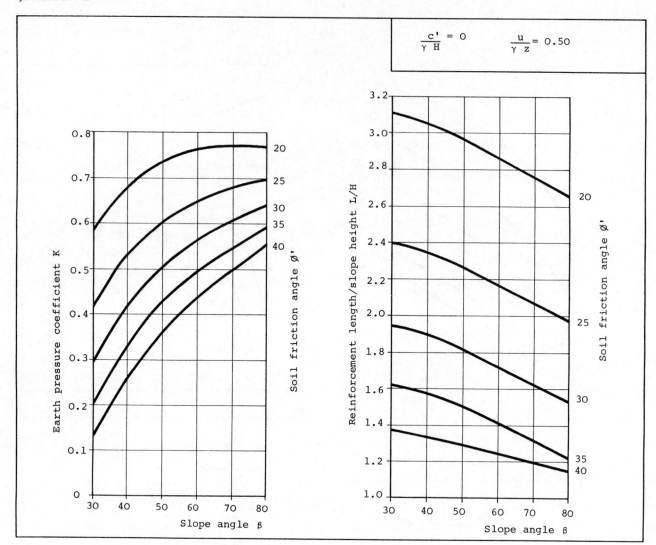

$$\frac{c'}{\gamma H} = 0 \qquad \frac{u}{\gamma z} = 0.50$$

CHART DESIGN PROCEDURE

1. Select the required embankment dimensions and surcharge loading

2. Select design values of soil properties and pore water pressures

3. Determine earth pressure coefficient K and length of reinforcement L from the charts

4. Choose "in soil" design strength properties for the reinforcement and an overall factor of safety

5. Obtain the factored reinforcement force P

6. Choose a minimum vertical reinforcement spacing and the spacing constant $Q=P/K\gamma v$

7. Perform tabular calculation for the number and spacing of reinforcement layers

8. Calculate the total horizontal force required for equilibrium $T=\tfrac{1}{2}K\gamma H^2$

9. Check $T/(\text{number of layers}) \leqslant P$

REFERENCES

Binnie & Partners (1982). WAGGLE - A computer program for the stability analysis of reinforced soil slopes and embankments. Program Manual.

Bishop, A.W. and Morgenstern, N (1960). Stability coefficients for earth slopes. Geotechnique, 10, 129-150.

Chen, W.F., (1975). Limit analysis and soil plasticity, 341-398. Elsevier, New York.

Department of Transport (UK) (1978). Reinforced earth retaining walls and bridge abutments. Technical memorandum (bridges) BE 3/78. London

Jewell, R.A., (1980). Some effects of reinforcement on the mechanical behaviour of soils. PhD Thesis, University of Cambridge.

Jewell, R.A., Milligan, G.W.E., Sarsby, R.W. and Dubois, D. (1984). Interaction between soil and geogrids. Proc. Symp. Polymer Grid Reinf. In Civ. Engng - London.

McGown, A., Andrawes, K.Z., Yeo, K.C. and Dubois, D. (1984). The load-strain-time behaviour of Tensar Geogrids. Proc.Symp. Polymer Grid Reinf. in Civ. Engng.. London

McGown, A., Paine, N. and Dubois, D. (1984). The use of geogrid properties in design. Proc.Symp. Polymer Grid Reinf. in Civ. Engng.. London

Ministere des Transports (1979). Les ouvrages en Terre Armee. Recommendations et regles de l'art. LCPC, Service d'etudes techniques des routes et autoroutes.

Schlosser, F., Jacobsen, H.M., and Juran, I. (1983) Soil reinforcement. General report. Proc. 8th Eur. Conf. Soil Mech. Fndn. Engng., Vol 3, Helsinki.

Schlosser, F. and Juran, I. (1979). Design parameters for artificially improved soils. General report. Proc. 7th Eur. Conf. Soil Mech. Fndn. Engng., Vol 5, Brighton.

Turnbull, W.N., and Hvorslev, M.J. (1967). Special problems in slope stability. ASCE J. Soil Mech. Fndns. Div., Vol 93, SM4, July, 499-528.

Whitman, R.V., and Bailey, W.A. (1967). The use of computers for slope stability analysis. ASCE J. Soil Mech. Fndns. Div, Vol 93, SM4, July, 475-498.

NOTATION

c'	effective cohesion of soil
f_δ	coefficient of inter-wedge roughness
H	height of slope
K	coefficient of earth pressure
L	length of reinforcement
N	minimum required number of reinforcement layers
n	integer number
P	factored reinforcement strength (design value)
P_R	reinforcement force
Q	spacing constant for reinforced slope
r_u	coefficient of pore water pressure
s_v	vertical spacing between reinforcement layers
T	gross horizontal force required for equilibrium
u	pore water pressure
v	minimum vertical spacing of reinforcement
w_s	vertical surcharge loading
β	slope angle measured from horizontal
γ	unit weight of soil
φ'	effective angle of friction of soil
σ	stress in soil
θ	angle of wedge surface from horizontal

Geogrid reinforced earth embankments with steep side slopes

Synthetic tensile elements to reinforce earth fills have been used successfully in the past in U.K. and Japan though not in Canada. For a Provincial highway project in Brampton, Ontario, property acquisition costs and other problems necessitated the design of retaining structures or steep side slopes for the 7 m high embankment required. Cost benefit studies showed that steep side slopes reinforced with synthetic tensile elements were considerably cheaper than other alternatives. This method was therefore selected. The paper describes the successful design and construction in 1983, of the reinforced earth slope using Tensar SR2 as the main reinforcement. Instrumentation to be installed for monitoring purposes is also described.

M. S. Devata, *Ontario Ministry of Transportation and Communications*

INTRODUCTION

This paper describes the design and construction of a 7 m high expressway embankment by the Ontario Ministry of Transportation and Communications, with 1:1 slopes using a technique hitherto not tried out in Canada. Since the highway is located in an urban area of Southern Ontario where land costs are extremely high, it was necessary to minimize property requirements in order to achieve an economical design. Various types of retaining structures were considered together with methods of constructing steeper-than-normal side slopes for the embankment. The paper provides cost comparisons between these various alternatives and from these it can be seen that the most economical method by far is one that utilizes synthetic tensile elements to reinforce the embankment material thus enabling very steep stable slopes to be constructed. This technique had been used frequently in the U.K. and Japan though not at all in Canada. This was the method selected for the east side of the highway and overall savings of about three quarters of a million dollars were thereby realized for an expressway length of 1.0 km. On the west side of the expressway standard 2 horizontal to 1 vertical side slopes were found to be economically advantageous. Construction of the project commenced in June, 1983 and about half of the work was completed by the end of August, 1983. The remainder of the project will be constructed in 1984 or later. The paper discusses the important aspects of the design considerations and construction techniques of the reinforced slopes and also the instrumentation programme developed to monitor future performance, which will be installed in the portion of the project yet to be built.

DESCRIPTION OF SITE

Topography: The site is located in the east portion of the City of Brampton, Ontario, Canada, where Highway 410 (Brampton By-Pass) has been under construction since 1982. At the location in question, the route traverses a newly developed heavily industrialized area in which the terrain is generally fairly level. The grade of the highway is such that embankments of about 7 m in height are required.

Subsurface Conditions: Subsurface conditions in this area are generally favourable from the foundation point of view. The overburden consists of a very stiff to hard glacial deposit of silty clay with sand and occasional gravel overlying shale bedrock. Groundwater is about 2 m below the ground surface.

Physical properties of the overburden are summarized as follows:

Grainsize distribution (%)	Gravel:	2- 5
	Sand:	16-31
	Silt:	48-55
	Clay:	19-25
Liquid Limit w_L (%)		23-30
Plastic Limit w_P (%)		13-19
Water Content w (%)		12-17
Unit Weight γ (kN/m^3)		19-22
SPT 'N' Values (blows/0.3m)		16-100

Table 1 - Physical Properties

DESIGN OPTIONS AND CONSTRAINTS

The proximity of the east side of the highway to recently constructed industrial buildings was such that standard side slopes of 2 horizontal to 1 vertical could not be constructed without the acquisition of private property at great expense which would probably have resulted in expropriation proceedings with consequent lengthy delays. The flattest slope which could be constructed without acquiring new property was 1:1. It was necessary therefore to consider the various options open to the Ministry in some detail to determine the most economical solution.

Polymer grid reinforcement. Thomas Telford Limited, London, 1985

The options were:

i)   to construct an earth retaining structure which could be built without acquiring new property.

ii) to construct a rockfill embankment with side slopes of 1-1/4 horizontal to 1 vertical which would require some additional property and

iii) to construct a 1:1 side slope of earth fill reinforced with synthetic tensile elements.

This latter method had never been used in Canada but had been used in the U.K. and Japan. The following table shows a comparison of the costs of the various alternatives considered including property costs for the east side of the highway for a total length of 1.0 km.

Construction Method	Construction Costs	Property Costs	Total Costs
Reinf. Conc. Wall	1.52	Nil	1.52
Bin Wall	1.42	Nil	1.42
Reinf. Earth Wall	1.38	Nil	1.38
1-1/4:1 Slope (Rockfill)	0.93	0.30	1.23
1:1 Slope (Reinforced with Tensar)	0.48	Nil	0.48
2:1 Slope (Earthfill)	0.30	0.90	1.20

Table 2 - Cost Comparisons (million dollars)

From the above table, it is readily apparent that a 1:1 earth slope reinforced with synthetic tensile elements is the most economically advantageous option and indicates savings of $720,000 over its nearest rival. Research and inquiries carried out by the Ontario Ministry of Transportation and Communications indicated that the reinforced earth slope method was technically feasible particularly since many actual cases are on record in the U.K. and Japan. There remained, however, one partially unanswered question, that of the performance where temperature extremes ranged from -35°C to +40°C such as is possible at the site under consideration. The synthetic material to be used as reinforcement is called Tensar SR2 which is manufactured by Netlon Limited, U.K. Laboratory tests performed by Netlon on this material indicated satisfactory performance in freeze-thaw cycle ranges of -20°C to +40°C however, we were not aware of any records detailing the actual performance of Tensar installed in soil subjected to these temperature variations. It was therefore decided that the frost penetration zone of soil to be reinforced would consist of well graded granular material (MTC Granular 'A') which would not be susceptible to frost action.

DESIGN DETAIL

For convenience the design of the reinforced slope is separated into the three most important aspects being:

1)   External Stability

2)   Internal Stability

3)   Surface Erosion Protection

Each aspect is discussed under the appropriate heading.

External Stability: Determination of the external stability of the slope requires that analyses be carried out to ensure adequate safety factors against deep seated base failure of the foundation soil or sliding failure along the surface of the foundation soil. As indicated previously, foundation conditions are excellent at this site and stability analyses resulted in satisfactory safety factors for the overall (or external) stability.

Internal Stability: The internal stability of a soil mass reinforced with tensile elements is governed by the physical properties of the soil together with the strength, spacing and length of the tensile elements. Lateral strains within the mass are resisted by friction between the soil and the reinforcing elements. Design is therefore a matter of determining the strength and spacing of these elements for the particular type of fill material to be used and its geometry. Required stability analyses are those in which local stability of the soil near a single reinforcement strip is considered and those which consider the overall stability of wedges of soil within the reinforced zone.

Reinforcement Requirements: Properties of the fill to be used in the reinforced zone of the embankment and the properties of the reinforcing elements are given below. As noted previously, the outer 1.2 m of soil within the frost penetration zone was to consist of well graded granular material (MTC Granular 'A'). The main fill was to be silty clay with sand and gravel (glacial till).

Main Fill:
Optimum Moisture (%)	11
Max. Dry Density (t/m^3)	2.1
Effective friction angle ϕ'	31°
Effective cohesion c'	0
Liquid Limit w_L (%)	18
Plastic Limit w_P (%)	14
Water Content w (%)	9
Friction Coefficient (Soil and SR2)	0.8 tan 31°

Outer Fill: Optimum Moisture (%)   8
(Gran. 'A') Max. Dry Density (t/m^3)   2.2
Effective friction angle ϕ' 35°
Effective cohesion c'   0
Friction Coefficient   0.8 tan 35°
(Soil and SR2)

Tensar SR2: Ultimate Tensile strength (kN/m) 79
Permissible tensile force (kN/m) 32
(40%) of Ultimate)

Geometry: East slope (H:V)   1:1
West slope (H:V)   2:1

Assuming the above mentioned properties of soil and reinforcing elements and a final slope of 1:1 for the design, it was first of all necessary to determine the limits of the zone to be reinforced. Since the reinforcement provides internal stability to this zone, it can be assumed to behave monolithically and should have an adequate overall factor of safety against failure. In this case a safety factor of 1.5 was deemed to be sufficient and a series of stability analyses showed the zone of reinforcement required to be 11 m back from the toe of slope for an effective fill height of 10 m. To determine the required vertical spacing between tensile elements, an analysis developed by Sarma (1973) was used to investigate two-part wedge failure surface within and around the reinforced zone. This work was carried out by Golder Associates who were retained by Tensar Incorporated. By this means the required spacing to achieve a factor of safety of 1.3 for internal stability was determined for various embankment heights. Details of this requirement are shown in Figure 1.

Finally the forces in the reinforcement were checked to ensure that they did not exceed the limit of 32 kN/m (40% of ultimate) also embedment lengths were checked to ensure a factor of safety of 2.0 against pull-out. To provide additional internal strength in the outer granular fill (within the frost penetration zone) additional reinforcing elements of Tensar SS1 were provided. These were 1.5 m in length and the vertical spacing was 250 mm.

Surface Erosion Protection: An essential part of the slope design was to provide a finished surface slope adequately resistant to erosive forces such as rain, snow and frost. It was decided that a 1:1 slope could support a grass vegetation cover provided that the surface soil could be stabilized until the formation of an adequate root mat. This Ministry has experimented with and cultivated a mixture of different types of grass which has a good appearance and which provides dense growth in a short period of time. It provides excellent protection against erosion and has the additional advantage of being relatively short in length thus not requiring periodic mowing (which would be difficult on the 1:1 slopes). To ensure stability of the surface soil until adequate vegetation growth took place, the surface of

the slope was covered by a mesh of Netlon CE111 placed directly on top of a thin layer of topsoil 50 mm thick. Further details of the seeding and mulching operations are given under the heading Construction Details below.

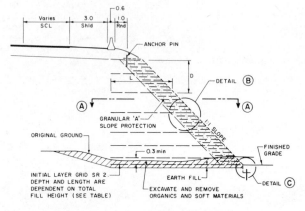

DIMENSION CHART FOR GRID SR 2			
L	D	L	D
2.5	0	8.0	5.5
3.5	1.0	9.0	6.5
5.0	2.5	10.0	7.5
6.5	4.0	11.0	8.5

NOTE:
ANCHOR PINS FOR CE III SHALL BE 300mm LONG, SPACED 1.0 m VERTICALLY AND 1.85 m HORIZONTALLY (ON OVERLAP). ANCHOR PINS SHALL HAVE A STAPLE TYPE HOOK TO HOLD THE MESH.

TYPICAL DETAIL— EARTH REINFORCEMENT OF 1:1 SLOPE

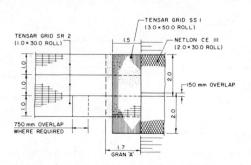

PLAN SECTION A-A

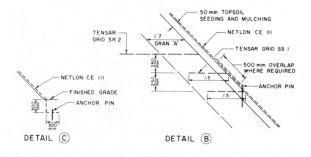

DETAIL C   DETAIL B

Fig. 1.

CONSTRUCTION DETAILS

As discussed earlier, three different tensile elements were incorporated in the construction of the reinforced embankment. Tensar SR2 was used in the fill to provide overall external stability. For internal strength, Tensar SS1 was utilized strictly in the granular frost protection zone. To protect against surface erosion, Netlon CE111 was placed directly on the embankment surface. The details of the reinforced earth embankment are illustrated in Figure 1.

During construction of the embankment, SR2 geogrids were placed horizontally in the silty clay fill. The design depths and lengths are included in the "Dimension Chart" in Figure 1.

The SR2 geogrids were spaced at 1.0 or 1.5 m vertically, while the SS1 elements were placed horizontally between the SR2 elements at a vertical spacing of 250 mm. Unlike the SR2 geogrids, however, the SS1 elements only extended 1.5 m in the 1.7 m granular embankment cover and were not continued into the silty clay.

Upon completion of the reinforced embankment, Netlon CE111 was placed directly over the topsoil as a control for slope erosion. The slope was then seeded.

Prior to the commencement of the embankment construction, the area beneath the reinforced fill was prepared. All organic and loose material was excavated by means of scrapers and removed from the site. The excavated area was then levelled transversely to the highway alignment to match the design elevation of the toe of the fill.

Tensar SR2 was the first geogrid placed on the prepared surface and also the last placed on the surface of the subgrade. The second and subsequent geogrid layers, either SR2 or SS1 were placed on well compacted earth and granular fill. No construction equipment was permitted to travel directly on the tensile elements until a layer of fill of minimum uncompacted thickness of 250 mm was placed.

It is to be noted that the silty clay fill and the granular fill cover were constructed simultaneously using scrapers for the main fill and a front-end loader for the outer 1.7 m granular zone. Both materials were compacted using vibratory rollers. Very little difficulty was encountered in maintaining the 1:1 slope of the granular zone since sufficient moisture was provided during the compaction operation.

Tensar SR2 geogrid was placed in 1.0 m abutting strips transversely in the fill. As indicated in the "Dimension Chart" in Figure 1, the geogrid extended horizontally from the face of the slope to a length varying with depth.
Tensar SS1 geogrid was placed horizontally between the SR2 elements at 250 mm vertical spacings. The geogrids were placed in such a way as to coincide with each second compacted

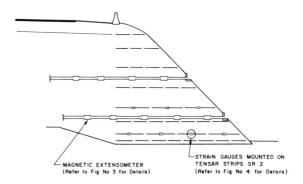

Fig. 2. Proposed instrumentation scheme

MAGNETIC EXTENSOMETER
(Refer to Fig No 3 for Details)

STRAIN GAUGES MOUNTED ON
TENSAR STRIPS SR 2
(Refer to Fig No 4 for Details)

lift. The 1.5 m SS1 tensile element was placed longitudinally in the granular zone. A small lip was formed of the SS1 at the face of the slope to provide direct contact with the overlying Netlon CE111.

Geogrid orientation and lapping requirements are shown in the "Plan Section" and "Detail B" in Figure 1.

Upon completion of the embankment fill, some difficulty was experienced in keeping the 50 mm loose topsoil cover on the 1:1 slope. This resulted in having to compact the topsoil on the slope - an action which is not often practised as it impedes the germination of grass seeds.

After the topsoil was applied, Netlon CE111 mesh was placed on the slope surface in strips of 2.0 m overlapping 150 mm with adjacent strips. The Netlon geogrid was anchored in place by using wooden pins at specified intervals. In addition, the mesh was anchored at the toe of slope as illustrated in "Detail C" in Figure 1.

The final operation in the construction of the reinforced embankment was carried out by the Ministry of Transportation's own forces who are experienced in seeding methods.

The slope was seeded utilizing a Ministry owned hydroseeder and at the same time was covered with a hydraulic mulch and tackifier to hold all of the ingredients in place. Successful germination was observed three weeks after the initial application and the area was totally covered with dense grass vegetation eight weeks after the initial seeding operation.

FUTURE MONITORING

The Ministry has decided to implement an instrumentation program for the portion of road still to be constructed. The intention is to measure deformations within the fill material by means of multipoint magnetic extensometers so that comparisons may be made between the reinforced 1:1 slope on the east side and the non-reinforced 2:1 slope on the west side. In

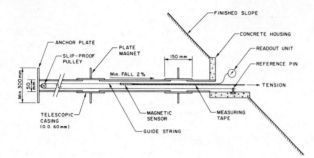

Fig. 3. Section of magnetic extensometer

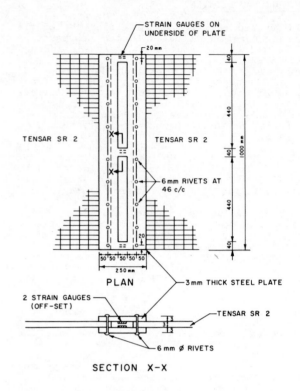

Fig. 4. Strain gauge assembly

addition, it is intended by means of strain gauges to determine the actual stresses developed in the reinforcing elements so that an evaluation may be made of the parameters assumed for the design. Figure 2 shows details of the instrumentation to be installed. A description of the extensometers and strain gauges to be used is as follows:

Multipoint Magnetic Extensometers: To compare deformation characteristics between the reinforced and non-reinforced section of the embankment, two multipoint magnetic extensometers will be installed along a selected section of each side of the embankment. To achieve the greatest accuracy with respect to horizontal movements, installation of the IRAD gauge FMC multipoint magnetic extensometer system has been decided. Basically the system consists of anchors containing magnets which can slide over an access tube. The access tube with up to twenty anchor points can be placed in the soil or attached to the geogrid. Readings are taken by inserting the flexible sensor in the access tube and connecting it to the read-out instrument. The distance either from the first anchor to each succeeding one, or the distances between each succeeding pair of anchors are obtained directly with the read-out instrument, reading to 0.025 mm. The system in the horizontal mode can be easily installed without delay to construction.

The detailed schematic section of the multipoint magnetic extensometer is shown in Figure 3.

Strain Gauge Assembly: To understand how the reinforcement works, it will be necessary to measure the strains occuring in the geogrid, as well as the strains occuring in the soil. It was decided that high elongation strain gauges, with adequate protection for long term buried conditions be used. The strain gauges will be mounted on slotted aluminum plates which will be connected to geogrid strips by means of rivets as shown in Figure 4. It is necessary prior to field installation that the strain gauges mounted to the slotted aluminum plates and attached to the geogrid strips be calibrated at different load and temperature ranges in the laboratory.

CONCLUSIONS

From our experience to-date with the reinforced earth project in Brampton, the Ministry has been able to draw the following conclusions:

(1) Where subsurface conditions are favourable and property acquisition is a major concern, considerable economy can be effected by using steep reinforced earth embankment slopes.

(2) The design methods used were partly empirical and partly analytical and presented no major difficulties to the designers.

(3) Construction was relatively straight forward, and required no special training on the part of contractor's staff.

(4) Performance to-date has fulfilled all expectation and no distortions of the slope surface have occurred. Protection from erosion using the method designed by the Ministry's Maintenance Branch has been excellent.

ACKNOWLEDGEMENTS

This paper is presented with the permission of Mr. H. F. Gilbert, Deputy Minister, Ontario

Ministry of Transportation and Communications. The author is indebted to Mr. D. R. Brohm, Manager, Engineering Materials Office for his encouragement and to Mr. K. G. Selby, of the Foundation Design Section for his valuable assistance and suggestions during the preparation of this paper.

The assistance of Mr. G. Petruzziello in the preparation of figures and Messrs. R. M. Dell, H. Szymanski and L. Politano in the preparation of construction and instrumentation details is also acknowledged.

The author is particularly grateful for the advice and suggestions given by Netlon Limited, The Tensar Corporation and their consultants Golder Associates in matters relating to the design.

REFERENCES

Dubois, D. (1982), The variation in the tensile strength of Tensar SR when subjected to temperature cycles over the range 40°C to -20°C. Netlon Report #16

Jones, C.J.E.P. (1977), Practical design considerations. Symposium of reinforced earth, T.R.R.L. and Heriot-Watt, T.R.R.L. #451 Edinburgh

Murray, R.T. (1981), Fabric reinforced earth walls: development of design equations, T.R.R.L. Supplementary Report #496

Sarma, S.K. (1973), Stability analyses of embankments and slopes. Geotechnique 23, 423-433

Reinforced embankments at the Great Yarmouth bypass

Construction of embankments up to 8m high is in progress along the route of the A47 Great Yarmouth Western Bypass over very soft organic alluvial deposits up to 22m depth.

Geogrids have been used in the base of the embankments to totally enclose a layer of coarse sand and gravel to form a freely draining reinforced granular mattress.

This paper describes the site preparation and the construction of the granular mattress. Considerations regarding the layout of the geogrids are discussed including the layouts at roundabouts. The jointing requirements and jointing methods are described and commented upon.

D. Williams, *C. H. Dobbie and Partners*

INTRODUCTION

This paper describes the construction of embankments reinforced with Tensar SS1 and SS2 geogrid on the A47 Great Yarmouth Western Bypass in Norfolk.

The route of the new road is shown in Fig.1 for which construction, entirely upon embankment, began in July 1983. On reaching their design height, following controlled filling, the embankments will be surcharged. They will then reach maximum heights of 8m at Acle Road Roundabout and 7.5m at Breydon Water. To the south of Mill Road they are generally 2.5m high.

The soft alluvial deposits on which the bypass is founded form the marshlands around Breydon Water. Much of the marshes have been reclaimed by earth bunding and drained to provide farmland but the ground level remains very low at around 0.0m AOD.

In many places the road embankment follows the line of a former railway embankment and spans either side onto the marsh. Very considerable geotechnical problems including failures were experienced during construction of the railway in the early 1900's. The track suffered from long term settlement until closure in 1953.

In view of this history, problems associated with settlement and stability of embankments in the area were foreseen. Site investigation results confirmed these initial conclusions and measures were incorporated into the design to minimise the problems.

The concept of the design is to form embankments which will exhibit a minimum of differential effects after opening of the new road. The principal agents to achieve this desired effect are:
(i)   the incorporation of reinforcement;
(ii) the installation of vertical wick drains, and
(iii) embankment surcharging.

The design concept was evolved partially as a result of the construction of a trial embankment. As a part of the trial, geotextiles and geogrids were considered. Geogrid reinforcement was selected for the main contract.

The reinforcement is installed near the base of the embankments and totally encloses a layer of coarse sand and gravel forming a freely draining reinforced granular mattress. The purpose of the mattress is to stiffen the base of the embankment, to reduce lateral spread and to regulate differential settlement of the finished road.

The construction of the mattress and its immediate post-construction performance have been satisfactory. Some problems associated with jointing and directional laying of the geogrid, have been successfully resolved. The full post – constructional performance of the mattress will be presented in a later paper.

SOIL CONDITIONS

The subsoil conditions along the route, shown in Fig.2, comprise up to 22m of soft and very soft organic clay and silt with some layers of peat.

Topsoil or railway embankment fill overlies soft and very soft dark grey clay and silt with numerous partings of fine sand up to 0.5mm thickness. Below this layer at around 7.0m depth is firm dark brown peat. This is underlain by very soft dark grey very organic clay, becoming less organic and more silty with depth. Below the silty clay is firm dark brown and grey peat interspersed with organic clay. At the base of the alluvial deposits is dense glacial sand and gravel, over Norwich Crag. These deposits have been proved to be in excess of 20m thickness.

SITE PREPARATION

Prior to construction of the granular mattress, site clearance including removal of topsoil was followed by spreading a layer of coarse sand and gravel to form a stable working surface up to 0.3m above the marsh level. Work commenced in an area where the former railway embankment had been removed by the early 1970's and thin topsoil had been re-established over the site.

Vertical wick drains were installed to within 3m of the basal sand and gravel at 2.3m centres. These consist of a corrugated polypropylene core contained within a non woven polypropylene fabric jacket.

Intensive geotechnical instrumentation was installed comprising hydraulic piezometers, inclinometers, magnet settlement gauges, hydrostatic profile monitors and rod settlement gauges. Results from the instrumentation will be used to control rates of construction and to assess the performance of the embankments.

At the time of writing the majority of the granular mattress has been constructed. The remaining area will be constructed during the spring and summer of 1984.

Polymer grid reinforcement. Thomas Telford Limited, London, 1985

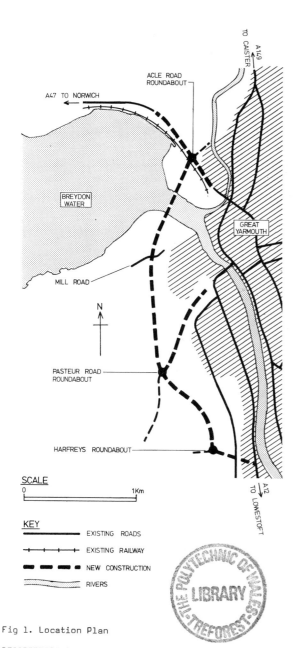

Fig 1. Location Plan

British Standard Sieve Size	Range of Grading Percentage by Mass Passing
100mm	100
75mm	100
37.5mm	85 – 100
10mm	45 – 100
5mm	25 – 85
600μm	8 – 45
212μm	0 – 25
75μm	0 – 5
Minimum Permeability	1×10^{-4} m/s

Table 1 Grading and permeability of the granular material

Material	Tensar SS1 Polypropylene	Tensar SS2 Polypropylene
Tensile strength per metre across roll width (kN/m²)*	20.9	36.2
Tensile strength per metre along roll length(kN/m²)*	12.6	17.0
Roll width (m)	3.0	3.0
Roll length (m)	50.0	50.0
Roll weight (kN)	0.3	0.47

Table 2 Properties of the geogrids
(As supplied by Netlon Ltd)

* Tensile strength measured on samples 3 apertures long by 1 aperture wide extended at a constant rate of 50mm/min at a temperature of $20 \pm 1^{\circ}$ C.

To improve the soil conditions in the areas of likely instability and provide some restraint against lateral movement, toe trenches 1m deep are installed along both sides of the embankment. These trenches are lined with geogrid and filled with granular material. The lower geogrid layer is completed by covering the area between the toe trenches.

The second geogrid layer is placed above 500mm of granular material placed directly on the lower layer. The two layers are joined together along either side throughout the embankment length. The third layer, where present, is not fixed to the lower layers and is placed 1m above the second layer, separated by suitable embankment fill.

It can be seen from Table 2 that the geogrids are considerably stronger across the roll width than longitudinally. For greatest economy the geogrid is, therefore, laid to orientate its greatest strength in the direction of greatest tensile stress in the embankment base. The maximum stress is believed to occur at the base of the embankment below the upper part of the side slope, i.e. generally corresponding with the edge of the former railway embankment and is bisected by

DESCRIPTION OF THE GRANULAR MATTRESS

The grading and permeability of the granular material enclosed in the mattress are given in Table 1. The properties of the geogrids are given in Table 2.

The higher embankments are reinforced with three layers of Tensar SS2 as shown in Fig.3, whereas the lower embankments are reinforced with two layers of Tensar SS1 as shown in Fig.4. The trial embankment was constructed between the area of Breydon Water and the low embankments. This section contained mainly geotextile reinforcement and will be incorporated into the final scheme without further reinforcement.

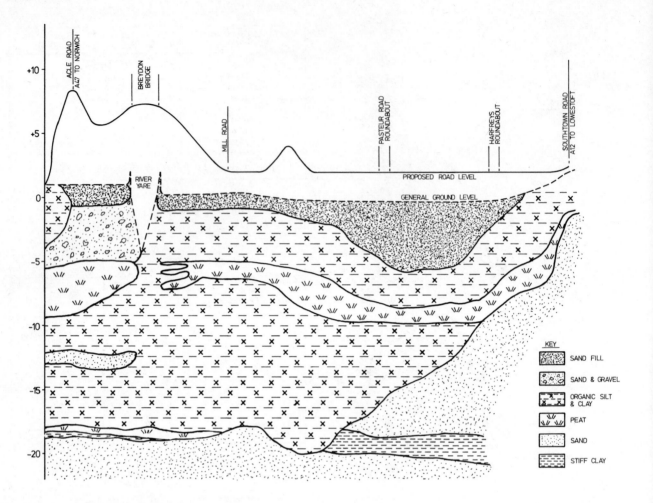

Fig. 2 Geological Long Section

the most critical predicted stability failure plane, as shown in Fig 5.

In order to orientate the geogrid so that its greatest strength corresponds to the direction of greatest stress, the grids have been laid longitudinally. The joints between the lengths of geogrid are therefore subjected to the maximum stress and are required to match the quoted tensile strength for the grid.

JOINTING

To achieve the required full strength continuity across joints either a reliable jointing system, or wide overlapping is required. The overlap width is dependent upon the subjected normal force developing friction between the geogrids and the interlocking soil to give a shearing resistance equal to the geogrid tensile strength. Results from laboratory tests indicate the required overlaps for the completed embankments are in the range 250 - 500mm. It was anticipated that during construction, when an embankment is less than full height, considerable movement on these overlaps could occur due to the low normal force. Hence the effective overlap could be considerably less than

intended, especially when differential movement developed at the edge of the former railway embankment. In consequence greater overlaps would be required, resulting in excessive wastage. A decision was therefore taken to stitch all joints using an interwoven high density polyethylene (HDPE) braid, of 2kN breaking load.

Two sewing techniques were tried on site. The first consisted of overlapping the adjacent geogrids by a single aperture and interweaving the braid through every aperture of the two geogrids using a crescent shaped needle. The second technique was more complex and consisted of an overlap of three apertures. The interweaving braid was fed through an aperture on one side of the joint and back through the following aperture. It was then fed through the corresponding aperture on the opposite side of the joint before returning to the first side of the joint, three apertures further along where the technique was repeated. This process was termed an overstitching technique.

Comparisons were made between the joint strengths provided by the two stitching techniques, together with the material wastage and the time per 50m run. As a result of these comparisons the first technique was preferred.

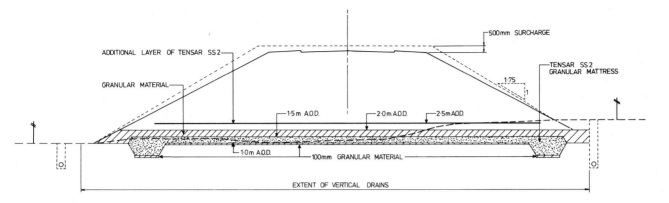

Fig. 3 Cross section of a typical high embankment

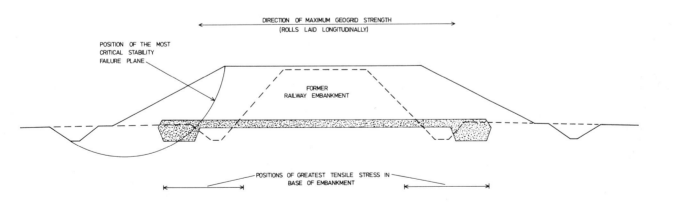

Fig. 4 Cross section of a typical low embankment

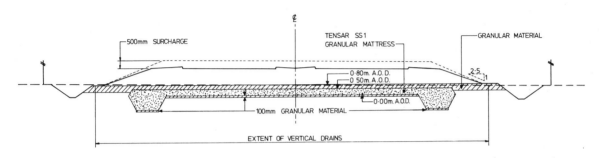

Fig. 5 Orientation of geogrid in relation to embankment stresses

JOINT STRENGTHS

Tests performed at Bolton Institute of Higher Technology in a 300mm shear box gave coefficients of interaction for an overlapped joint of 0.83 for Tensar SS1 and 0.78 for Tensar SS2. For full strength joint continuity the required overlaps became 250 - 500mm. As mentioned above this was considered uneconomic and stitching was preferred.

Pull out tests were carried out by Ground Engineering at Hatfield Polytechnic on geogrid samples 250mm wide with a 300mm length of embedment. The results are presented in Table 3.

	Unjointed	Every Aperture Stitching Technique	Overstitching Technique
Tensar SS1			
Tensile strength kN/m	57.9	57.3	60.0
% strain at failure	4.5	11.0	7.0
rate of displacement mm/min	1.92	1.0	1.0
Tensar SS2			
Tensile strength kN/m	76.0	62.7	76.0
% strain at failure	4.3	4.7	5.0
rate of displacement mm/min	1.0	1.0	1.0
Confining Material			
Borehamwood pit sand			
Coarse - medium sand with some gravel			
D50 size = 500 U			
Coefficient of uniformity = 3			
Maximum dry density 1.88 Mg/m^3			
Relative density 90%			

Table 3 Details of laboratory pull out tests.

All tests show strengths approximately 100% greater than the quoted strengths from simple extension tests. The stitching technique adopted for the works of stitching every aperture surprisingly gave a lower test result than the unjointed specimen. This result nevertheless meets the design requirements for the geogrid.

GRANULAR MATTRESS CONSTRUCTION SEQUENCE

The toe trenches were excavated using a tracked excavator; a specially fabricated bucket forming the 60° side slopes. They were then lined with the geogrid and backfilled with granular material. The geogrid edges, either side of the toe trench, were left exposed.

Lengths of geogrid were then placed longitudinally between the toe trenches, stitched together and to the exposed inside geogrid edges at the toe trench to complete the lower geogrid layer.

The 500mm layer of granular material was spread, with roughly formed 60° side slopes, covering the entire width of the geogrid. The upper layer of geogrid was

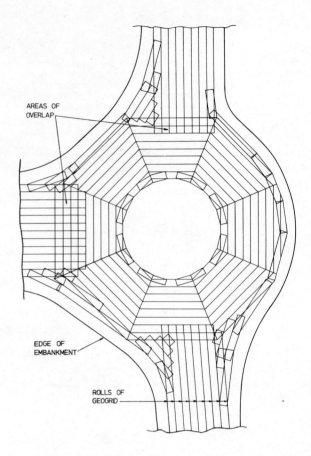

Fig.6 Layout of geogrid at Harfreys Roundabout

then laid and stitched sandwiching the fill. The mattress edges were formed by pulling up the exposed geogrid from the outer edge of the toe trench and stitching this to the upper grid, keeping the grids as taut as possible. Fill was spread and compacted over the mattress commencing above the toe trenches. The action of the rolling and tracking over the toes, further tensioned the geogrid edges. The lower embankments were then raised 0.5m in height per week. The higher embankments were raised at 0.5m height per week for the first metre at which point a third geogrid layer was placed and stitched. The fill above this layer was placed at 0.5m per week with periods of cessation called for when porewater pressures reached critical values.

LAYOUT OF THE GEOGRIDS

The requirement for the geogrids to be laid longitudinally has already been discussed. Where mattress widths varied at junctions and at roundabouts, detailed layout plans were prepared. In order to prepare these layouts detailed consideration was given to the following:
a) Maximum geogrid strength corresponding with the direction of the maximum embankment stress.
b) Wastage
c) Quantity of cutting and stitching
d) Difficulty laying geogrid around curved toe trenches.

e) Feasibility of checking the layout on site.

The typical layout of the geogrid around a roundabout is shown in Fig 6. The need to achieve continuity of geogrid in the areas of principal stress resulted in large areas of overlapping geogrids. Nevertheless the joints were still all stitched. The toe trenches around the roundabout were constructed in a series of straights. The length of each straight was maximised for ease of construction but controlled by the need to maintain the geogrid strength corresponding to the direction of maximum stress.

Consideration was given to using a higher strength geogrid for the roundabouts such that the minimum strength of this geogrid would be equal to the maximum strength required. This would permit a more simplified layout, but further consideration regarding the maximum stress being taken diagonally by the geogrid would be necessary. This option was discarded because of the excessive cost of a suitable geogrid.

CONSTRUCTION RATES

Direct comparisons of output between the two stitching techniques were made. To stitch every aperture with a single aperture overlap the times varied between $\frac{1}{2}$ hour and $1\frac{3}{4}$ hour per 50m. An average of 1 hour per 50m appeared normal. For the overstitching technique a time of 2 hours was achieved during trials. This could probably be reduced with familiarisation but was considered by all parties to be a considerably more time consuming method. The former stitching method was adopted throughout the scheme.

During the summer and autumn of 1983 approximately 100,000 m of stitching was completed by up to 25 operatives over a three month period. In this time the operatives placed, cut and stitched up to 200,000 square metres of geogrid.

The operatives tried various methods of prefabrication in attempts to expedite the work. The most successful of these consisted of pre-stitching 2 rolls of geogrid and stocking these until they could be laid in the mattress. This reduced the periods of time when access along the site was restricted by the stitching process and when trafficking over the geogrid was not permitted. Attempts at prestitching more than 2 rolls were less successful due to the increased handling difficulties.

The operatives found a single geogrid roll was readily handled by one person. The prefabricated rolls were, handled and laid by two people without undue difficulty.

All operations benefitted from the generally good weather which prevailed during summer and autumn of 1983. On colder days the geogrid was noticeably stiffer and stitching was correspondingly slower. The operatives often expressed the view that the work would be very undesirable and slow in winter months. An important point to note with respect to using the material is that the grid did cause numerous cuts and abrasions to the hands and knees. It is therefore considered necessary that gloves and kneepads are used by the operatives laying and stitching the grid.

End tipping and spreading of the fill over the grid was liable to cause intolerable rucking in the grid. This problem was overcome by placing small heaps of fill, with an excavator, onto the grid in advance of the filling. It was found that this provided sufficient frictional resistance to prevent movement of the grid over the underlying soil.

The most time consuming single operation was stitching. Construction rates and hence the cost of laying the grids, would be vastly improved if the geogrid was available in rolls, wider than the 3m maximum currently available. It would be useful if 6m

wide rolls could be produced, either as single rolls or by factory stitching of two 3m wide rolls.

The development of a faster jointing technique would be of great benefit. A hand operated gun releasing small plastic clips around the geogrid strands and possible heat bonding of these clips should be worth investigating.

WASTAGE

A total of approximately 230,000m^2 of geogrid is included in the contract and, as stated, all joints are stitched to avoid the excessive material wastage at overlaps. Overlap joints would incur an additional 15% material requirement. The stitching method adopted resulted in an effective wastage of 2% while the overstitching technique would create a 7% wastage.

The granular mattress construction involved recutting and patching in of geogrid to form the irregular shapes required, particularly at the toe beams and roundabouts. These operations created off-cuts of geogrid which were generally reused as patches over any damaged material or around the geotechnical instrumentation which passed vertically through the geogrid. In the event the actual wastage was very small.

Wastage caused by carrying out repairs to damaged geogrid was minimal. Very little damage resulted from transporting and handling of the geogrid but damage would easily be inflicted if full protection was not provided by overfilling prior to trafficking.

QUALITY CONTROL

In general the quality of the geogrids delivered to site was very good. Very occasional damaged sections were found consisting of split strands. These were detected, either during laying or later inspection and patched over with sound material.

A large variation was detected in the size of aperture in the stretched direction. The dimension between the strand centres, quoted on the manufacturers' data sheets is 39.5mm and 39.4mm for SS1 and SS2 respectively. This dimension was found to vary between 55mm and 30mm. This is a function of the manufacturing process and was not considered to, or detected as, generating weak spots.

INITIAL PERFORMANCE

The immediate effects of the first geogrid layer was very noticeable in soft areas where rutting was reduced from 200 - 300mm to less than 50mm. In one exceptional area, prior to placement of the geogrid, the ground failed to support a track mounted crane. This area became perfectly trackable subsequent to laying of the geogrid and granular cover.

The settlement profiles across the embankments during filling indicate slightly less differential settlement than predicted. Further analysis is required of the actual and predicted settlements and lateral movements of the embankments over a longer period of time.

CONCLUSIONS

Joints between the geogrids will be subjected to the maximum stresses due to the orientation of the grid. The joints must therefore be capable of achieving full strength continuity.

Stitch jointing was considered more economic and reliable than overlapping.

All tests performed on the geogrids have indicated strengths in excess of the quoted values for free air extension tests.

Detailed layout plans are necessary to ensure the grids are correctly orientated around curves and at junctions.

Stitching rates achieved on site averaged 1 hour per 50m length per man. This would reduce significantly in inclement weather conditions.

In a 3 month period approximately 200,000m² of geogrid was placed, with 100,000m of stitching by up to 25 men.

Construction rates were improved by prestitching pairs of rolls for later use. Development of an improved jointing system could be worthwhile.

The manufacture of wider rolls should be considered.

Very little wastage of geogrid was evident although the stitching method causes an inherent 2% effective material wastage.

The quality of the geogrids was very good and very little damage is caused if sensible protective measures are taken.

The initial performance of the geogrids is very encouraging.

ACKNOWLEDGEMENTS

The scheme was designed and construction supervised by C.H.Dobbie and Partners, Consulting Engineers to The Department of Transport, Eastern Regional Office.

The works were constructed by May,Gurney and Co.Ltd.

The author wishes to thank all persons involved in the design and construction of the scheme and all persons who assisted in the preparation of this paper.

This paper is given by permission of Mr.D.I.Evans BScTech MSE CEng Director (Tp) Eastern Region,Department of Transport.

REFERENCES

1) Engineers reports on the construction of the Lowestoft Junction Railway for the Midland and Great Northern Railways Joint Committee. 1901 – 1904.

2) WILLIAMS, D.Construction and performance of a trial embankment. To be published 1984.

3) WILLIAMS,D. Design and performance of reinforced embankments over very soft organic marshland. To be published 1985.

Design methods for embankments over weak soils

V. Milligan, *Golder Associates,* and P. La Rochelle, *Laval University*

The paper examines existing analytical methods to determine the stability of embankment fills over weak soils. Present methods generally involve establishing the limited equilibrium of a free body failure mass and are markedly influenced by the assumed mode of failure and the shearing resistance of the foundation soil and fill. Most of the existing methods assume total or effective stress conditions to express the shearing resistance. However, whether or not such conditions are justified is dependent, inter alia, on the character of the foundation soil, the form of construction and on the definition of factor of safety. Further, the mode of failure may be strongly related to the presence of an upper foundation crust or strong layer which also may influence the mode of embankment deformation. Design methods are proposed to allow for these factors. It is suggested that, for consistently reliable design, the analysis of fills over weak soils should be based on the lower limit of the foundation strength available in situ, or on the large strain strength.

INTRODUCTION

Current methods of predicting stability of embankments on weak foundations basically involve establishing the limiting equilibrium of a free body failure mass as illustrated on Figure 1. Various trial failure surfaces are analyzed and the failure surface which provides the minimum factor of safety is taken as the critical surface. Although many other factors are involved, the key components of the limit equilibrium method of analysis are the shearing resistance of the soil mass along the failure surface and the failure mode (i.e. nature of the failure surface itself). These basic components are considered separately in the following sections.

METHOD OF ANALYSIS

Common practice in assessing the stability of embankments on weak foundations has involved the use of both effective stress analyses and total stress ($\phi = o$) analyses. The advantages and disadvantages of both these methods have been examined in detail (Tavenas et al, 1980 and Pilot et al, 1982). Stability analyses based on a total stress approach avoid some of the problems in the effective stress approach involving an estimation or even assumption of pore pressures and the degree of mobilization of the shear strength of the foundation soils, but problems still remain concerning the mobilized strength of the embankment fill and, in particular, on whether the undrained shearing resistance of a foundation soil, as measured on intact samples or by in situ vane tests, is representative of the strength actually mobilized in situ, (Bjerrum, 1973).

SHEARING RESISTANCE OF THE FOUNDATION SOIL

The critical period with respect to fill stability on weak foundations is generally immediately following construction. In this period, the soil is essentially undrained and in its weakest state. With time, by consolidation, water is expelled from the pores of the soil and its strength increases. It is common to measure soil shearing resistance in an undrained condition and to use this minimum design (undrained) strength in stability analyses. It is now recognized that many factors affect the design undrained strength and that undrained strength is not, in fact, a unique property of the soil deposit. Some of the major factors which affect undrained strength are as follows:

o Variation with depth - As shown on Figure 2A, the undrained strength of a soft clay deposit typically varies with depth. In the surface crust,

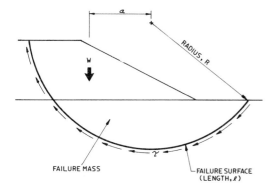

FACTOR OF SAFETY, F.S. $= \dfrac{\gamma \ell R}{W a}$ $\dfrac{\text{(RESISTING MOMENT)}}{\text{(DISTURBING MOMENT)}}$

Fig. 1. Stability analysis - basic elements

Polymer grid reinforcement. Thomas Telford Limited, London, 1985

95

which has been altered by natural weathering processes, the undrained strength is high but decreases with depth to a minimum below the base of the crust. Below this depth, the undrained strength generally increases in a uniform manner, with the rate of increase in strength dependent on the stress history and the soil type. This type of variation in strength with depth is common in soft clay deposits.

o Weak layers - Because of the nature of deposition, it is not uncommon for soil or weak deposits to incorporate layers of weaker soils, (Figure 2A). These weak layers, if present, are difficult to detect and can control stability (Leonards, 1982). Even in cases where discrete weak layers are not present, the zone of minimum shear strength below the crust represents a weak zone which is found in most deposits.

o Rate of shearing - Laboratory and field strength tests are often carried out at rates which are much more rapid than actual field loading (i.e. rate of fill construction). For example, field vane tests are carried out at rates which are up to 10,000 times as fast as actual fill construction. The undrained strength measured in common laboratory strength tests is dependent on strain rate and typically exhibits a 10% - 15% increase in strength for each order of magnitude increase in strain rate.

o Anisotropy - As shown on Figure 2B, the undrained strength of soft clays depends on the direction of shearing (i.e. the undrained strength is anisotropic). The degree of anisotropy decreases with increasing plasticity. The complex mode of failure associated with the common field vane tests results in a trend which does not reflect other tests in which the boundary conditions are better defined.

o Strain Dependence - Many soft clays are strain-softening (Figure 3). In most situations, fills are designed such that the factor of safety against failure is low, (~ 1.3). In this condition, a large portion of the foundation clay deposit is overstressed (i.e. the imposed shear stress exceeds the shear strength of the soil). Therefore, if the soil is strain softening, large portions of the foundation soil will have reached at failure a lower "large strain" strength so that the maximum shearing resistance does not apply throughout the deposit. Further, progressive failure will occur along the failure surface as load is transferred from the over-stressed zones to adjacent "under-stressed" zones.

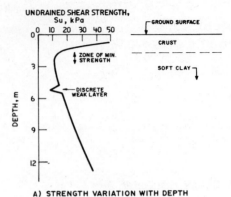

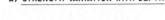

A) STRENGTH VARIATION WITH DEPTH

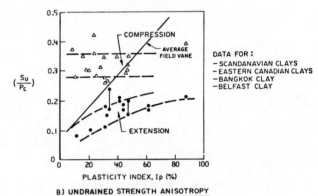

B) UNDRAINED STRENGTH ANISOTROPY

Fig. 2. Factors affecting undrained shear strength

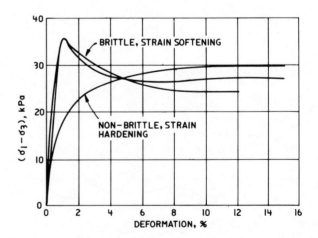

(AFTER LA ROCHELLE ET AL, 1974)

Fig. 3. Types of stress-strain behaviour

Although significant research has been carried out on the strength behaviour of weak soils and, in particular, on soft clays, there are still unresolved problems associated with the prediction of stability of fills on soft clays. This situation is well illustrated by the predictions made by various researchers since 1960, (Figure 4). Bishop and Bjerrum (1960) established the validity of the $\phi = o$ (short term-undrained strength) approach by back-analyzing a number of failures on clays; as shown on Figure 4A, this approach appeared to provide a reasonable solution for each of the cases studied and became a common method of analysis in geotechnical engineering. By the early 1970's, it became apparent that this approach did not always provide the correct prediction, as indicated on Figure 4B (Bjerrum, 1973). The magnitude of error in computed factor of safety increased with increasing plasticity when the undrained strength was measured using the field vane. This trend led Bjerrum to propose an empirical correction factor to be applied to the measured vane strength. This modification was obviously compelling and was adopted widely. However, exceptions which did not conform with Bjerrum's correction factor were reported, (Figure 4C) and questions arise as to the advisability of using Bjerrum's approach, based on both fundamental and practical considerations.

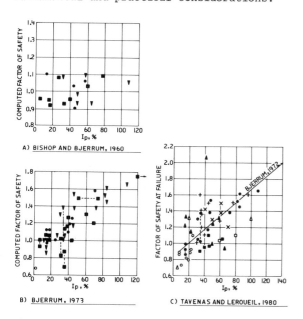

Fig. 4. Undrained strength stability analyses

Based on the above, it is obvious that the selection of undrained strength to be used in stability analysis is a complex issue. What then is the "correct" undrained strength? This question has been addressed by a number of researchers using different approaches and surprisingly, there is a reasonable consensus of opinion. Three of the more significant

approaches are described below and are similar in one respect - the undrained strength proposed for use in analysis is related to the maximum past vertical effective stress (p_c) on the soil.

o Bjerrum/Mesri - Mesri (1975) reinterpreted Bjerrum's original data related to the strength at failure of soft clay deposits (Figure 5A) and pointed out that when Bjerrum's correction factor was applied to s_u and p_c plots, there was a unique ratio of s_u/p_c of 0.22, independent of plasticity.

o Ladd and Foott (1974) have recognized that the normalized direct simple shear undrained strength as determined on destructured clay is the average strength mobilized in the foundation of an embankment on soft clay and have successfully used this strength value to explain failures (Figure 5B). As observed by Trak et al (1980), for slightly overconsolidated cohesive soils with an overconsolidation ratio between 1 and 2, the ratio s_u (DSS)/p_c varies between 0.18 and 0.23.

o La Rochelle et al (1973) reasoning from the premise that at failure, the strength mobilized in the soft clay foundation would everywhere be equivalent to the strength at large strain, have shown that the undrained strength obtained at large strain (USALS) in laboratory compression tests, could successfully be used in stability analysis of embankments on soft clays. It was later shown by Trak et al (1980) that the ratio s_u (USALS)/p_c varies between 0.2 and 0.25.

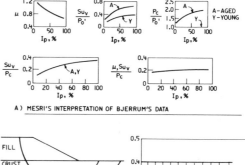

A) MESRI'S INTERPRETATION OF BJERRUM'S DATA

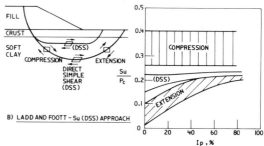

B) LADD AND FOOTT - Su (DSS) APPROACH

Fig. 5. Definitions of undrained shear strength at failure

Thus, it would appear that the undrained strength at failure can be taken to lie in the range of $(0.2 - 0.25)$ p_c. However, the geotechnical profession as a whole has not as yet adopted this or any other unique definition of failure strength for defining the strength to be used in the analysis of embankment stability on soft clay. Rather, strength values measured in a variety of ways are used, sometimes indiscriminately, and which in some cases, do not reflect the actual field strength.

FAILURE MODE

In the majority of stability analyses and back analyses of failures, it is assumed that the failure surface is circular which is a kinematically admissible failure mode (Figure 6A). Possible errors in back analyses related to three-dimensional effects are discussed by Azzouz et al (1981); errors related to the form of expression of the factor of safety as related to the different balance of forces in varying methods of limit equilibrium analyses have been examined by Fredlund et al (1981). However, there are many instances in which the assumption of a circular surface does not account for other important factors such as the presence of a weak layer, or a zone of minimum shear strength located immediately below the crust. Attempts to fit circular failure surfaces passing through this zone can lead to unrealistic failure geometries within the fill and beyond the toe of the fill slope. The assumption that failure surfaces are circular often appears to be consistent with surface and subsurface evidence of failure (i.e. back scarp, toe heave extent and depth of significant deformation in the soft clay foundation material); however, as shown on Figure 6B, other modes of failure, and in particular, predominantly translational movement, are also consistent with this physical evidence. In some cases monitoring of lateral deformations in the soft clay clearly indicates concentration of deformation within the zone of minimum strength below the crust.

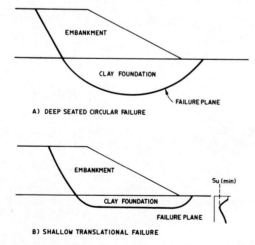

A) DEEP SEATED CIRCULAR FAILURE

B) SHALLOW TRANSLATIONAL FAILURE

Fig. 6. Modes of failure

It is considered that translational failure modes are most likely to occur when the crust is thin and there is a pronounced zone of minimum shear strength below the crust. On the other hand, circular failure modes would be associated with thicker crusts and for uniform strength-depth profiles.

Analyses incorporating non-circular and translational failure modes do exist (e.g. wedge analyses) but are used less frequently than circular arc analyses. Further, many of the assumptions incorporated in the method of slices, which are common in both circular and non-circular analyses, are not necessarily applicable to analysis of translational failure modes.

THE EFFECT OF REINFORCEMENT TO FILLS

Based on limited prior experience, reinforcement at the base of fills improves their stability on soft clay foundations. An early example of reinforcement of fills on soft ground is the "corduroy" road in peat bogs where cut trees are laid side by side across the intended route and fill is placed on top of this "corduroy" arrangement. However, caution should be exercised in terms of inference drawn from this example, since the mode of failure associated with low road fills on peat is generally related to excessive deformation as opposed to rupture of the foundation soils - the case with soft clay foundations. The predominant effect of the corduroy is to keep the fill intact and to prevent distress to the road pavement as a result of spreading (lateral displacement) of the fill, in association with deformation of the underlying peat. It should also be appreciated that while corduroy reinforcement of low road fills on peat may help to keep the fill intact, it will not reduce (and in some instances could increase) overall deformations.

There are a number of reasons why reinforcement of the base of fills enhances their stability on soft clay. For example, reinforcement tends to reduce "splitting" of the fill and therefore increases whatever contribution the fill strength makes to overall stability; however, it is considered that the major effect of reinforcement is to cause a beneficial change in stress distribution in soft clay foundation.

The hypothesis is as follows: Construction of fills on a flat ground surface induces movement and associated shear stresses are significant, as indicated by the data obtained during construction of the Gloucester test fill, (Figure 7). The effect of incorporating reinforcement at the base of the fill is to reduce the magnitude of these shear stresses at the interface. Thus, as the fill tends to spread, the reinforcement is put into tension and, if sufficiently stiff, will reduce the magnitude of displacement, and, therefore, result in a lower shear stress at the fill/ground surface interface.

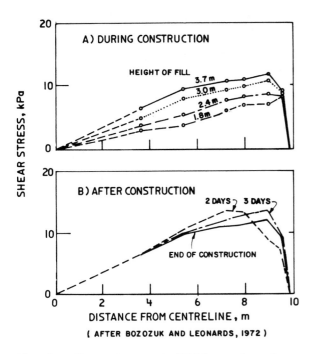

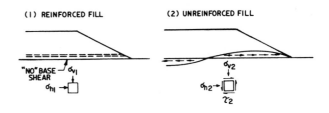

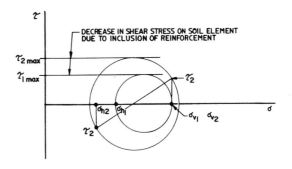

Fig. 7. Shear stress at fill/ground surface
interface
(Gloucester Test Fill)

Fig. 8. Illustration of effect of Tensar
reinforcement on shear stress
in foundation soil

This change in external boundary stresses can
result in a significant change in the stress
distribution in the foundation material by
introducing a horizontal tensile stress component and by preventing rotation of the
principal stress directions at the ground
surface. It should be noted, however, that
the vertical stress distribution will be
greatly influenced by the rigidity of the
reinforcement and may remain practically
unchanged for reinforcement of low rigidity
(e.g. conventional geotextiles).   To illustrate
the influence of reinforcement, Figure 8 shows
typical stress conditions on an element of soil
below reinforced and unreinforced fills for the
case of perfectly rigid reinforcement.   The
combination of increased total horizontal
stress and reduced boundary shear stress
causes a reduction in the maximum shear stress
cn the soil element.   However, the magnitude
of the reduction in shear stress associated
with reinforcement decreases with depth.   For
a site where the surface crust is thin and a
typical strength profile with depth exists,
the consequence of the decrease in shear
stress magnitude at shallow depth is important,
since the zone of highest shear stress is no
longer coincident with the zone of minimum
shear strength.   This results in a deeper
failure surface through more competent
materials and, as a result, a higher factor of
safety (more stability) for a given fill
height.   The corollary is that on a given site,
a higher fill can be constructed before
failure (F = 1) occurs.

EXAMINATION OF EXISTING DESIGN METHODS FOR
REINFORCED FILLS

A variety of methods of analysis of reinforced
embankments on weak foundations have been
reported in the recent literature.   The majority of these methods employ total stress
(ϕ = o) limit equilibrium concepts to assumed
circular arc failure surfaces (Brakel et al,
1982;   Fowler, 1982);   however, some assume
three part wedge modes of failure (Enka, 1978).
The force applied by the reinforcement has
been variously assumed, either to act only on
the failure surface or to act tangentially to
it.   Jewell (1982) separates in his circular
arc analysis the distribution of the reinforcement force required to provide equilibrium
with a specified factor of safety (F) on the
mobilized soil shear strength and the distribution of mobilized force along the
reinforcement at that value of F available to
provide equilibrium.   If the required forces
are everywhere less than the available then
the embankment has as a minimum, the value of
F presumed.

It is evident that there are a variety of
assumptions used in practice for limit
equilibrium analyses of reinforced embankment
stability on weak foundations.   We can query
if detailed assessments of the amount and
direction of the reinforcement force are
justified if the assumptions made with respect
to undr_ined strength and failure mode can be
in error.   Therefore, it is considered that
since the stability of unreinforced fills on

soft clay cannot as yet be predicted with absolute confidence using conventional limit equilibrium approaches, modification of these existing methods to include the effects of reinforcement can be subject to the same errors and uncertainties.

EXAMINATION OF A PROPOSED DESIGN METHOD

Constitutive models to define stress redistribution effects in the fill and in the foundation soil attributable to reinforcement, have not as yet been developed. At present, there does not exist an agreed reliable method of analysis which will properly account for the effect of reinforcement although solutions will undoubtedly develop as a result of ongoing research. At this stage of our knowledge, in using reinforcement to enhance embankment performance, we therefore have to rely on instinct and growing experience. As a guide, it appears that the best available method is to adopt a modified limit equilibrium method, despite the uncertainties discussed previously.

Such a method must recognize the overwhelming importance of a rational assessment of the foundation strength. It assumes that firstly, the total stress (ϕ = o) approach is applicable for first stage construction; secondly that the lower limit of undrained strength is adopted, as defined by an S_{u/p_c} value of about 0.2 to 0.25; thirdly, that the mode of failure should fit, wherever possible, to known or suspected weak layers within the foundation. The factor of safety assessed by this model should then consistently represent a lower bound and give a safe design model.

Simple limit equilibrium equations suitable for expression in a dimensionless form can be developed as shown in Figures 9 and 10. The equations assume a circular arc or translational failure through the clay foundation and a Coulomb failure surface in the fill. The assumption in the fill allows the failure wedge to be replaced by a single active force Pa and with this configuration the centre of the critical circular arc can be shown to be on the vertical line through the mid point of the slope. A computer search is necessary to locate the vertical distance from the ground surface to the centre of the critical arc.

Figure 9 presents a general form of the equation in which the clay foundation has a crust thickness t, and average shear strength C_t. Under the crust, the clay is assumed to be near normally consolidated with its shear strength increasing linearly with depth. The lower bound of S_{u/p_c} is likely to be 0.2.

For a shallow depth of crust or of a clay foundation in relation to the embankment dimensions the failure mode will tend to be translational (refer to Figure 6). The dimensionless equation for translational failure satisfying force equilibrium and assuming constant shear strength in the foundation is

shown in Figure 10. A comparison of the two alternative failure modes (circular arc and translational) for unreinforced embankment stability, with the more rigorous method developed by Sarma (1973) indicated good agreement for the circular arc mode at D/H = 1.0, with the translational mode being more conservative at all values of D/H.

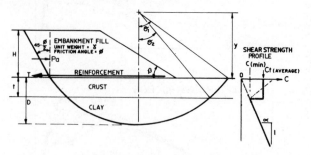

FOR MOMENT EQUILIBRIUM THE STABILITY EQUATION CAN BE PRESENTED IN A DIMENSIONLESS FORM :

$$\frac{2T}{\delta H^2} Y = 2 YZ - \frac{\cot^2\beta}{12} + K_A (Y - \frac{1}{3}) + Z^2 - \frac{4C_t}{\delta H} (Y+Z)^2 (\sigma_1 - \sigma_2) - \frac{2\alpha}{\delta} Z [(Y+D)^3 \sin\sigma_2 - \sigma_2 Y(Y+D)^2]$$

WHERE ; $Y = \frac{y}{H}$, $Z = \frac{D}{H}$

$$\tan\sigma_1 = \frac{1}{2\frac{Z}{Y} + (\frac{Z}{Y})^2} \quad , \quad \tan\sigma_2 = \frac{\sqrt{(Y+Z)^2}}{(Y+X)^2} - 1$$

WHERE $X = \frac{t}{H}$

Fig. 9. Circular arc failure: general analysis

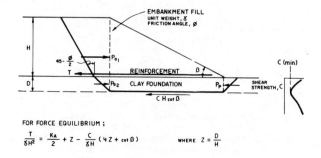

FOR FORCE EQUILIBRIUM :

$$\frac{T}{\delta H^2} = \frac{K_A}{2} + Z - \frac{C}{\delta H} (4Z + \cot\beta) \quad \text{WHERE } Z = \frac{D}{H}$$

Fig. 10. Shallow translational failure

It can be recognized that the method proposed "idealizes" to some extent both soil and reinforcement behaviour and ignores possible strain dependence for either the soil or the reinforcement element; this presumption is however essential for the limit equilibrium approach and, as discussed below, for most cases is reasonable.

Strain Compatibility of the Reinforcement and Foundation Soil

The strength and stiffness of Tensar geogrids, as with all other polymerics, varies with temperature and strain rate.
Strength and stiffness increase as

temperature decreases. Freezing conditions therefore do not present a problem while the standard reference temperature of 20°C is unlikely to be exceeded in temperate climates where the material is buried.

Both strength and stiffness decrease with decreasing strain rate. In addition, such material exhibits a tendency for brittle fracture at some extension beyond the point of ductile yield. The strain, and therefore the stress, induced in an element of reinforcement is dependent on the strains induced in the soil in the working condition. Limiting the design stress to some fraction of the yield stress does not, in itself, ensure that brittle fracture will not occur; if strains are sufficient, the stress level could be high enough to cause continuing creep and eventually brittle failure, within the life of the structure. It is therefore a necessary part of stability calculations of Tensar reinforced embankments to ensure that strains in the zone of the reinforcement are such that brittle fracture will not occur during the life of the structure. Currently available data suggest that the maximum strain allowable, "the performance limit strain" for Tensar SR2, for 120 year life, is 10 per cent.

Two factors relate to the effect of creep in the reinforcement:·

i) As the height of the embankment reaches its final value, the resulting load will produce creep both in the reinforcement and the foundation soil. If the creep rate in the reinforcement is larger than in the foundation soil, some load will be shed from the reinforcement into the foundation soil, and vice versa, where a higher creep rate would be developed in the soil. The load transfer in either case will cease when the stress levels in the reinforcement and the soil will be such that both creep rates will have the same value.

ii) The strength of the foundation soil and therefore its capacity to resist the stresses imposed by the embankment, tends to increase with time due to consolidation. As a consequence, the stress level and the creep rate in the foundation soil will decrease, with a resulting increase in the load transfer from the reinforcement to the soil.

Available data for embankments on soft clay foundations indicate that for factors of safety (ϕ = o) in the range of 1.2 to 1.3, the maximum lateral strains in the subsoil are several per cent where the maximum vertical deformation is about 10 per cent of the embankment height. (The average lateral deformation is 18 per cent of the maximum settlement for sensitive soils (Tavenas, 1979)). The performance limit strain on Tensar SR2 is unlikely to be exceeded in these

applications and limit equilibrium methods of analysis can thus be used safely to calculate stability; however, where settlement is anticipated to exceed about 10 per cent of the embankment height, the design should include a detailed assessment of strain, and limit equilibrium methods must be used with caution.

In many applications, it is probable that strains will be insufficient to mobilize the full design capacity of the reinforcement. This situation is consistent with most applications of limit equilibrium analysis in soil mechanics (e.g. the use of active and passive pressures to calculate the stability of rigid retaining walls) and the unmobilized capacity of the reinforcement can be considered as providing the required security against failure.

ACKNOWLEDGEMENTS

The authors thank J. H. A. Crooks and J. R. Busbridge for their critical comment on this paper and for their significant input to the review of existing practice in the analysis of fills on soft clays.

REFERENCES

Azzouz, A.S., Balich, M.M., and C.C. Ladd (1981). "Three-Dimensional Stability Analyses of Four Embankment Failures". Proc. 10th Int. Conf. ISSMFE. Vol 3, Stockholm.

Bishop, A.W. and Bjerrum, L. (1960). "The Relevance of the Triaxial Test to the Solution of Stability Problems." N.G.I. No. 34, Oslo.

Brakel, J., Coppens, M., Maagdenberg, A.C., and Risseuw, P. (1982). "Stability of Slopes Constructed with Polyester Reinforcing Fabric, Test Section at Almere Holland 1979." Proc. 2nd Int. Conf. on Geotextiles, Vol 3, Las Vegas.

Bjerrum, L. (1973). "Problems of Soil Mechanics and Construction on Soft Clays." State-of-the-Art Report, Session 4, Proc. 8th Int. Conf. ISSMFE. Vol 3, Moscow.

Bozozuk, M. and G.A. Leonards, (1972). "The Gloucester Test Fill." Performance of Earth and Earth-Supported Structures, Vol 1, Part 1, ASCE.

Enka, B.V. (1978). "Design of the Suez Breakwater." Unpublished Report.

Fowler, J. (1982). "Theoretical Design Considerations for Fabric Reinforced Embankments." Proc. 2nd Int. Conf. on Geotextiles, Vol 3, Las Vegas.

Fredlund, D.G., Krahn, J., and D.E. Pufahl,

(1981). "The Relationship Between Limit Equilibrium Slope Stability Methods." Proc. 10th Int. Conf. ISSMFE, Vol 3, Stockholm.

Jewell, R.A. (1982). "A Limit Equilibrium Design Method for Reinforced Embankments on Soft Foundations." Proc. 2nd Int. Conf. on Geotextiles, Vol 3, Las Vegas.

La Rochelle, P., Trak, B., Tavenas, F., Roy, M. (1974). "Failure of a Test Embankment on a Sensitive Champlain Clay Deposit." Cdn. Geot. Jour., Vol II, No. 1.

Leonards, G.A. (1982). "Investigation of Failures." ASCE Jour., Geot. Eng. Div., Vol 108, G.T.2.

Mesri, G. (L975). Discussion on "New Design Procedure for Stability of Soft Clays" by C.C. Ladd and R. Foott. ASCE Jour., Geot. Eng. Div., Vol 101, G.T.4.

Pilot, G., Trak, B. and P. La Rochelle (1982). "Effective Stress Analyses of the Stability of Embankments on Soft Soils." Cdn. Geot. Jour., Vol 19, No. 4.

Raymond, G.P. (1972). "Prediction of Undrained Deformations and Pore Pressure in Weak Clay Under Two Embankments." Geotechnique, Vol 22, No. 3.

Sarma, S.K. (1973). "Stability Analysis of Embankment and Slopes." Geotechnique, Vol 23, No. 3.

Tavenas, F., Leroueil, S. (1980). "The Behaviour of Embankments on Clay Foundations." Cdn. Geot. Jour., Vol 17, No. 2.

Tavenas, F., Mieussens, C., Bourges, F., (1979). "Lateral Displacements in Clay Foundations Under Embankments." Cdn. Geot. Jour., Vol 16, No. 3.

Tavenas, F., Trak, B. and S. Leroueil (1980). "Remarks on the Validity of Stability Analyses." Cdn. Geot. Jour. Vol 17, No. 1.

Trak, B., La Rochelle, P., Tavenas, F., Leroueil, S., Roy, M. (1980). "A New Approach to the Stability Analysis of Embankments on Sensitive Clays." Cdn. Geot. Jour., Vol 17, No. 4.

The use of a high tensile polymer grid mattress on the Musselburgh and Portobello bypass

S. Edgar, *Lothian Regional Council Highways Department*

The Musselburgh and Portobello Bypass, when completed will link the A1 trunk road and the proposed Edinburgh City Bypass with the eastern industrial areas of Edinburgh and the port of Leith, eliminating the present heavily congested areas of Musselburgh and Portobello.

During the design of the Bypass the problem of ensuring the stability of a 15 metre high embankment over weak ground was encountered. Two alternatives were considered:- the excavation of the unsuitable material and its replacement with a suitable rock fill, or the construction of a polymer grid mattress over the weak material at ground level. Contractors were invited to price both these alternatives with the successful contractor electing to construct the polymer grid mattress which he considered the more economic and satisfactory solution.

INTRODUCTION

From its commencement at the Old Craighall Road, the bypass runs in a westerly direction across essentially level arable farmland, crossing two railway lines and one road. After crossing the Millerhill Marshalling Yard goods line which is carried on a 6 m high embankment, the bypass runs north westerly across an area of marshy low lying ground drained by several small streams, before crossing arable farmland to the A6095 Newcraighall Road. After this road crossing the bypass crosses the eastern section of the existing colliery spoil heap (Newcraighall Bing) from the disused Newcraighall Colliery before passing over derelict land and the railway lines from Niddrie Junction. The bypass then continues northwards crossing the Edinburgh Suburban railway line and the valley of the Niddrie Burn, before joining the previously constructed

roundabout adjacent to the Asda Superstore.

As can be seen from the above route description, and from Fig 1, the bypass crosses several roads and railways and as a result the majority of the road will be carried on embankments up to a maximum height of approximately 15 metres.

The site investigation carried out revealed the line of the road to be generally underlain by glacial clay with a localised covering of course granular raised beach deposits in the east. Laboratory tests showed that the strength of the above materials was adequate to support the embankments proposed. However in the area of the marshy ground adjacent to the Millerhill Marshalling Yard, where the road embankment reached its maximum height of 15 metres boreholes and subsequent laboratory testing revealed strata of low strength to depths of 6 metres below existing ground level. Calculations based upon the results of

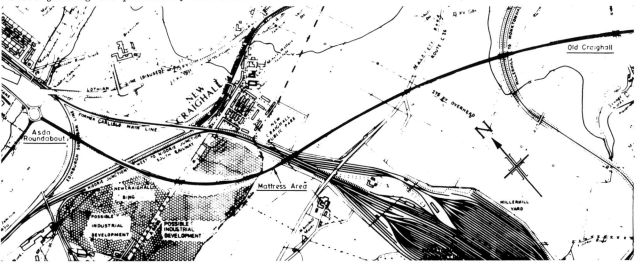

Fig 1 - Location Plan

Polymer grid reinforcement. Thomas Telford Limited, London, 1985

103

laboratory tests carried out on the above low strength materials produced factors of safety against foundation failure of the order of 1.0 - 2.0. This was not considered acceptable and it was decided to investigate alternative solutions, to increase this factor of safety.

SOILS INVESTIGATION:-

(i) Site Description:-

The area of bad marshy ground over which the embankment is to be constructed is bordered to the north by the embankment of the Millerhill Marshalling Yard, and opens to the south, east and west on to gently sloping arable farmland. The marshy ground is drained by a stream running in an easterly direction which is fed by several small streams ('spreads') towards the north of the site. At the northern end of the site, an existing culvert carries the stream below the railway embankment.

(ii) Ground Conditions:-

Boreholes were sunk by shell and auger and rotary open hole and core drilling methods together with two trial pits excavated mechanically. The position of all these boreholes and trial pits are shown on the site plan (Fig 2).

Generally beneath 0.20 m to 1.10 m of topsoil and/or clay and ash fill, the boreholes and trial pits (S5-1 to S5-4 inclusive, R22, R23A, R23E, R23F, R23G and R23H and S5-TP1 and S5-TP2) encountered glacial till, with a local covering of alluvial deposits, resting on sedimentary bedrock.

Boreholes R22, R23E and trial pit S5-TP2 penetrated up to 5.3 m of alluvial materials, generally comprising very soft organic clay, soft to firm, firm and firm to stiff sandy silty clay and clayey sandy silt containing gravel, firm and firm to stiff, often laminated, silty clay with

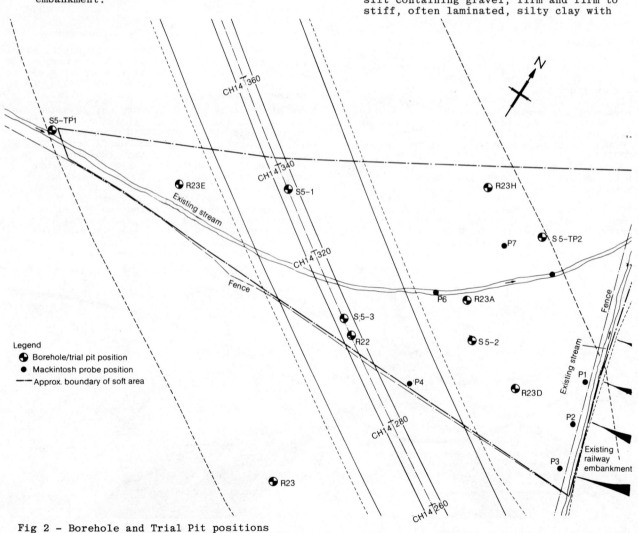

Fig 2 - Borehole and Trial Pit positions

sand partings and a thin layer of clayey sand with gravel at the base of the deposit. Trial pit S5-TP2 terminated within the alluvium at a depth of 3.20 m.

Boreholes S5-1, R23A, R23F, R23G and R23H and S5-TP1 penetrated between 1.65 m and 3.10 m of weathered glacial till comprising soft, soft to firm, firm and firm to stiff mottled very sandy silty clay, or clayey sandy silt containing gravel with local layers of medium dense silty sand.

The weathered glacial till and the alluvium were both underlain by unweathered glacial till, consisting of stiff and very stiff sandy silty clay or locally clayey sandy silt, containing gravel, cobbles and boulders and occasional thin layers or lens of sand and gravel.

Boreholes R23B, R23C and R23D penetrated 2.30 m of alluvial material comprising soft to firm sandy clay and firm to stiff clayey silt with gravel, rootlets and decaying vegetable material overlying soft to firm sandy weathered glacial clay and gravel. The weathered material was again underlain at a depth of 3.10 m by firm to stiff becoming very stiff and hard unweathered glacial clay with gravel.

Ground-water was encountered during boring and excavation in boreholes R22, R23A, R23B, R23C, R23E, R23G and trial pit S5-TP2 at depths ranging from 0.60 m to 5.30 m. Standpipes inserted in boreholes R22 and R23A indicated a standing water level of the order of 1.50 m and 1.00 m, respectively, below ground surface.

Boreholes S5-1 to S5-4, R23D, R23F, R23H and trial pit S5-TP1 did not encounter any ground water

(iii) Summary

From inspection of the boreholes logs and analyses of the laboratory results obtained from the site investigation, the underlying strata below the proposed embankment can be summarised as follows:-

(a) A hard crust of topsoil/vegetation to a maximum thickness of 400 mm.

(b) A soft layer of alluvial materials ranging from soft sandy silty clay to soft clayey sandy silt, with a maximum thickness of approximately 6 metres.

(c) Stiff to hard unweathered glacial clay with gravel recorded to a maximum thickness of approximately 10 metres in borehole S5-2.

(d) Sedimentary bedrock:- sandstone, mudstone.

BOREHOLE NO	SAMPLE DEPTH (m.)	NATURAL MOISTURE CONTENT (%)	NATURAL WET DENSITY (Mg/m³)	APPARENT COHESION (kN/m²)	ANGLE OF SHEARING RESISTANCE (ϕ°)
R22	0·4	22	1·90	82	13
	2·4	28	1·90	33	0
R23D	1·1	24	1·86	50	8
	2·1	20	2·08	47	0
R23E	5·5	13	2·22	70	8
R23F	0·4	28	1·78	37	0
R23H	1·5	17	2·07	46	7
	2·5	17	2·16	45	0
R25	1·2	·13	2·13	119	13
S5-1	2·0	15	2·13	95	0
	3·8	14	2·26	53	6
	5·5	11	2·25	204	11

TABLE NO 1 - Results of Shear Strength Tests

BOREHOLE NO	SAMPLE DEPTH (m)	EFFECTIVE SHEAR STRENGTH	
		COHESION INTERCEPT C' (kN/m²)	ANGLE OF SHEARING RESISTANCE (ϕ')
R23E	1·5	5	32·5
	3·0	5	33·0
	4·0	9	25·0

TABLE NO 2 - Shear Strength Parameters in terms of Effective Stresses

Test results for the alluvial and glacial clay materials are summarised in Tables 1 and 2 above. Using these test results, it was found that the factor of safety against foundation failure of the alluvial materials carrying a 15 metre high embankment ranged from 1 to 2.

Expected consolidation settlements in the alluvial material and the unweathered glacial clay were calculated to be 275 mm and 85 mm respectively, with the majority of the settlement occurring within the first 4 years.

DESIGN ALTERNATIVES:-

(i) The excavation of the soft material and its replacement with an imported suitable rockfill material.

(ii) The partial excavation of the soft material and the displacement of the remainder by deposition of suitable rockfill until finally stable.

(iii) Drainage of the soft material, together with a controlled filling of the embankment to permit dissipation of pore pressure and thus a resultant gain in strength of the weak layer.

(iv) The construction of a high strength geogrid mattress to carry the embankment over the weak ground, with the granular fill of the mattress acting as a drainage blanket.

The second alternative was considered unsatisfactory since it would require extremely careful control on the placing of the rockfill with no real indication of the exact volume of materials being displaced. The third alternative it was felt was unworkable due to the period of time required to construct the 15 metre high embankment. This method would also have required strict monitoring to ensure that the rate of filling did not lead to excessive build up of pore pressures.

In order to increase the rate of settlement of the alluvial material below the embankment and prevent the build up of excess pore pressure during its construction, the geogrid mattress option included provision for the installation of vertical band drains driven through the mattress into the alluvial material below. Calculations indicated that the six metre thick layer of alluvial material would settle approximately 275 mm over a period of four years, but by installing a system of vertical band drains with a triangular equidistant spacing of 2.5 metres, 70% of this settlement could be achieved in two years.

It was therefore finally decided after consideration to include both alternatives (i) and (iv) in the contract documents, giving the contractor the choice of construction which he considered the most economic and applicable to his programme.

DESIGN PHILOSOPHY FOR GEOGRID MATTRESS:-

(i) Introduction

From a study of the site investigation results and the proposed cross-sectional geometry of the embankment, the problem to be analysed is as shown in Fig 3.

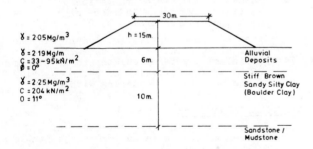

Fig 3 - Embankment Cross-section and Foundation soils

From Terzaghi's expression for the nett ultimate bearing capacity of long footings, at or below the surface of any soil conforming to general shear:-

$$q_{nett} = cN_c + \gamma d (N_q-1) + \gamma \frac{b}{2} N_\gamma \quad ... \quad (1)$$

Using the chosen maximum and minimum values of cohesion and angle of shearing resistance, the safe bearing capacity is the above expression divided by the

required factor of safety, plus the original overburden pressure, γZ.

Therefore for design stability

$$F \text{ of } S = \frac{q_{nett}}{w - \gamma Z}$$ where w is the applied load intensity .... (2)

The calculated factors of safety ranged from approximately 1 to 2. These were well below the target figure of 3 and the use of a geogrid foundation mattress as a means of distributing the embankment load and thereby increasing the factor of safety against foundation failure was considered.

(ii) Principles of Geogrid Mattress

Netlon were approached by Lothian Regional Council Highways Department for advice on the use and design of a suitable mattress for increasing foundation stability.

Advice on the analysis of a high embankment on a thin layer of soft foundation material was obtained by Netlon from Dr Richard Bassett of King's College, London.

The use of a rigid geogrid mattress alters the direction of the normal slip circle failure plane by forcing it to pass vertically through the mattress and as a result, the slip plane is forced deeper into the stiffer layers of the underlying stiff brown sandy silty clay. Before this stiff underlying clay could shear, a plastic failure would be initiated in the soft alluvial layer above, so therefore the plastic failure in the soft layer is the critical design condition.

The method of analysis relies on the following characteristics of the foundation mattress:-

(a) A high tensile strength to ensure that the full shear strength of the soil is mobilised on the base.

(b) Rigidity to ensure an even distribution of the load onto the foundation material. This is provided by the high tensile strength of the geogrids and the adoption of a cellular construction.

(c) High friction on the base of the mattress. The geogrid base of the mattress allows the granular infill material to partially penetrate through the apertures, creating a rough underside to the mattress.

(iii) Design of Mattress

Reference is made to Chapter 12.5 of 'Engineering Pasticity': W Johnson and P B Mellor (VNR) which deals with the plastic failure of material between two rough, rigid, parallel platens, when the platen width exceeds the material thickness.

Using the above reference, the pressure diagram for half the embankment, after adding twice the cohesion for the effect of passive pressure beyond the toe of the embankment and neglecting the effect of upthrust within 0.45 x 2h of the toe, is shown in Fig 4.

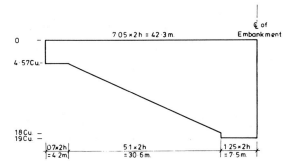

Fig 4 - Pressure Distribution Diagram

Adopting an average value of apparent cohesion of 53 kN/m² for the soft layer, the load to failure for half the embankment is:-

4.57 x 53 x 4.2   =   1017.3
(13.43/2+4.7) x 53 x 30.6 = 18302.0
19 x 53 x 7.5   =   7552.5
                            _____
                            26871.8 KN

Imposed loading from half the embankment is:-

30 x 15 x 20   =   9000   KN

This gives a factor of safety against foundation failure of 2.986

For a factor of safety reduced to 1.0 the equivalent apparent cohesion value of foundation material would be 17.73 kN/m². From a Mohr Circle construction it can be shown that the horizontal load to be resisted by the geogrid mattress is:-

$$T = \frac{Cu}{\sin \phi^1} \quad \ldots\ldots\ldots\ldots (3)$$

From Table No 2 an average value of the internal friction in terms of effective pressures ϕ^1 is 29°.

Therefore the horizontal load for Cu = 17.73 kN/m², with a factor of safety = 1 is:-

$$T_1 = \frac{17.73}{\sin 29°} = 36.57 \text{ kN/m. run}$$

This is the minimum condition and applies over the centre portion of the embankment.

The worst likely condition is when the factor of safety drops to 1.5. For the embankment under consideration, this would require an equivalent apparent cohesion of 26.6 kN/m², which is below the minimum values obtained in the laboratory.

The embankment load foundation support pressure diagram for this condition is shown in Fig 5.

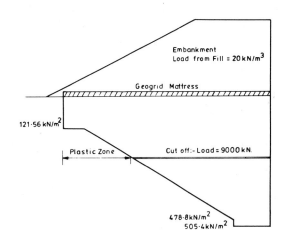

Fig 5

The extent of the plastic zone of failure is shown by the cut-off line at the foundation support value of 9000 kN.

Over this plastic zone, extending to approximately 14.0 metres from the edge of the mattress, the horizontal load in the geogrid mattress is:-

$$T_{1.5} = \frac{26.6}{\sin 29°} = 54.87 \text{ kN/m.run}$$

Assuming a working load in the SR- 2 Geogrid of 23.7 kN/m. width,(approximately 30% of the ultimate load) the strengths of 1 metre and 0.5 metre mattress cells are:-

1 metre cells $T = \frac{23.7 + 23.7}{\sqrt{2}} = 40.46$ kN/m

0.5 metre cells $T = \frac{2 \times (23.7+23.7)}{\sqrt{2}} = 80.92$ kN/m

Therefore 0.5 metre cells are required within 14 metres of the edge of the mattress ie within the plastic zone and 1.0 metre cells will suffice over the remainder of the mattress.

<u>CONSTRUCTION</u>

The foundation mattress was constructed using a base of Tensar SS-2 geogrids, with Tensar SR-2 geogrids forming the sides and diaphragms. The vertical diaphragms divided the mattress into a cellular structure consisting of 0.5 metre cells within 14 metres of both

embankment edges and 1.0 metre cells over the centre portion, as shown in Fig 6 below.
A selected filling material was then placed in all the cells.

(i) Specification

(a) <u>Geogrids</u>

	Tensar SS-2	Tensar SR-2
Structural Characteristics:		
Roll Length (m)	50.0	-
Roll Width (m)	3.0	1.0
Grid Wt(g/m^2)	320.0	938.0
Colour	Black	Black
Mechanical Properties:		
Tensile Strength(kN/m)		
– along Roll	17.0	70.0
– across Roll	36.2	-
Polymer Characteristics:		
Raw Material	Polypropylene	High Density Polyethylene

Chemical Resistance — Resistant to all natural occurring alkaline and acidic soil conditions.

Biological Resistance — Resistant to attack by bacteria, fungi and vermin.

Sunlight Resistance — Resistant to Ultra Violet attack

(b) <u>Infill Material</u>

Selected filling material was crushed rock or sand/gravel mixture, well graded from a maximum size of 50mm down to a 75 um sieve, with not more than 10% passing the 75 um sieve. It had a uniformity coefficient exceeding 10 prior to compaction.

(ii) <u>Programme of the Works</u>

Construction work in the area of the embankment mattress commenced at the beginning of September 1983, with the construction of a new culvert to carry the existing stream which then flowed

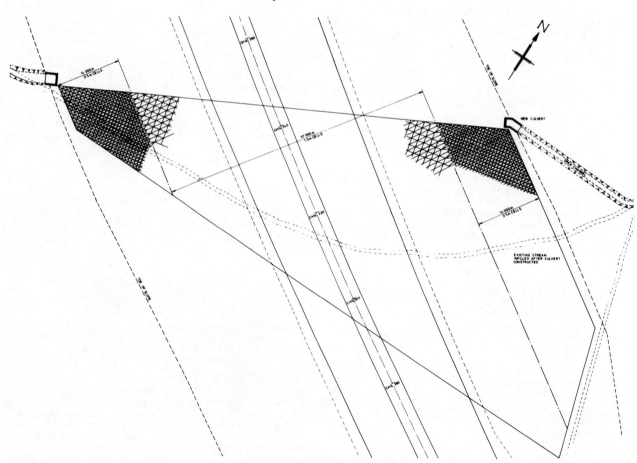

Fig 6 - Mattress Layout

across the area over which the mattress was to be constructed. While this work was proceeding a commencement was made on the mattress to the south of the stream. This necessitated a temporary crossing of the stream as a means of access for plant and materials. This access was also required for the construction of the adjacent railway crossing.

Once the culvert was finished and the stream diverted, work on the mattress was now able to proceed uninterrupted towards completion.

The above programme of the works is shown diagramatically in Fig 7 below.

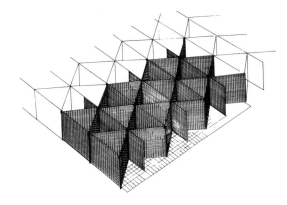

Fig 8 - Typical Mattress Layout

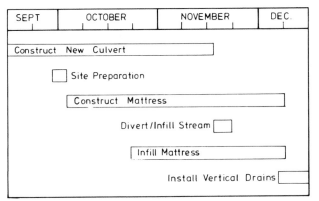

Fig 7 - Programme of Works

(iii) <u>On Site Construction</u>

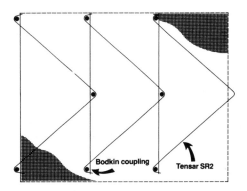

Fig 9 - Triangular Cell Arrangement

    (a) Since it was intended to construct the mattress over the existing ground without removal of topsoil or vegetation, the only preliminary work undertaken was the general levelling of the site, and the removal of any large obstructions such as boulders and the like.

    (b) The base geogrid of Tensar SS-2 in 3m wide strips was laid out over the site commencing from the southern boundary. These strips overlapped by 150 mm and were stitched together with P.V.C. binding. Transverse diaphragms of Tensar SR-2 were then attached to the base grid by similar binding and the intermediate triangular diaphragms of Tensar SR-2 were inserted and connected to the transverse diaphragms by steel bodkins.

    See Figs 8 to 10.

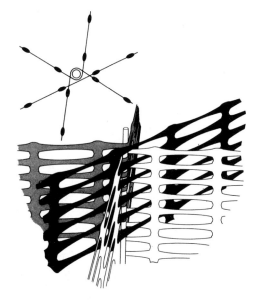

Fig 10 - Coupling of Diaphragms

Initial assembly of the mattress cells was slow and rather haphazard, until a recognised sequence of fabrication was adopted and the work force gained in competence at erection. No specific labour skills were required for the jointing and assembly work, but it was found that it was essential that the sequence of construction was adhered to and consequently a fair degree of supervision was required.

After the initial stages where progress was slow it was found that the triangular diaphragms could be more easily inserted if all the transverse diaphragms were tensioned. A simple hand winch was used for this operation. This greatly helped to maintain the shape of the mattress and to develop sound joints between the diaphragms, thus increasing the rigidity of the mattress prior to infilling.

(c) Filling of the mattress commenced from the southern boundary once sufficient cells had been constructed, and progressed northwards with the previously filled area of mattress forming a working platform for plant and machinery. A full circle, slewing back acting crawler excavator was used to fill the mattress cells, with the material being shaken and not dropped into the cells from the excavator's 0.75 cu m. bucket.

The cells were gradually filled with the specified material, taking care to fill the first two rows of cells to half height and then the first row to full height. Filling was then continued by this method, ensuring that the leading row was always half filled before the trailing row was fully filled.

No compaction of the infill material was specified, and it was found that with the well graded granular material used, no large voids were formed. The only compaction the material did receive was from the plant operating over it, and to prevent any damage to the vertical diaphragms from this plant, and to compensate for any settlement of the material within the mattress, the cells were overfilled by approximately 100 - 150 mm.

(d) On completion of the mattress, vertical band drains were installed over a period of five days. Trials previously undertaken had shown no difficulty in penetrating the base geogrid of the mattress with the mandrel used for inserting the drains. This was borne out by the ease of installation experienced on site by the contractor. All the drains were driven to refusal with the maximum penetration being approximately six metres.

COMPARISON OF COSTS

Eight reputable, experienced contractors were invited to submit tenders for the estimated £5 million advance bridge and earthworks contract. Within the earthworks section of the bill of quantities, contractors were asked to price the following two options:-

(a)   The excavation of 12,000 cu metres of unsuitable material and its replacement with a suitable imported rockfill.

(b)   The construction of a 1 metre thick geogrid mattress, covering an area of approximately 4120 sq metres, infilled with suitable granular free draining material.

The three lowest tenderers for the whole of the works, priced the geogrid mattress option cheapest. The average costs based upon these three lowest tenders were £87,000 for the mattress option and £116,000 for the excavation option. However the average costs based on all eight tenderers prices submitted were £99,000 and £96,000 respectively for the mattress and excavation options.

Advantages and Disadvantages of using a Geogrid Mattress

(i)   No excavation and minimal site preparation.

(ii)   No tips required for disposal of unsuitable excavated material.

(iii)   No difficulty working at or below the water table.

(iv)   Less prone to delays due to inclement weather.

(v)   No difficulties due to having to excavate unsuitable material adjacent to existing structures, embankments etc.

(vi)   Some initial training to familiarise the workforce with construction procedures is required.

(vii)   Labour intensive if high rate of progress to be maintained.

(viii) Supervision and strict pattern of working required.

(ix)   In order to maintain the rigidity of the mattress during construction and infilling a large degree of tensioning of the transverse vertical diaphragms is necessary.

From experience on site, this last point regarding the amount of tensioning, appeared to have been underestimated by the successful contractor.

The main disadvantage of the excavation option is the high cost of the imported rockfill, which accounted for approximately 85% of the total cost of all the tenderers prices for this option. If however a local ready supply of rockfill was available, or material was being quarried or excavated elsewhere on site during the course of the works, both of which were not available on this contract, the excavation option would obviously become more competitive.

CONCLUSIONS

By adopting the design of a geogrid mattress to carry the proposed embankment over the area of weak ground, it has been possible to increase the factor of safety against foundation failure of the weak material to an acceptable level. Careful choice of the mattress infill material has enabled the mattress to act as an effective drainage blanket for the vertical band drains which were installed through the mattress into the weak material below to accelerate settlement and alleviate any excess pore pressure developed during construction of the embankment.

Being a new innovation, progress in constructing the geogrid mattress was slow initially until the contractor's workforce had familiarised themselves with the construction procedure Once this had been achieved and a recognised sequence of construction had been adopted progress was greatly increased. This was especially noticeable after the first half of the mattress had been completed.

With regard to costs each possible application of a geogrid mattress needs to be considered on its own merits. In many instances, where the thickness of the weak layer is less than three metres, or where there is a ready supply of suitable infill material close by, the obvious solution from both design and cost considerations will be to excavate and substitute the weak material with imported fill material. However, a ready supply of infill material is not always available near the works, and excavation may prove difficult due to the presence of a high water table, site conditions or other restraints and it is in such circumstances that the use of a geogrid mattress becomes cost effective.

ACKNOWLEDGEMENTS

The author wishes to thank Mr P J Mason, Director of Highways, Lothian Regional Council for his permission to produce this paper. He would point out that the views and opinions expressed are his own and not necessarily those of the Regional Council.

Thanks are also due to Netlon Limited, Tarmac Construction, Cementation Ground Engineering Limited and W A Fairhurst, Consulting Engineers for their advice and assistance provided during the preparation of this paper.

REFERENCES

1) Capper P L, Cassie W F, Geddes J D - Problems in Engineering Soils, London E and F N Spon Ltd (1968)

2) Johnson W, Mellor P B (VNR) - Engineering Pasticity

3) Netlon Limited (1982) - Design Suggestions for the Use of High Tensile Strength Polymer Grids in the Construction of the Musselburgh Bypass Rural Section.

4) Smith G N - Elements of Soil Mechanics for Civil and Mining Engineers London, Crosby Lockwood (1971) Second Edition

5) Taylor D W - Fundamentals of Soil Mechanics (Wiley International Edition) John Wiley and Sons (London)

Embankments: report on discussion

R. H. Bassett, *King's College, London*

When the design methods described in paper 3.1 are compared with the alternative approach used by The Reinforced Earth Company a number of discrepancies and difficulties are highlighted;

a) In order to generate stresses in the reinforcement the strain necessary to mobilise the reinforcement has to be assessed.

b) If the tensile force mobilised is over-estimated then the safety factor is over-estimated.

c) The mechanism being strain-controlled, could give rise to the strain corresponding to the application of the safety factor being larger than the strain which would cause failure of the structure.

d) The linear variation of tensile forces in the reinforcement with depth, as suggested by the author, did not correspond to the distribution of maximum tensile forces either in the theoretical case for inextensible reinforcement or in his practical experience.

Mr Paine pointed out that his proposals were for application to slopes and he was not aware of how the techniques used by The Reinforced Earth Company could actually be applied to slopes. He suggested that one major difference was that in slopes there was no effect on the facing units. Application of the author's design analysis to some real structures, showed the analysis fitted well, and the Dewsbury Wall was quoted as an example.

The main problem in obtaining data from real structures was one of interpreting forces in reinforcement via the measurement of strains, particularly when the polymer grids in question were a non-uniform material.

Comparison of the author's design methods with the work of others was being undertaken as an independent study.

Experiments conducted on scaled down Geocell mattresses at the Royal Military College, Kingston, Ontario were described by Prof. Jarrett.

Vertical reference rods on the foundation passed up through the gravel filled mattresses and were used to measure vertical movement at various locations of the mattress.

Displacement of the gravel filled mattress under a beam load was compared with displacement of unreinforced gravel.

Initially there was little difference in response but by the time the beam had displaced 50mm a 50% additional load was required in the reinforced Geocell case compared to the unreinforced gravel.

The data showed that the vertical movement in the foundation was spread over a much wider area when using the mattress indicating that the Geocell mattress has an ability to spread load laterally, and hence must possess a measure of bending stiffness.

Experience in Norway on soft soils, marshes and bogs, frequently shows a shallow type of failure. It was considered that the Tensar Geocell would prevent this type of failure, not because it behaved as a rigid plate or foundation - but because it took up tensile forces in the base.

The weak soil didn't respond to mathematics, but only to creation of weak spots.

The failure in tension, rather than shear was explained by taking the example of a cube under compression from 4 sides, and if compressed one millimetre, will increase in height by 2 millimetres, whereas if compressed one millimetre vertically downward, it will extend only half a millimetre to retain the volume.

Only half the energy was therefore required to cause failure in tension rather than in shear. Supplying additional tension at the base therefore allowed the Tensar Geogrid to do a good job.

PARTICIPANTS:

Dr Juran

Mr Paine

Professor Jarrett

Mr Knutson

Polymer grid reinforcement. Thomas Telford Limited, London, 1985

Embankments: written discussion contributions

The contours shown in Fig. 1 represent the stress level within the embankment. The cellular mattress to be placed across the width of the embankment would be underlain by soft alluvial clay and by sands and gravels. It immediately became apparent that the geogrid would fail before construction was completed - in fact after about the second lift. At that stage it was evident that a single layer of cellular geogrid would not be sufficient to prevent instability. However, it was felt that it could still be of benefit because we found some rather interesting secondary results.

The first relates to the settlements. We ran a comparison with geogrid and without geogrid to see how the settlements are affected. From Fig. 2 it is evident that firstly the geogrid was not acting as a rigid plate and secondly the geogrid made virtually no difference to the settlement. So we had considerable difficulty in deciding where the benefit was originating from. That there was a benefit there was no doubt because the embankment stood up for longer than it did without the geogrid.

The next thing we looked at was the lateral displacements. In Fig. 3 it is evident that the geogrid makes a considerable reduction in the lateral displacements near the base of the embankment which is, we presume, giving the benefit in strength and allowing us to construct the embankment slightly more quickly than we could otherwise have done.

On the results of these analyses, we modified the design to reduce the stress concentrations that would snap the geogrid at an early stage (Fig. 1) by flattening the side slopes and putting a 10 m berm at the toe. The principal message is that a cellular geogrid does not produce rigid plate deformations and that the theory seems to be slightly inappropriate for this case, because the benefit seems to be related more to lateral rather than vertical stiffness.

Dr M. F. Symes, R. Travers Morgan & Partners
I have recently been involved in trying to design some embankments which are 9 m high on some very soft alluvial clay. A quick visit to the site revealed that a meandering river zigzagged its way across it and so we decided to use some sort of cellular earth reinforcement to span the channels. We then realised that we had little idea as to how to assess the benefits on the strength and at what rate we could construct the embankment. Under normal circumstances it would require three-stage construction. We decided to contact Netlon and see what they could offer in the way of theories. The only theory currently available is based on rigid plate analysis. To investigate the problem from a different point of view we decided to conduct some finite element analysis using an elastoplastic formulation. This was performed for us by the Geotechnical Consulting Group.

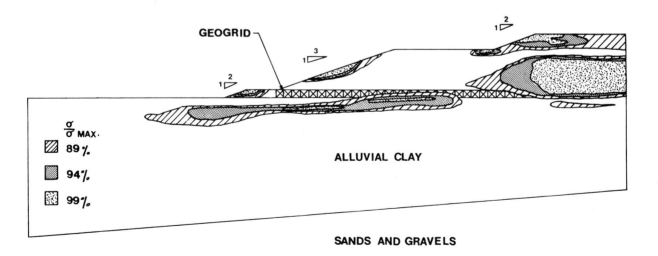

Fig. 1. Contours of stress level

Polymer grid reinforcement. Thomas Telford Limited, London, 1985

113

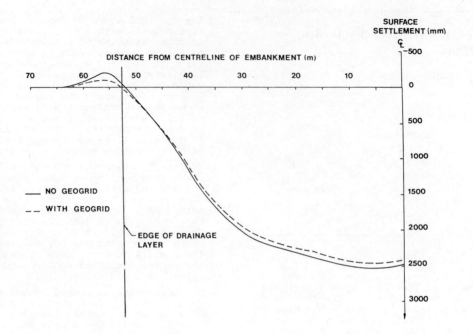

Fig. 2. Comparison of settlements with and without cellular geogrid

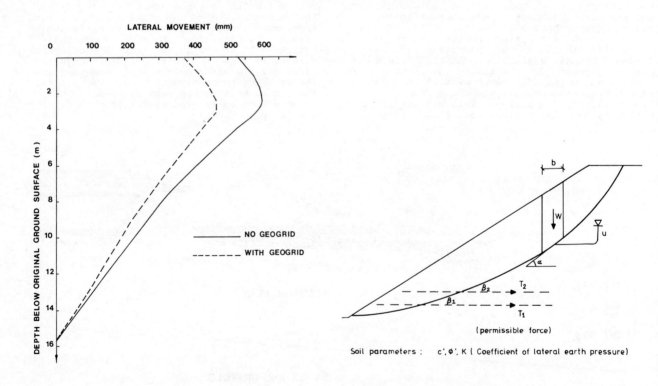

Fig. 3. Comparison of lateral movement with and without cellular geogrid

Fig. 4. Consideration of a potential slip surface in a slope with geogrid reinforcement

Mr J. R. Greenwood, Department of Transport

The papers by Murray (2.1) and Jewell et al. (3.1) present methods for analysing slope repairs and steep embankments where geogrid reinforcement is proposed. These analytical methods are somewhat complex and require considerable computer power. To assist the designer the authors produced a series of design charts.

However, the practising design engineer is often uneasy about the use of design charts, particularly where layered soil systems are present. The use of complex computer programs may leave doubts about the validity of the base assumptions and computations unless the results can be checked against a simple routine method or hand calculation.

The following simple method of analysis may be applied to check the stability of any slope or foundation and may be readily adapted to include the effects of geogrid reinforcement.

Let us consider a potential slip surface within a slope where geogrid reinforcement is proposed (Fig. 4). The factor of safety F may be defined most simply as suggested by Lambe & Whitman (1969) as

$$F = \frac{\text{shear resistance along the slip surface}}{\text{shear stress along the slip surface}}$$

The simple stability equation (1) is derived by consideration of vertical soil slices and the resolution of effective stresses to determine the effective normal force of the base of each slice. The analysis may be applied to both circular and non-circular slip surfaces (Greenwood, 1983).

$$F = \frac{\Sigma[c'\, b \sec \alpha + (W - ub) \cos \alpha \tan \phi']}{\Sigma W \sin \alpha} \qquad (1)$$

The addition of geogrid reinforcement layers may then be considered by resolving the appropriate permissible force (T) along the slip surface.

$$F = \frac{\Sigma[c'\, b \sec \alpha + (W - ub) \cos \alpha \tan \phi'] + \Sigma T \cos \beta}{\Sigma W \sin \alpha} \qquad (2)$$

As for all limit analysis solutions the strain compatibility of the geogrid and soil must be considered together with the necessary embedment lengths to prevent pull out.

Equation (2) is equally applicable to the circular and non-circular undrained ($\phi = 0$) analysis preferred by Milligan et al. in paper 3.4 (C' becomes Su and the tan ϕ' term becomes zero).

The basic simple equation (1) gives satisfactory results for all stability analyses but it may be conservative in that it ignores the contribution of lateral soil stresses to the shear resistance available. In a normal slope the contribution of lateral stresses may well be insignificant because of the proximity of the slip surface to the edge of the slope. However, the reinforced slope may benefit considerably from the available lateral stresses and it is suggested that this effect may be studied by consideration of the full stability equation based on Mohr's circle of stress (Greenwood, 1983).

$$F = \frac{\Sigma[c'\, b \sec \alpha + (W - ub)(1 + K \tan^2 \alpha) \cos \alpha \tan \phi'] + \Sigma T \cos \beta}{\Sigma W \sin \alpha} \qquad (3)$$

Considerable research and monitoring is required before the appropriate K values for use in this equation can be determined and for the present state of knowledge it is suggested that K is assumed to be zero.

References
Greenwood, J. R. (1983). A simple approach to slope stability. Ground Engineering 16, No. 4, 45-48.

Lambe, T. W. & Whitman, R. V. (1969). Soil mechanics, pp. 363-365. New York:Wiley.

gn of unpaved roads and
cked areas with geogrids

J. P. Giroud, *GeoServices Consulting Engineers,* C. Ah-Line, *Woodward-Clyde Consultants,* and R. Bonaparte, *The Tensar Corporation*

An increasing number of unpaved roads and trafficked areas, such as parking lots and drilling pads, are built using geogrids. These unpaved structures are composed of a base layer placed on a subgrade soil and reinforced with one or several layers of geogrids. At the present time, no design method for these structures is available. This paper presents the initial development of a design method for geogrid-reinforced unpaved structures. The paper begins by reviewing the mechanisms through which geogrids can improve the performance of an unpaved structure. An analysis of published full-scale unpaved road test data is presented next. The analytical development of the design method is then described, as is the preparation of design charts. The charts give base layer thickness as a function of the subgrade soil undrained shear strength, geogrid tensile stiffness, traffic, and allowable rut depth of the unpaved structure. The use of the charts is illustrated by a design example. It is pointed out that the design method is still undergoing development and that field and/or laboratory verification of the method is required. When fully developed, the method will allow designers to evaluate the cost effectiveness of geogrids as compared with other methods for constructing unpaved roads and trafficked areas.

INTRODUCTION

Geogrids (Fig. 1) are increasingly used to reinforce geotechnical structures, including unpaved roads and trafficked areas. Unpaved roads and trafficked areas ("unpaved structures"), consisting of a base layer (usually made of aggregate) placed on a subgrade soil, can be reinforced by geogrids placed within the base layer and/or, more often, at the base layer/subgrade interface. Such reinforcement improves the performance of the structure: for a given base layer thickness and a given allowable rut depth, the traffic can be increased, in comparison with the allowable traffic on the same thickness of unreinforced base layer; or, for a given base layer thickness and a given traffic, the rut depth is smaller than with an unreinforced base layer. Alternatively, for the same traffic and allowable rut depth, the use of geogrid reinforcement allows: a reduction in base layer thickness in comparison with the thickness required when the base layer is unreinforced; or construction of a base layer with material of lesser quality than usually required.

The improved unpaved structure performance associated with geogrid usage was recognized by first users. At present, however, no design method specifically developed for geogrid-reinforced unpaved structures is available. Design methods developed for geotextile-reinforced unpaved structures are not suitable because they do not take into account the mechanism of geogrid/base layer material interlocking. However, certain other mechanisms by which geogrids and geotextiles reinforce unpaved structures are similar, and consequently design methods for geogrid-reinforced unpaved structures may utilize features from existing geotextile design methods.

Starting from the design method published by Giroud and Noiray (1981) for geotextile-reinforced unpaved roads, a design method for geogrid-reinforced unpaved structures is currently being developed. This paper presents the initial analytical development of the method based on the authors' understanding of the behavior of unreinforced and reinforced unpaved structures. A preliminary design chart is presented with a design example. The design method is discussed and recommendations for additional work are presented .

CONCEPT AND BEHAVIOR OF UNPAVED STRUCTURES

Concept of Unpaved Structure

When a given traffic cannot be supported by a soil ("subgrade soil") because of insufficient bearing capacity, the traffic load should be distributed so that the stress on the subgrade soil becomes less than the subgrade soil bearing capacity. In unpaved structures, load distribution can be achieved by placing a layer of material ("base layer") between the load and the subgrade soil.

To achieve its load distribution function, the base layer should have: (i) adequate mechanical properties; and (ii) sufficient thickness.

To maintain its load distribution function under repeated loading, the base layer must remain intact, requiring that: (i) deformations of the subgrade soil be limited to prevent lateral or vertical displacements at the bottom of the base layer; and (ii) the mechanical properties of the base layer be such that it is not sheared or degraded by traffic loads.

In comparison to paved structures where only small deformations can be accepted, rather large deformations are often acceptable in unpaved structures. This is the case with construction site access roads where earthwork equipment is available to maintain the base layer (fill ruts and regrade). Since failure mechanisms and magnitude of deformations for paved and unpaved roads are usually different, it is not recommended that methods developed for paved structures be used to design unpaved structures and vice versa. For instance, using unpaved road design methods to evaluate the effect of geotextiles or geogrids in paved roads is inadequate.

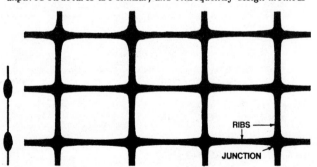

Fig. 1   Typical geogrid used in unpaved structures.

Polymer grid reinforcement. Thomas Telford Limited, London, 1985

Performance of an Unreinforced Unpaved Structure

The performance of an unreinforced unpaved structure is considered satisfactory until failure occurs. A road or trafficked area is considered failed when surface deformations are such that conditions for traffic become unacceptable. (The criterion usually considered is the rut depth, defined subsequently.) Unacceptable deformations can occur after one axle passage if the axle load exceeds the bearing capacity of the structure. Unacceptable deformations can also occur after several passages, as a result of:

o   accumulation of small permanent deformations occuring at each axle passage and accelerated by progressive deterioration of the structure; and/or

o   large deformation associated with shear failure of the structure which occurs when the bearing capacity becomes smaller than the axle load as a result of progressive deterioration of the structure.

Deterioration of the structure results from deterioration of the subgrade and deterioration of the base layer. Progressive deterioration of the subgrade results from the decrease of shear strength of the subgrade soil due to fatigue generated by repeated loading. Progressive deterioration of the base layer occurs through one or a combination of the following mechanisms affecting its thickness and/or mechanical properties:

o   lateral displacement of base layer material resulting from tensile and shear strains related to bending and low confining stresses at the bottom of the base layer;

o   contamination of base layer by fine particles moving upward from subgrade;

o   sinking of base layer aggregate into subgrade soil; and

o   breakdown of base layer aggregate due to repeated loading and abrasion.

Progressive deterioration of the structure has three consequences:

o   acceleration of the occurence of unacceptable surface deformations resulting from accumulation of small permanent deformations;

o   decreased shear strength of both base layer material and subgrade soil and increased risk of shear failure; and

o   decreased ability of the base layer to distribute loads (because of thickness reduction and decrease in mechanical properties of the base layer), resulting in increased applied stress to the subgrade.

Behavior of a Reinforced Unpaved Structure

Since failure of an unpaved structure results from base layer and/or subgrade failure, either or both should be prevented or delayed to improve the performance of the structure. Geogrids can be placed at the base layer/subgrade interface or at various levels within the base layer. They can improve performance and prevent or delay progressive deterioration of the unpaved structure as discussed below.

Influence of a Geogrid on Base Layer Behavior

Geogrids can improve the performance of the base layer as a result of interlocking between geogrid and base layer material (especially if this material is aggregate). By interlocking with the base layer aggregate, geogrids provide reinforcement which can prevent shear failure and reduce permanent deformations of the base layer. According to McGown and Andrawes (1977), optimum reinforcement depth is approximately 0.3 times the width of the load. Raymond and Hayden (1983) found the optimum depth to be in the range of 0.3 to 0.6 times the load width. The width of a typical dual wheel contact area is approximately 0.3m (as discussed later in this paper). Consequently, the optimum depth of geogrid would be of the order of 0.1m to 0.2m. It is also important to consider any possible detrimental effects of reinforcement on the behavior of base layers. For instance, it is known that a smooth slippery surface at a depth smaller than the width of the load decreases the shear strength of the base layer. This would be the case of a smooth surface geotextile placed at a depth less than approximately 0.3m under a typical dual wheel. Geogrids do not act as slip surfaces because of their interlocking with aggregate, and, therefore, the considered detrimental effect is not relevant.

Geogrids can prevent or delay progressive deterioration of the base layer (i.e., base layer thickness reduction and decrease in base layer material properties) as a result of interlocking between geogrid and base layer material. The influence of geogrids on the actions leading to progressive deterioration of the base layer aggregate is discussed below:

o   Lateral Displacement of Base Layer Aggregate. - By interlocking with the base layer aggregate, geogrids reduce permanent lateral displacements which accumulate with increasing numbers of load repetitions. Optimum placement of the geogrid is as described above. Reduced displacements result in reduced deterioration of base layer material properties while preserving the effective thickness of the base layer.

o   Contamination. - By interlocking with the base layer aggregate, a geogrid, placed at the bottom of the base layer, prevents lateral movements of aggregate, thus preventing development of openings at the bottom of the base layer and limiting the rapidity and extent of penetration of fine subgrade soil particles into the aggregate. The effectiveness of geogrids in limiting contamination of base layer material by fine soil particles depends on several parameters including : subgrade soil consistency, water content and shear strength; and, particle size distribution of aggregate.

o   Aggregate Sinking. - By interlocking with the base layer aggregate, a geogrid prevents aggregate stones from sinking into the subgrade soil. Experience indicates that aggregate sinking is minimized through use of well-graded materials with maximum particle size at least one-half the width of the grid aperture.

o   Aggregate Breakdown. - By interlocking with the base layer aggregate and reducing movements, geogrids can decrease the amount of aggregate breakdown. This is particularly important when weak aggregate is used.

Influence of Geogrids on Subgrade Soil Behavior

Geogrids can improve performance of the subgrade soil in a way that is different from the way they improve the performance of the base layer. Unlike the base layer, subgrade soil properties cannot be improved or maintained through reinforcement. The performance of the subgrade can be improved only through changes in the boundary conditions, i.e., applied stresses and deformations. Geogrids can improve the performance of the subgrade soil through three mechanisms: confinement (which reduces deformations); and improved load distribution through the base layer and tensioned membrane effect (which reduces stresses).

o   Confinement. - When the vertical stress on the subgrade soil of an unreinforced unpaved structure exceeds the elastic limit (assuming elasto-plastic behavior), local shear failure and large deformations ensue.   These large deformations result in accelerated deterioration of the base layer and fatigue of the subgrade soil, causing the subgrade soil to be subjected to ever higher stress levels (i.e., an increasing ratio between applied and allowable stress).   Consequently, after a relatively small number of vehicle passages the plastic limit (ultimate bearing capacity) of the subgrade soil is exceeded and general shear failure occurs.   Experience has shown that if the subgrade soil is confined, deformations resulting from local shear failure do not become large and the subgrade soil can support a vertical stress close to its plastic limit (which is significantly higher that its elastic limit).   Confinement requires continuity of the reinforcing element.   Geotextiles provide confinement because they have small openings. Geogrids provide confinement if the quantity of subgrade soil escaping through geogrid openings is not sufficient to cause heave of the base layer.

o   Load Distribution. -   As explained previously, load distribution results from the mechanical properties and thickness of the base layer.   Geotextiles, by separating base layer and subgrade soil, prevent aggregate contamination and aggregate sinking.   Separation, therefore, reduces degradation of the mechanical properties of the aggregate and helps maintain the thickness of the base layer.   However, as the unpaved road structure deforms, the base layer aggregate spreads laterally.   Due to the absence of interlocking, a geotextile placed at the base layer/subgrade interface, does little to prevent lateral displacement of the base layer aggregate. In contrast, a geogrid is expected to reduce lateral displacement due to interlocking with aggregate.   Also, the geogrid tensile stiffness adds to the stiffness of the base layer.   The result is a stiffer base layer with reduced deterioration.   The base layer is therefore able to provide better applied load distribution than possible with unreinforced, or geotextile-reinforced, base layers.

o   Tensioned Membrane Effect. -   If the subgrade soil is incompressible (such as saturated clay), deformation of the subgrade soil under the wheels causes heave between and beyond the wheels.   Therefore, the geogrid exhibits a wavy shape; consequently, it is stretched.   When a stretched flexible material has a curved shape, normal stress against its concave face is higher than normal stress against its convex face.   This is known as the "tensioned membrane effect".   Therefore: (i) between the wheels and to a lesser extent beyond the wheels, the normal stress applied by the geogrid on the subgrade is higher than the normal stress applied by the base layer on the geogrid; and (ii) under the wheels, the normal stress applied by the geogrid on the subgrade is smaller than the normal stress applied by the wheels plus the base layer on the geogrid.   This action provides two beneficial effects, confinement of the subgrade soil between and beyond the wheels (as discussed above), and reduction of the stress applied by the wheels on the subgrade.   The stress reduction resulting from the tensioned membrane effect is effective if traffic loads are repeated at exactly the same location (i.e., if the traffic is channelized), and if the tensile stress in the reinforcement does not decrease with time (i.e., if the reinforcement has low creep susceptibility at working loads).

Geogrids can reduce subgrade soil deterioration by reducing the magnitude of repeated deformation of the structure, thereby reducing disturbance of the subgrade soil and, consequently, fatigue resulting in a progressive decrease in shear strength of the subgrade soil.

DESIGN PARAMETERS

Geometry of Unpaved Structure

The thickness of base layer is termed h_0 when there is no reinforcement and h when there is reinforcement.   The subscript s is affixed to h_0 and h when the axle load is the standard axle load (defined subsequently).   The ratio, R, between h and h_0 for a same axle load is called the thickness ratio.   If there is only one geogrid, it is assumed to be at the base layer/subgrade interface.   If there are two geogrids, it is assumed that one is at the base layer/subgrade interface and that the other is in the base layer; the exact location of this second geogrid is not specified because this parameter is not considered in the method presented hereafter.   The optimum depth of the second geogrid layer may be obtained from the previous discussion on the behavior of reinforced unpaved structures.   The subgrade soil is assumed to be homogeneous, at least over a thickness, H, sufficient to allow development of subgrade soil failure.   The thickness H can be estimated using a classical bearing capacity analysis which shows that H is usually less than 1.5m.   Therefore, the assumption of homogeneous subgrade soil is often valid.

Traffic

Traffic can be channelized, as is the case on a road, or unrestricted over an area.   A channelized traffic is characterized by the number of passages, N, of a given axle during the design life of the structure.   A traffic unrestricted over an area is less easy to characterize and some judgement on the designer's part is required.   However, in most trafficked areas there are locations, such as entrances and exits, where traffic is channelized and can therefore be characterized by a number of passages.

Axles and Loads

The geometry of axle is depicted in Fig. 2a in which e is the distance between the midpoints of the two sets of wheels. Dual wheels are considered because they are more common than single wheels for trucks using unpaved structures.

The axle load, P, is considered to be evenly distributed between the four wheels:

$$P = 4A_c p_c \qquad (1)$$

where: P = axle load (N); A_c = contact area of a tire (m^2); and p_c = tire inflation pressure (N/m^2), assumed to be equal to the average value of the actual contact pressure (non-uniformly distributed) between each tire and the base layer.

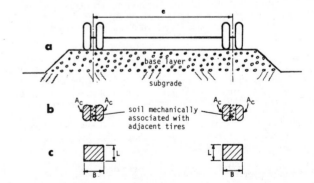

Fig. 2   Vehicle axle and contact area: (a) Geometry of vehicle axle with dual wheels; (b) Tire contact areas; and (c) Equivalent contact area used in analysis.

The soil between the tires of a dual wheel is mechanically associated with these tires (Fig. 2b). Since no failure of the base layer material and subgrade soil is expected to occur between the tires, each double contact area $2A_c$ is replaced in the theoretical study by a rectangle L x B of larger area (Fig. 2c). By examining several dual tire prints, the following value appears reasonable:

$$LB = 2A_c\sqrt{2} \qquad (2)$$

The actual contact pressure (nonuniformly distributed) between each tire and the base layer induces the same mechanical effects in the subgrade soil as an "equivalent contact pressure", p_{ec}, (assumed uniformly distributed) between the rectangle L x B and the base layer; therefore:

$$P = 2LBp_{ec} \qquad (3)$$

The relationship between the equivalent contact pressure, p_{ec}, and the tire inflation pressure, p_c, is deduced from Eqs. 1, 2, and 3:

$$p_{ec} = p_c /\sqrt{2} \qquad (4)$$

Examination of typical dual tire prints leads to the following approximate value for the length L, in the case of on-highway trucks:

$$L = B /\sqrt{2} \qquad (5)$$

Eliminating L from Eqs. 3 and 5 and using Eq. 4 yield:

$$B = \sqrt{P/p_c} \qquad (6)$$

Eq. 6 is useful for the subsequent analysis because vehicles are usually characterized by axle load, P, and tire inflation pressure, p_c, which is almost equal to the actual contact pressure as previously stated. For the American-British standard axle load ($P = P_s = 80$ kN) and a tire inflation pressure of 620 kN/m^2, L = 0.25m and B = 0.36m.

Different relationships would be necessary for other types of trucks such as off-highway trucks, and construction equipment.

Also, replacement of the actual contact area by the rectangle LxB is valid only for analyzing effects on the subgrade (the only failure mechanism explicitly considered in the proposed design method). It may not be appropriate for evaluating shear failure of the base layer.

Rut Depth

When the traffic is channelized, ruts develop at the surface of the base layer. The rut depth, r, is the vertical distance between the highest point of the base layer surface between the wheels and the lowest point of the rut. In the case of trafficked areas, where the traffic is not channelized, an erratic pattern of ruts develop. The rut depth can then be defined as the vertical distance between high spots and low spots of the base layer surface in the considered area.

Properties of Base Layer

In this study, the base layer consists of an aggregate which is assumed to have the properties usually required to ensure adequate distribution of the applied load. In a practical sense, this means that the California Bearing Ratio (CBR) of the aggregate is larger than 80.

No other base layer material (such as sand, silt, etc.) is considered in this study.

Properties of Subgrade Soil

The subgrade soil is assumed to be saturated and to have a low permeability (silt, clay). Therefore, under quick loading (such as traffic loading), the subgrade soil behaves in an undrained manner. Practically, this means that the subgrade soil is incompressible and frictionless. Consequently its shear strength is equal to its undrained cohesion, c_u.

The value of c_u is measured in the laboratory using unconsolidated undrained or unconfined triaxial tests, or quick direct shear tests; and in the field using vane shear test. The value of c_u can also be approximated using a cone penetrometer or deduced from CBR value (less than 5) using one of the following relationships:

$$c_u = q_c/10 \qquad (7)$$

$$c_u \ (N/m^2) = 30\ 000\ CBR \qquad (8)$$

where: q_c = cone resistance (N/m^2); and CBR = California Bearing Ratio (dimensionless). Equations 7 and 8 should be replaced by site-specific strength correlations where available.

Properties of Geogrids

The geogrid mechanical properties relevant to this study are described by the geogrid behavior in a uniaxial tensile test giving the force per unit width, F/w, as a function of the elongation, ε.

Force per unit width-elongation curves related to the longitudinal and the transverse direction of the geogrid Tensar SS2 are presented in Fig. 3. The definition of the secant tensile stiffness, K, is shown on one of these curves.

The curves presented in Fig. 3 have been obtained for a rate of elongation of 25% per minute, at 20°C. This rate of elongation corresponds approximately to the quick loading by a moving vehicle. Values of the tensile stiffness obtained for a 2% elongation are 400 kN/m for the longitudinal direction and 700 kN/m for the transverse direction. An average value of 500

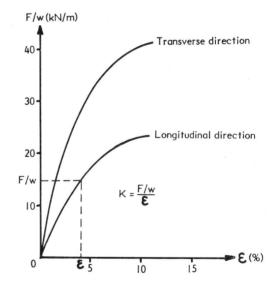

Fig. 3   Typical force per unit width - elongation curve of geogrid. K is the secant tensile stiffness of the geogrid at elongation ε.

kN/m has been considered in the development of the design method. The average value for the tensile stiffness of the geogrid Tensar SS1 is estimated at 300 kN/m, based on limited test data.

The tensile behavior of polymeric material depends on the rate of elongation. Smaller values of tensile stiffness and tensile strength, and larger values of elongation at failure would be obtained for lower rates of elongation.

In addition to the tensile behavior, it is necessary to address the friction characteristics of the geogrid/base layer aggregate interface. The limited available data leads to the conclusion that for the types of geogrids and base layer aggregate considered herein, the frictional resistance at the geogrid/base layer aggregate interface is approximately equal to the frictional resistance of the base layer aggregate alone. Thus, geogrids, through their interlocking with the base layer material, have adequate friction characteristics to prevent failure by sliding along the geogrid/base layer interface.

DESIGN OF UNREINFORCED UNPAVED STRUCTURES

The proposed method for designing geogrid-reinforced unpaved structures has been established using the design of unreinforced unpaved structures as a starting point.

Test Results on Unreinforced Unpaved Roads

An extensive test program on unreinforced unpaved roads has been conducted by the Corps of Engineers (Hammit, 1970). The failure criterion for this study was a rut depth of 0.075 m. From the test results, Webster and Alford (1978) established a chart giving the thickness of the base layer as a function of the number of passages, N, and the CBR of the subgrade soil for a standard axle load, P_s = 80 kN. Giroud and Noiray (1981) found the following formula to be in good agreement with Webster and Alford's chart:

$$h_{os} = 0.19 \log N / (CBR)^{0.63} \qquad (9)$$

where: h_{os} = base layer thickness (subscript o refers to unreinforced and s to standard axle) (m); N = number of passages of standard axle (load P_s = 80 kN); and CBR = California Bearing Ratio of subgrade soil. The use of Eq. 9 is restricted to standard axle load and 0.075 m rut depth.

Combining Eqs. 8 and 9 yields:

$$h_{os} = 125 \log N / c_u^{0.63} \qquad (10)$$

where: c_u = undrained shear strength (cohesion) of the subgrade soil (h_{os} in m and c_u in N/m^2 exclusively).

To extend Eq. 10 to rut depths r other than 0.075 m, Giroud and Noiray have proposed replacing log N by $[\log N - 2.34 (r - 0.075)]$. This expression has been empirically deduced from test results presented by Webster and Watkins (1977), showing that the increase of rut depth with number of passages is much more marked as soon as the rut depth exceeds 0.075 m. Eq. 10 thus becomes:

$$h_{os} = \left[125 \log N - 294 (r-0.075)\right] / c_u^{0.63} \qquad (11)$$

(h_{os} and r in m and c_u in N/m^2 exclusively)

This equation has been established by extrapolation and should not be used when the number of passages exceeds 10 000.

A chart for the design of unreinforced unpaved roads, established using Eq. 11 is presented in Fig. 4.

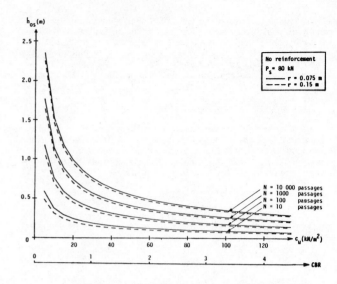

Fig. 4   Unreinforced base layer thickness, h_{os}, versus subgrade soil shear strength, c_u, for a standard axle load (P_s = 80kN) and two values of rut depth r.

Interpretation of Test Results on Unreinforced Unpaved Roads

As explained earlier in this paper, progressive deterioration of the subgrade soil and the base layer are major factors leading to the failure of unpaved roads. Therefore, it is important to identify the relative influence of these two mechanisms in the above mentioned test data.

Progressive Deterioration of the Subgrade Soil

The mechanical property which governs the behavior of the subgrade soil is its undrained shear strength. Therefore, the progressive deterioration of the subgrade soil can be expressed by a decrease of its undrained shear strength when the number of passages increases:

$$c_{uN} = \lambda c_u \qquad (12)$$

where: c_{uN} = undrained shear strength (cohesion) of the subgrade soil at the Nth passage of the axle (N/m^2); and c_u = undrained shear strength of the subgrade soil before or at the first passage of the axle (N/m^2).

The coefficient λ represents the progressive deterioration or fatigue of the subgrade soil generated by repeated loading due to traffic. The following conditions should be met by λ :

$\lambda = 1$ if N = 1   ($c_{uN} = c_u$)

$\lambda = 1$ if $c_u = 0$ (a liquid is not susceptible to fatigue)

$\lambda < 1$ if $c_u > 0$ and $N > 1$ ($c_{uN} < c_u$)

These conditions can be met using various empirical equations, including:

$$\lambda = c_{uN}/c_u = 1/\left[1 + (\log N)^{3/2} c_u/1000\right] \qquad (13)$$

(with c_u in kN/m^2 exclusively)

Values of λ calculated with Eq. 13 are presented in Fig. 5.

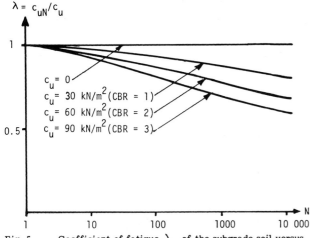

Fig. 5   Coefficient of fatigue, λ, of the subgrade soil versus number of passages, N, calculated using Eq. 13.

Progressive Deterioration of the Base Layer

Failure of the subgrade soil occurs where applied stresses exceed the bearing capacity of the subgrade soil. As discussed earlier in this paper, progressive deterioration of the base layer gradually increases stresses on the subgrade soil. These stresses could be estimated using linear or nonlinear elastic or elasto-plastic analyses. Such analyses would necessitate detailed numerical calculations. A simpler approach, used herein, consists of assuming that the base layer provides a pyramidal distribution of the applied wheel loads (Fig. 6). The pyramidal distribution is a classic assumption in soil mechanics as discussed by Giroud (1982), and is suitable for the comparison of peak vertical stress magnitudes, as used herein. The vertical stress, p_{os}, on the subgrade, calculated using the pyramidal distribution, is.:

$$p_{os} = (P_s/2)/\left[(B+2\,h_{os}\,\tan\alpha_o)(L+2\,h_{os}\,\tan\alpha_o)\right]+\gamma h_{os} \qquad (14)$$

where: P_s = standard axle load (80 kN); $P_s/2$ = dual-wheel load (40 kN); B and L = width and length respectively of the rectangular area replacing the actual contact area between a dual wheel and the surface of the base layer (m); h_{os} = base layer thickness (m); α_o = load distribution angle for unreinforced base layer; γ = unit weight of soil (N/m³); and γh_{os} = gravity stress at depth h_{os} (i.e., at the bottom of the base layer) (N/m²) (Note: subscript o refers to unreinforced and subscript s to standard axle).

In Eq. 14, it is assumed that the axle is long enough that the pyramids related to the two dual wheels do not interfere. This assumption is valid for most values of α_o and h_{os} typically considered.

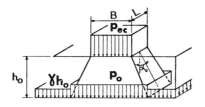

Fig. 6   Concept of pyramidal load distribution.

Progressive deterioration of the base layer may be modeled by a decrease of load distribution angle and/or effective base layer thickness when the number of passages increases. A decrease in the load distribution angle is considered in this study and is used to evaluate the progressive increase of applied subgrade stress due to base layer deterioration. To evaluate the assumed decrease of this parameter in the test data presented in the previous section, the value of p_{os} obtained using Eq. 14 should be compared with the maximum stress that can be allowed on the subgrade soil.

As discussed previously for unreinforced unpaved roads, the deformation of the surface of the subgrade soil, and therefore the rut depth, becomes large if the vertical stress on the subgrade soil exceeds the elastic limit, given by:

$$p_e = \pi c_{uN} + \gamma h_{os} \qquad (15)$$

where: c_{uN} = undrained shear strength (cohesion) of the subgrade soil at the Nth passage of the axle (N/m²); and γh_{os} = gravity stress at depth h_{os} (i.e., at the bottom of the base layer) (N/m²).

Combining Eqs. 14 and 15 yields:

$$\lambda c_u = c_{uN} = (P_s/2)/\left[\pi(B+2h_{os}\,\tan\alpha_o)(L+2h_{os}\,\tan\alpha_o)\right] \qquad (16)$$

where: $\lambda = c_{uN}/c_u$ = ratio between undrained shear strength of subgrade soil at the Nth passage and at the first passage.

The load distribution angle, α_o, can be calculated by combining Eqs. 5, 6, 9, 13 and 16, and solving the resulting quadratic equation:

$$\tan\alpha_o = \frac{\sqrt{(\sqrt{2}-1)^2 P_s/(2p_c)+2P_s/(\lambda\pi c_u)} - (\sqrt{2}+1)\sqrt{P_s/(2p_c)}}{6.5\log N/c_u^{0.63}} \qquad (17)$$

where: P_s = standard axle load (80 kN); p_c = tire inflation pressure (kN/m²); and c_u = undrained shear strength (cohesion) of the subgrade soil (kN/m²).

Values of $\tan\alpha_o$ calculated using Eq. 17, with a tire inflation pressure $p_c = 620$ kN/m², are presented in Fig. 7 which shows that $\tan\alpha_o$ decreases as the number of passages increases. This illustrates that the ability of the base layer to distribute

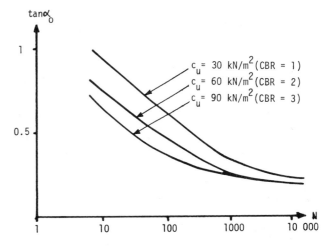

Fig. 7   Load distribution angle, $\tan\alpha_o$, versus number of pasages, N, calculated using Eq. 17.

121

load decreases as the base layer progressively deteriorates due to repeated loading. This effect is more marked when the shear strength of the subgrade soil is smaller.

Although Eq. 13 is arbitrary, it is qualitatively valid because it meets the conditions for representing the progressive deterioration of the subgrade shear strength as explained above. Trends derived from Eq. 13 and shown in Fig. 5 and 7 can therefore be considered to qualitatively represent real behavior. These trends can be summarized as follows:

o   Traffic causes progressive deterioration of the subgrade soil, expressed as a decrease of the subgrade soil shear strength when the number of passages increases. This deterioration is more marked for subgrade soils with a higher strength (Fig. 5).

o   Traffic causes progressive deterioration of the base layer. This deterioration is more marked when the subgrade soil is softer (Fig. 7) because the softer the subgrade soil, the larger the magnitude and/or rate of action of the mechanisms for progressive deterioration cited previously (lateral displacement of base layer aggregate, contamination by fine subgrade soil, sinking of aggregate into subgrade soil, and breakdown of aggregate particles).

These findings, resulting from a theoretical interpretation of full scale tests, are in good agreement with generally accepted modes of failure in roads. From an immediate and practical view point, the above discussion provides the understanding and equations necessary to extend the design method for unreinforced unpaved structures with standard axle loads to unreinforced unpaved structures with any axle load and, more importantly, to the case of geogrid-reinforced unpaved structures.

Design of Unreinforced Unpaved Structures for Any Axle Load

Eqs. 10 and 11, established from test data, are valid only for standard axle loads (i.e., $P_s = 80$ kN). However, similar equations can be derived for other axle loads using Eq. 16, if it is assumed that base layer and subgrade behave in a similar manner as in the above discussed full scale tests, and that this behavior is dependent on the subgrade stress level.

Eq. 16 can be written for any axle load P, corresponding to a thickness h_o, by replacing P_s and h_{os} by P and h_o, respectively. Combining the two versions of Eq. 16 and solving the resulting quadratic equation leads to:

$$h_o = \frac{\sqrt{(B-L)^2 + 4(B+2h_{os}\tan\alpha_o)(L+2h_{os}\tan\alpha_o)(P/P_s)} - (B+L)}{4\tan\alpha_o} \quad (18)$$

where: h_o = base layer thickness corresponding to an axle load P (m); h_{os} = base layer thickness corresponding to the standard axle load $P_s = 80$ kN (m); α_o = load distribution angle for unreinforced base layer; and B and L = width and length respectively of the rectangular area replacing the actual contact area between a dual wheel and the surface of the base layer (m). (Note: for $P = P_s$, it is easy to verify that Eq. 18 gives $h_o = h_{os}$.)

Replacing h_{os} in Eq. 18 by its value given in Eq. 10 or 11 enables one to determine the thickness of an unreinforced unpaved base layer, h_o, corresponding to any axle load P, and to prepare charts valid for any axle load similar to the charts presented in Fig. 4 for a standard axle load. Eq. 18 is valid only if the wheel print (represented by the rectangle B x L) is the same, or approximately the same, for the considered axle and the standard axle. A more complex equation would be necessary if the geometry of the two dual wheels were significantly different.

DEVELOPMENT OF A DESIGN METHOD FOR GEOGRID REINFORCED UNPAVED STRUCTURES

A discussion of the mechanisms through which a geogrid can improve the behavior of an unpaved structure has been presented earlier in this paper. The design method presented hereafter takes into account only three mechanisms: confinement of the subgrade soil, improved load distribution and tensioned membrane effect.

Influence of Confinement of the Subgrade Soil

As discussed previously, the vertical stress on the subgrade soil can be as large as the ultimate bearing capacity of the soil when confinement is provided. The ultimate bearing capacity (i.e., plastic limit) of the subgrade soil is expressed by:

$$p_{lim} = (\pi+2)\,c_{uN} + \gamma h \quad (19)$$

where: c_{uN} = undrained shear strength (cohesion) of the subgrade soil at the Nth passage of the axle (N/m^2); and γh = gravity stress at depth h (i.e., at the bottom of the base layer) (N/m^2).

A comparison between Eqs. 15 and 19 outlines the benefit resulting from subgrade soil confinement by a geogrid; the term $\pi\,c_{uN}$ in Eq. 15 referring to elastic limit (local shear) is replaced in Eq. 19 by the term $(\pi+2)c_{uN}$ referring to plastic limit (general shear).

It has been assumed that the value of the undrained shear strength, c_{uN}, at the Nth passage for the confined subgrade is the same as for the unconfined subgrade, if the ratio between confined and unconfined subgrade stress is equal to the ratio between the plastic and elastic limit of the subgrade soil. The rationale behind this assumption is that fatigue of subgrade soil results from remolding caused by repeated deformations; and since deformations are of equivalent magnitudes in an unconfined soil at the elastic limit and a confined soil at the plastic limit, fatigue will be the same in both cases.

Influence of Load Distribution

The improvement in load distribution capability of the reinforced base layer relative to the unreinforced base layer can be quantified by replacing the angle α_o used in Eq. 14 by a larger angle α. The vertical stress transmitted by the base layer to the upper face of the geogrid becomes:

$$p' = (P/2)/\left[(B+2h\tan\alpha)(L+2h\tan\alpha)\right] + \gamma h \quad (20)$$

where: P = axle load (N); P/2 = dual-wheel load (N); B and L = width and length respectively of the rectangular area replacing the actual contact area between a dual wheel and the surface of the base layer (m); h = base layer thickness (m); α = load distribution angle for reinforced base layer; γ = unit weight of soil (N/m^3); and γh = gravity stress at depth h (i.e., at the bottom of the base layer) (N/m^2).

In Eq. 20, it is assumed that the axle is long enough that the pyramids related to the two dual wheels do not interfere. This assumption is valid for most values of α and h typically considered.

The determination of the load distribution angle of a geogrid-reinforced base layer is discussed later.

Influence of the Tensioned Membrane Effect

As explained previously the normal stress is not the same on both sides of a reinforcing element exhibiting a tensioned membrane effect. Consequently, the vertical stress on the lower (convex) side of the geogrid under the wheels is:

$$p = p' - p_m \qquad (21)$$

where: p' = vertical stress on the upper (concave) side of the geogrid given by Eq. 20 (N/m^2); and p_m = normal stress difference between the two sides of the geogrid resulting from the tensioned membrane effect (N/m^2).

The magnitude of the tensioned membrane normal stress, p_m, has been evaluated by Giroud and Noiray (1981) as a function of the tensile stiffness and elongation of the reinforcement and the shape of the deformed surface of the subgrade soil.

Combined Influence of the Three Effects

The equations presented above for the three effects, confinement, load distribution and tensioned membrane effect, are combined to obtain the ratio $R = h/h_o$ ("thickness ratio") between the thicknesses of the base layer with and without reinforcement respectively.

Eliminating γ between Eqs. 14 and 15 written for a load P (i.e., replacing P_S and h_{os} by P and h_o, respectively), and Eqs. 19 and 20, p' between Eqs. 20 and 21, and c_{uN} between Eqs. 15 and 19, and solving the resulting quadratic equation lead to:

$$R = h/h_o = \left[\sqrt{(B-L)^2+4Y} - (B+L)\right]/(4h_o \tan\alpha) \qquad (22)$$

where: R = thickness ratio (dimensionless); h and h_o = thicknesses (m) of the base layer with and without reinforcement, respectively, for an axle load P; B and L = width and length respectively of the rectangular area replacing the actual contact area between a dual wheel and the surface of the base layer (m); α = load distribution angle of the reinforced base layer; and Y = term given by Eq. 23 (m^2):

$$Y=1/\left[(1+2/\pi)/(B+2h_o\tan\alpha_o)(L+2h_o\tan\alpha_o)+2p_m/P\right] \qquad (23)$$

where: P = axle load (N); B and L = width and length, respectively, of the rectangular area replacing the actual contact area between a dual wheel and the surface of the base layer (m); and h_o and α_o = thickness and load distribution angle, respectively, of the unreinforced base layer; and p_m = normal stress difference between the two sides of the geogrid (N/m^2).

As noted previously, the value of p_m is obtained from several lengthy equations, making it impractical to prepare a limited number of simple charts if p_m is taken into account. However, Y (and consequently R) are simplified if p_m is neglected. Systematic comparisons of values of R calculated with and without p_m have shown that: (i) if the rut depth is 0.075m the effect of p_m on R (hence on the design thickness of the base layer) is negligible; and (ii) if the rut depth is 0.15m, the value of R calculated with p_m is approximately 10% smaller than the value of R calculated neglecting p_m, regardless of the other parameters.

Consequently, in the design method presented hereafter, the normal stress difference, p_m, resulting from the tensioned membrane effect is neglected, and when it is not negligible a lump reduction of 10% of the design thickness of the base layer is recommended.

Values of the thickness ratio, R, calculated using Eqs. 22 and 23 with $\tan\alpha_o = 0.6$ are presented in Fig. 8. These values were obtained using values of B and L corresponding to a tire inflation pressure of 620 kN/m^2. However, almost identical values are obtained for a wide range of tire inflation pressures, provided the thickness of the base layer is at least 0.15m.

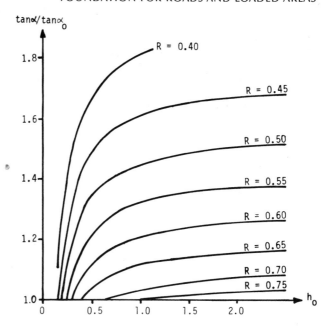

Fig. 8   Thickness ratio, R, versus load distribution improvement ratio, $\tan\alpha/\tan\alpha_o$, and thickness of unreinforced base layer, h_o.

For simplicity, a single value of $\tan\alpha_o$ (0.6) was used for calculation of R. A parametric study using a range of values for $\tan\alpha_o$ (0.4 to 0.8) showed that R was only slightly influenced by $\tan\alpha_o$, particularly for h_o values greater than 0.3m. Thus, while it is recognized that a value of $\tan\alpha_o$ of 0.6 may not represent the actual stress distribution, it has little influence on the ratio $\tan\alpha/\tan\alpha_o$ and therefore little influence on the calculated thickness of aggregate.

To use Fig. 8 to determine the thickness ratio, R, for design of a reinforced unpaved structure, the load distribution angle, α, corresponding to the considered reinforcement must be established. The purpose of the next section is to present an approach for determining α.

STUDY OF THE LOAD DISTRIBUTION ANGLE

Purpose and Approach

The purpose of reinforcing a base layer is to increase its load distribution capability by improving its mechanical properties. Improving the mechanical properties of the base layer does not necessarily mean improving the properties of the base layer as constructed. It means ensuring that the properties of the base layer after a given number of passages are superior to the properties that the base layer would have had without reinforcement after the same number of passages. This can be achieved if the reinforcement prevents or delays deterioration of the base layer in addition to, or instead of, improving the properties of the base layer as constructed.

The analyses presented below have been conducted to evaluate the effectiveness of geogrids in improving the mechanical properties of base layers and preventing their degradation.

123

Scope of Elastic Analyses

Linear elastic theory has been used to describe the behavior of all materials involved: base layer material subgrade soil and geogrid. Although in reality the subgrade is loaded at the plastic limit, elastic analyses were conducted because elastic theory allows the vertical stresses to be calculated with reasonable accuracy and because it is the simplest way to evaluate the effect of a variety of parameters such as:

o   Material properties: elastic moduli of base layer material and subgrade soil; tensile stiffness of the reinforcement; effect of contamination and/or very low confining stress on base layer material .

o   Geometry: thickness of base layer, thickness of deteriorated base layer, location of reinforcement.

To simplify the analyses, a dual-wheel load of 40 kN was represented by a pressure uniformly distributed over a circular area, 0.3m in diameter. This load was applied at the surface of a multilayer soil system representing a base layer on a subgrade (Fig. 9). Several cases were considered in order to evaluate the effect of reinforcement on the load distribution angle. For all cases the subgrade soil was the same: a saturated clay with a Young's modulus of 10 MN/m^2 and a Poisson's ratio of 0.5.

Two cases of unreinforced base layer were considered (A and B in Fig. 9). In both cases, the bottom part of the base layer consisted of a contaminated aggregate under zero confining stress. The top part of the base layer consisted of clean aggregate in case A and contaminated aggregate in case B.

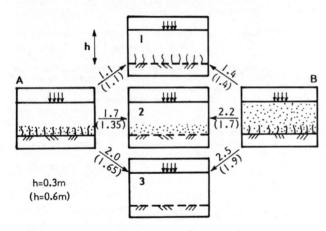

h=0.3m
(h=0.6m)

▢	Clean aggregate	$E_h = E_v = 30$ MN/m^2
▨	Contaminated aggregate	$E_h = E_v = 10$ MN/m^2
▧	Clean aggregate with $\sigma_c = 0$	$E_h = 0, E_v = 30$ MN/m^2
▧	Contaminated aggregate with $\sigma_c = 0$	$E_h = 0, E_v = 10$ MN/m^2
▨	Subgrade clay	$E_h = E_v = 10$ MN/m^2

Fig. 9   Summary of elastic analyses showing the considered cases (A and B = unreinforced; 1, 2 and 3 = reinforced), values of the moduli (E$_h$, horizontal, and E$_v$, vertical), and values of the load distribution improvement ratio tanα/tanα_o (written above the arrow for h = 0.3m and under the arrow, in parenthesis, for h = 0.6m) (h is the thickness of the base layer; σ_c = 0 means zero confining stress.)

Three cases of reinforced base layer were considered (1,2, and 3 in Fig. 9):

o   In case 1, the reinforcing element provides separation, thus preventing contamination, but negligible reinforcement, thereby permitting a large reduction in confining stress as the base layer deforms vertically and laterally.

o   In case 2, the reinforcing element provides reinforcement of the base layer. Aggregate/geogrid interlocking causes the confining stress on the aggregate to increase, relative to the unreinforced case, as the reinforcing element is put into tension. This prevents lateral displacement of the base layer material. In case 2, the reinforcing element is not assumed to provide separation. However, since a high confining stress in the base layer is developed, aggregate sinking and lateral displacement are limited, thereby reducing contamination.

o   In case 3, the reinforcing element provides reinforcement and separation.

Values of Young's moduli used in the elastic analyses are given in Fig. 9. These values were selected as follows: 30 MN/m^2 for a clean base layer aggregate; 10 MN/m^2 for the undrained Young's modulus of a saturated clay and for the aggregate contaminated by the clay; zero for the horizontal Young's modulus of the aggregate subjected to a zero confining stress. The zero Young's modulus represents complete dissociation of the base layer aggregate due to lateral displacement.

Two base layer thicknesses have been considered in the elastic analyses: 0.3m and 0.6m. Several values for the depth of the base layer affected by contamination and zero confining stresses have been considered. In addition, a number of intermediate cases, falling within the range of conditions presented in Fig. 9, have been considered, (i.e., cases where the depth of partial contamination was different from the depth of the zone where confining stress is zero). These additional cases will not be presented in detail.

Method of Interpretation of the Elastic Analyses

The results of the elastic analyses have been interpreted as follows:

o   For each case (reinforced or unreinforced), a finite element computer program is used to obtain the distribution of vertical stresses on the subgrade soil.

o   For a given unreinforced case (i.e., A or B in Fig. 9), the fraction of the load encompassed by a load distribution pyramid with an angle of $\alpha_o = \tan^{-1} 0.6$ is calculated by integrating the vertical stresses at the base layer/subgrade interface, over the width of the load distribution pyramid.

o   For each corresponding reinforced case (i.e., 1, 2, or 3 in Fig. 9, with the same thickness of base layer and depth of zone where confining stress is zero, if any, as in the given unreinforced case), the angle of the load distribution pyramid, α, encompassing the same fraction of the load at the base layer/subgrade interface as for the unreinforced case, is determined. The improvement of load distribution in the reinforced case compared to the unreinforced case is characterized by the ratio tanα/tanα_o, called load distribution improvement ratio.

A parametric study has shown that the use of 0.6 for tanα_o has only a minor effect on the calculated results.

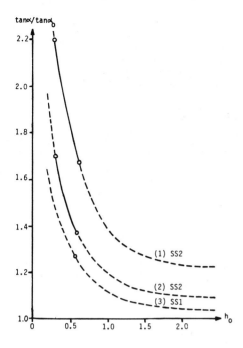

Results of the Elastic Analyses

Results obtained for Tensar SS2 geogrid (quick loading tensile stiffness K=500 kN/m) are summarized in Fig. 9 where values of $\tan\alpha/\tan\alpha_0$ obtained as described above, with $\tan\alpha_0$ = 0.6, are indicated. The following conclusions can be drawn:

o   If the reinforcing element provides separation, but only negligible reinforcement, the load distribution improvement ratio ($\tan\alpha/\tan\alpha_0$) is 1.1. to 1.4 depending on the expected degree of contamination of the base layer without reinforcement. These values of the ratio appear practically independent of the thickness of the base layer.

o   If the reinforcing element provides significant reinforcement, preventing a very low or zero confining stress (and consequently limiting the extent of contamination as explained above), the load distribution improvement ratio is 1.7 to 2.2, depending on the expected degree of contamination of the base layer without reinforcement. This range of values of the load distribution improvement ratio is for a base layer thickness of 0.3m. For a thickness of 0.6m, the range of values is 1.35 to 1.7. These results indicate that the effectiveness of a given reinforcing element (i.e., a reinforcing element with a given tensile stiffness) decreases as the thickness of the base layer increases.

o   If the reinforcing element provides separation and reinforcement, the load distribution improvement ratio becomes large (2.0 to 2.5 for a thickness of 0.3m and 1.65 to 1.9 for a thickness of 0.6m, as shown in Fig. 9).

In addition, calculations made with a base layer thickness of 0.6m result in a multiplier of 1.25 for Tensar SS1 geogrid (tensile stiffness K = 300 kN/m) instead of 1.35 for Tensar SS2 geogrid (tensile stiffness K = 500 kN/m) for the case A2 (reinforcement without separation).

From these results, three curves can be tentatively drawn as shown in Fig. 1: curve (1) related to Tensar SS2 geogrid for the case of a large number of passages (e.g., N larger than 1000) for which significant contamination would be expected without reinforcement; curve (2) related to Tensar SS2 geogrid and curve (3) related to Tensar SS1 geogrid for the case of a small number of passages for which no significant contamination would be expected without reinforcement.

Fig. 10   Curves giving the load distribution improvement ratio, $\tan\alpha/\tan\alpha_0$, as a function of the thickness of the unreinforced base layer, h_0. Curve (1) is related to the case of a large number of passages (e.g. N larger than 1000) where significant contamination would be expected without reinforcement; and curves (2) and (3) are related to the case of a small number of passages where no significant contamination would be expected without reinforcement. Circles represent the calculated points. Dashed curves have been extrapolated.

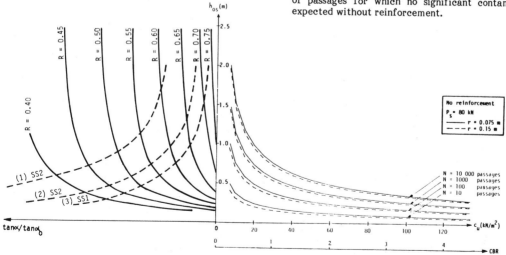

Fig. 11   Chart for the design of geogrid-reinforced unpaved structures. (Fig.11 combines Fig.4,8 and 10.)

DESIGN METHOD AND EXAMPLE

Proposed Tentative Design Method

Geogrid reinforced unpaved roads or areas can be designed using the chart presented in Fig. 11, valid for a standard axle load (P_s = 80 kN). Similar charts have been developed for other axle loads. Four steps are necessary:

o   <u>First step</u> - Given the subgrade undrained shear strength, c_u, the number of passages, N, and the allowable rut depth, r, (0.075 m or 0.15 m), determine the thickness, h_o, of the unreinforced base layer using the right part of Fig. 11.

o   <u>Second step</u> - Select in the left part of Fig. 11 the dashed curve related to the considered type of geogrid: Tensar SS1 or Tensar SS2.

o   <u>Third step</u> - At the intersection of the selected dashed curve and the thickness h_o, interpolate for the thickness ratio, R, between two solid curves.

o   <u>Fourth step</u> - Determine the thickness of the Tensar reinforced aggregate layer as follows:

$$h = R\,h_o \qquad\qquad (23)$$
(if rut depth is less than 0.15m and/or traffic is not channelized).

$$h = 0.9\,R\,h_o \qquad\qquad (24)$$
(if rut depth is 0.15m or more and if traffic is channelized; i.e., 0.9 accounts for tensioned membrane effect)

Design Example

The use of the proposed tentative design method is illustrated by the following design example.

An unpaved road consisting of an aggregate base layer is to be built on a subgrade soil with a CBR value of 0.3 . The expected traffic during the design life of the unpaved road is 1000 passages of trucks whose front axle load is negligible and rear axle load is the American-British standard axle load. The allowable rut depth is 0.075m.

The undrained cohesion of the subgrade soil can be derived from the CBR using Eq. 8:

$$c_u = 0.3 \times 30\ 000 = 9\ 000\ \text{N/m}^2 = 9\ \text{kN/m}^2$$

Using the right part of Fig. 11 for a rut depth r = 0.075m and a number of passages N = 1000, gives the design thickness of a unreinforced unpaved road for the American-British standard axle load (P_s = 80 kN): h_{os} = 1.20m.

Using the left part of Fig. 11 for h_{os} = 1.20m, and the curve (1) related to Tensar SS2 geogrid for a large number of passages, gives the following value of the thickness ratio: R = 0.55.

Since the allowable rut depth is less than 0.15m, Eq. 23 should be used to determine the thickness of the geogrid-reinforced unpaved road:

$$h = 0.55 \times 1.20 = 0.66\text{m}$$

If the geogrid had not improved the load distribution capability of the base layer, and had only provided confinement, the value of the thickness ratio would have been read on the axis of Fig. 11 for $\tan\alpha/\tan\alpha_o$ = 1: R = 0.76. Hence:

$$h = 0.76 \times 1.20 = 0.91\text{m}$$

CONCLUSIONS

Understanding of the Behavior of Unpaved Structures

The behavior of unpaved structures is complex because of the variety of phenomena involved. Little help can be provided by existing technical literature due to a lack of well documented observations on the behavior of unpaved structures. Observations of unpaved structure behavior are difficult because of the large number of parameters and, because unpaved structures, which are often temporary, are seldom instrumented and/or monitored, in contrast to permanent structures such as high vertical retaining walls.

Because of the lack of data from observations, conceptual models constitute a useful attempt at understanding the behavior of unpaved structures. From that viewpoint, this paper provides a detailed discussion on the behavior of unreinforced and reinforced unpaved structures, intended to identify and describe the mechanisms involved and their relationships. The paper also presents analyses to tentatively quantify the influence of some of the identified mechanisms. For instance : (i) elastic analyses were used to provide an evaluation of the influence of contamination and low confining stresses on the load distribution capability of a base layer; (ii) calculations have been carried out to compare the influence of subgrade soil deterioration and base layer deterioration on full-scale test unpaved road behavior; and (iii) the relative importance of the various mechanisms by which reinforcement improves the behavior of an unpaved structure, such as confinement of the subgrade soil, load distribution and tensioned membrane effect have been evaluated.

Relevance to Geogrids of the Proposed Tentative Method

The design method presented in this paper includes several mechanisms by which geogrids can improve unpaved structure behavior, in particular the improvement in load distribution capability of the base layer that geogrids are expected to provide. Calculations presented in this paper show that :

o   Improvement of load distribution capability of the base layer is one of the two mechanisms that provide most improvement to unpaved structures, among the identified mechanisms (it is therefore important to use reinforcing elements improving the base layer load distribution capability), the other mechanism being subgrade soil confinement.

o   Through interlocking with base layer material (especially if this material is aggregate) geogrids can provide significant improvement in base layer load distribution capability. Such improvement does not significantly result from the increased stiffness provided by the geogrid to the base layer as constructed, but results mostly from the fact that a geogrid, through interlocking with base layer material (especially if this material is aggregate) reduces or delays the mechanisms of deterioration of unpaved structures (i.e. decrease of effective thickness and decrease of properties of base layer, associated with increasing number of vehicle passages).

A conclusion that can be tentatively drawn from the proposed tentative design method is that the savings in base layer thickness is not proportional to the tensile stiffness of the geogrid. Such a conclusion should not be surprising since in an unpaved structure, base layer material/geogrid interaction is at least as important as geogrid tensile stiffness.

Discussion of the Results

Systematic calculations similar to the above presented design example show that the thickness of unreinforced unpaved structures can be reduced by 30 to 50% by using existing geogrids.

As is usual for the design of unpaved structures, base layer thickness values do not include a factor of safety. In addition, these values should be considered with caution because the method has not yet been calibrated with either small-scale or full-scale test data.

The calculations also show that approximately half of the thickness reduction resulting from geogrid reinforcement is due to subgrade confinement and approximately half of the improvement is from improved load distribution resulting from geogrid-base layer material interlocking.

Suggested Developments

The following developments and additional research work are suggested:

o Systematic elastic analyses, using the approach described in this paper, encompassing a wide range of parameters having an influence on the base layer load distribution capability (especially the base layer thickness). (The elastic analyses presented in this paper should be considered as an attempt at defining an approach and obtaining preliminary results.)

o Elastic analyses similar to the above with several layers of geogrids in the base layer.

o Calibration of elastic analyses with data from small-scale model tests.

o Review of laboratory test data on fatigue of clay caused by repeated loadings to refine the evaluation of the relative importance of base failure versus subgrade failure.

o Detailed study of full scale tests that have been conducted on unpaved structures to evaluate the relative importance of the various mechanisms of failure.

o Evaluation of field installations and field tests on unreinforced and geogrid-reinforced unpaved roads.

o Adaptation of the proposed tentative method to a variety of subgrade soils, such as peat, loose sand and frozen soils, and a variety of base layer materials.

ACKNOWLEDGEMENTS

The research into unpaved road behavior and the development of a design method for geogrid-reinforced structures was started when all of the authors were members of the Geotextiles and Geomembranes Group of Woodward-Clyde Consultants. The authors are indebted to J. Dixon for valuable comments, and S. Hartmaier and N. Stuart for assistance in the preparation of this paper. The authors thank Gulf Canada Ltd. for their support throughout the course of this study and Netlon Ltd. for the opportunity to participate in the Soil Reinforcement Design Group which, under the chairmanship of Sir Hugh Ford, provided a catalyst for the development of design procedures for geogrid-reinforced structures.

REFERENCES

Giroud, J.P., "Discussion and closure on Geotextile - Reinforced Unpaved Road Design", Journal of the Geotechnical Division, ASCE, Vol. 108, No. GT 12, (December 1982), 1654-1670.

Giroud, J. P., and Noiray, L., "Geotextile-Reinforced Unpaved Road Design", Journal of the Geotechnical Division, ASCE, Vol. 107, No GT9, Proc. paper 16489, (September 1981), 1233-1254.

Hammit, G., "Thickness Requirements for Unsurfaced Roads and Airfield Bare Base Support," Technical Report S-70-5, United States Army Engineer Waterways Experiment Station, Vicksburg, Miss., (July, 1970).

McGown, A., and Andrawes, K.Z., "The Influences of Nonwoven Fabric Inclusions on the Stress Strain Behaviour of a Soil Mass", Proceedings of the International Conference on the Use of Fabrics in Geotechnics, (Paris, 1977), 161-166.

Raymond, G. P., and Hayden, F.B., "Effect of Reinforcement on Sand Overlying Bases of Different Compressibilities Subject to Repeated Loading," C.E. Research Report No. 79, Department of Civil Engineering, Queens University, Kingston, Ontario, (Nov., 1983).

Webster, S. L., and Alford, S. J., "Investigation of Construction Concepts for Pavements Across Soft Ground," Technical Report S-78-6, United States Army Engineer Waterways Experiment Station, Vicksburg, Miss., (July, 1978).

Webster, S. L., and Watkins, J. E., "Investigation of Construction Techniques for Tactical Bridge Approach Roads Across Soft Ground," Technical Report S-77-1, United States Army Engineer Waterways Experiment Station, Vicksburg, Miss., (Feb., 1977).

NOTATION

Symbols used in the design method are as follows:

CBR = California Bearing Ratio of the subgrade soil (dimensionless);

c_u = undrained shear strength (cohesion) of the subgrade soil (N/m^2);

h = reinforced base layer thickness (m) corresponding to an axle load P;

h_o = unreinforced base layer thickness (m) corresponding to an axle load P;

h_{os} = unreinforced base layer thickness (m) corresponding to a standard axle load P_s;

N = number of passages of axle during design life of the structure (dimensionless);

P = axle load (N);

P_s = American-British standard axle load (80 kN);

R = h/h_o = thickness ratio (dimensionless);

r = rut depth (m);

α = load distribution angle for reinforced base layer;

α_o = load distribution angle for unreinforced base layer; and

$\tan \alpha / \tan \alpha_o$ = load distribution improvement ratio.

Model testing of geogrids under an aggregate layer on soft ground

The mechanisms by which the inclusion of a geogrid may improve the performance of unpaved roads and similar constructions are being studied at reduced scale by laboratory model tests. Tests so far have been conducted under plane strain conditions by applying monotonic loading from a rigid footing to reinforced and unreinforced soil-aggregate systems, using a range of fill thicknesses and subgrade strengths. Performance of the reinforced systems is significantly better, primarily because the grid effectively resists the tensile strains which develop in the base of the aggregate layer.

G. W. E. Milligan and J. P. Love, *Oxford University*

INTRODUCTION

It is a common construction technique to place a layer of coarse granular material on the surface of weak and compressible ground, for the formation of unpaved roads, working areas, parking lots, storage areas and the like. The design problems for such constructions vary with their purpose; they may be concerned with foundation failure under local concentrated loading, or with trafficking problems due to rutting. What they generally have in common is that they involve structures with a relatively limited life, in which larger-than-usual deformations are expected and acceptable, and on which the loading is often localised, variable, repeated, or moving, or some combination of all of these.

Such construction is difficult to design precisely, yet the use of very conservative methods may have severe financial implications; on the other hand, failures may also prove very expensive in terms of loss of access, machine down-time, or time spent on repair work. One method of improving the construction which has found rapidly-increasing use is to incorporate either a geotextile or a geogrid at the base of the aggregate by laying it on the surface of the ground before placing the aggregate. Not only may this allow a reduced thickness of fill to be used but should improve the reliability of performance by increasing substantially the load required to cause a complete failure of the soil-aggregate system. The performance limit for the system is then even more likely to be one of excessive deformation or rutting than of ultimate failure; the life of the structure may then be readily extendible as required by simple maintenance such as filling of ruts or regrading.

There is now much field evidence of the benefits of using a geotextile in such a system, but there is still no generally accepted design method. It is recognised that the fabric has three main functions:

(i) Separation of the aggregate fill from the soft ground.

(ii) Membrane action.

(iii) Reinforcement.

It seemed possible that the latter two functions at least might be performed better by a Tensar geogrid than a geotextile as it is generally stiffer and forms a positive interlock with the aggregate (Jewell et al 1984). The primary aim of the work reported in this paper was to investigate the performance of a Tensar geogrid in such a situation. By observing carefully the mechanisms of deformation and failure it was also intended that the relative merits of proposed general methods of design of soil-aggregate systems could be assessed and some progress made towards a generally acceptable method.

MODEL TESTING

As already noted above this is a difficult problem, because loading conditions are complex and pre-failure deformations, and hence the stiffness as well as strength of all the materials involved, are highly important. Even if attention is restricted to the access road, in which vehicles generally follow a single path and the problem is reduced to the rate at which ruts develop, there are many variables to consider: vehicle weight, axle configuration and tyre pressure; fill type and thickness; reinforcement type, strength and stiffness; and the geotechnical properties of the subgrade. Under site conditions the effects of construction technique such as the method of laying the reinforcement and compacting the fill, may also be highly significant. As a result much of the information obtained to date has been specific to a particular site and method of construction.

Polymer grid reinforcement. Thomas Telford Limited, London, 1985

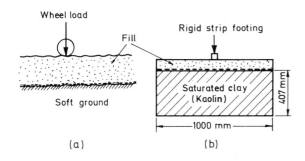

Fig 1.

Field problem   Idealised model

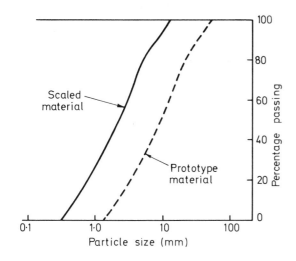

Fig 2.   Grading of fill material

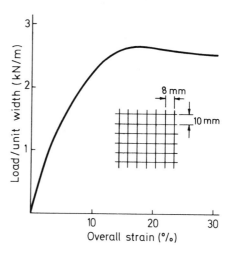

Fig 3.   Properties of grid material

Plate 1. View through side of apparatus
         (end of test)

Plate 2. Apparatus during consolidation stage

129

A decision was therefore made to work in the laboratory, performing model tests at approximately one quarter scale, to allow variables to be controlled more precisely and more accurate measurements to be made. The problem was also simplified initially as far as possible by investigating the behaviour under plane strain conditions of soil-aggregate and soil-fabric-aggregate systems loaded monotonically by a rigid footing (see Figure 1). A single type of subgrade soil (saturated kaolin clay), a single type of reinforcement (Tensar SS-type geogrid), and a single type of fill (type 2 sub-base material as specified by the Department of Transport), have been used. The principal variables investigated to date have been the depth of the aggregate layer and the strength of the subgrade soil.

Working at small scale introduces problems in the correct physical modelling of all aspects of the problem. Modelling need not in all cases be perfect; a common use of small scale physical modelling is to investigate general mechanisms of behaviour and hence suggest an appropriate analytical model which can then be applied to the full-scale problem, and this has been one purpose of the tests reported here. However, the analytical model must take proper account of all important parameters if it is to respond correctly to changes in behaviour from small to full scale. In this case the relative importance of various parameters was not known in advance; in addition, in the absence of a design method, it might prove possible to extrapolate results directly to full scale. An attempt has therefore been made to model the various components of the system as accurately as possible.

Dimensional analysis indicated that for correct modelling all dimensions, loadings, material strengths and material stiffnesses should be reduced by the geometrical scale factor of four. To meet the first condition the depth of fill was reduced, the range of 50 to 100mm corresponding to a realistic range of 200 to 400mm in the field, and the particle sizes in the aggregate scaled down as shown in Fig. 2. A miniature Tensar geogrid was produced by Netlon Limited to model the grid; the scale dimensions were in practice between those of SS1 and SS2 grids, and some difficulty was experienced in obtaining a regular grid with repeatable properties. Average dimensions and properties of the grid used are shown in Fig. 3. Particle size in the clay subgrade was considered to be of no significance since it was much smaller than the fill particle size and the grid aperture size at both full and model scale.

The stiffness and strength of the subgrade was taken to be characterized by a single parameter, the undrained shear strength. This is certainly appropriate for the strength, since the loading considered was rapid and any failure would take place under undrained conditions. For normally-consolidated clays the undrained stiffness is approximately proportional to the undrained shear strength. In these tests the method of preparation of the clay was such that it was overconsolidated to a greater or lesser extent

and it is probable that the stiffness varied less than in proportion to the shear strength, being relatively higher for the softer than for the stiffer clays. Subgrades with nominal strengths of 6, 10 and 16 kN/m^2 were used, corresponding to field strengths of 24, 40 and 64 kN/m^2. Ideally an even lower strength of say 10 - 12 kN/m^2 at full scale should also have been used, but the resulting scaled subgrades were impossibly soft to work with.

The material properties of the Tensar grid were a major difficulty. The method of manufacture was such that the properties of the aligned polymer could only be modified to a relatively small degree by the use of different polymers and variations in the draw ratio. Several types of grid, both oriented and non-oriented, were investigated. Finally use was made of the variation of Tensar properties with strain-rate to produce the scaling effect. The miniature grid was tested at a wide range of different strain-rates; both initial stiffness and peak strength decrease with reducing strain rate. An estimate was made of the typical strain-rate in a layer of fabric under a moving wheel load, and a rate of loading in the tests adopted which would give approximately the right reduction in fabric stiffness.

Modelling of the properties of the fill also caused some difficulty. In a frictional material the strength is automatically reduced in proportion to the ambient stresses, except that at low stress levels the angle of internal friction generally increases. This can be counteracted by using a slightly less dense material, which also helps to make the modelling of the shear modulus more accurate since this has been found to reduce less rapidly than in direct proportion to the stress level (Wroth et al. 1979). In practice however, it was found to be difficult to compact the fill on top of the soft clay and the degree of compaction achieved was probably somewhat less than would have been theoretically ideal.

APPARATUS AND METHOD OF TESTING

The main part of the apparatus is a strong box with internal dimensions of 1000 mm in length, 300 mm width and 600 mm depth. The ends and base are of aluminium plate and the sides, of 25 mm thick perspex, are supported by steel frames. The clay subgrade is formed as a block of kaolin approximately 400 mm deep by consolidation from a slurry. The initial volume of kaolin is about double the final volume so an extension box fits on top of the test box for the consolidation stage. The complete box fits inside a loading frame and the clay is consolidated by a rigid platen driven by three hydraulic cylinders operated by compressed nitrogen. Additional supports to the sides of the tank are provided during the consolidation stage to help resist the high lateral pressures. Drainage is allowed from the top and bottom but not the sides of the sample.

Once the kaolin has been consolidated to the required extent the platen is removed and the clay allowed to swell back under water until an

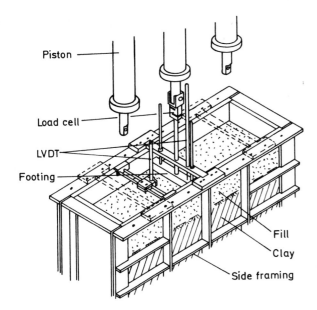

Fig 5. Apparatus for tests

equilibrium state is achieved. One side of the tank may also briefly be removed to allow a grid of markers to be placed on the side surface of the clay. These are visible through the perspex side wall and allow deformations in the clay to be observed during a test. The resulting sample of clay is not uniform and one full sample of each of the nominal strengths of clay used has been set aside for a thorough investigation of the variation of moisture content and undrained strength throughout the block. Typical results for one such investigation are given in Fig. 4 and Table 1.

Immediately before the test the surface of the clay is scraped to produce a flat, level surface. If a Tensar grid is to be incorporated it is then laid on the surface of the clay, without pretensioning. Fill is then placed in layers to the required depth with careful compaction by hand tamping. The fill is placed at about optimum moisture content; capillary suctions in a damp unsaturated fill will obviously be more significant at the model scale than at full scale, but dry fill was found to draw up moisture from the clay to a variable degree while attempts to work with the fill inundated were not successful.

Spot levels on the surface of the fill are taken at a grid of points in plan to obtain an accurate measurement of the thickness of the fill at the start of the test. This procedure is repeated at the end of the test to check the surface heave.

The tests themselves consist of driving a rigid footing into the fill at a constant rate using one of the hydraulic rams, as shown in Fig. 5. Footings of various widths have been used but for the main series of tests the width was 75mm, equivalent to 300mm at full scale. The rate of descent is controlled by having the low-pressure

Depth, mm	Undrained Shear Strength, Cu (kN/m^2)			Moisture Content (%)	
	SV1	SV2	SV3	MC1	MC2
70	7.4(10)*	6.8(7)*	6.6(7)*	60.1	60.3
170	9.0	8.2	7.2	59.0	59.6
270	9.4	9.2	9.3	58.0	58.3
370	9.4	10.0	10.1	57.6	58.3

* corresponding values of Cu from triaxial tests

Table 1. "Site investigation" results

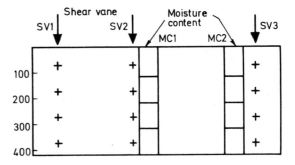

a) Location of tests

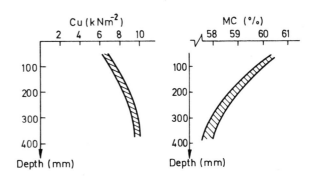

b) Profiles of shear strength and moisture content

Fig 4. Typical "site investigation" (for Cu 6 kN/m^2)

H (mm)	50	75	100	Reinforcement
Cu (KN/m^2)				(a) without grid (b) with grid
6	17P	8P	10P	(a)
	17TM	8TM	10TM	(b)
10	9TM	7P,14TM	12TM,15TM	(a)
	9P	14P	12P,15P	(b)
16	13TM	16P	11P	(a)
	13P	16TM	11TM	(b)

Table 2. Test numbers

side of the cylinders filled with hydraulic fluid which is allowed to flow out through a constant-flow-rate device, the high pressure side again being supplied from compressed nitrogen cylinders. The load on the footing is measured by an electrical load cell and its displacement by an LVDT. The test is stopped automatically when the footing has descended 50mm, equivalent to 200mm at full scale.

The primary test position is obviously in the centre of the box using the middle hydraulic cylinder, but secondary tests may subsequently be performed at either end using the outer cylinders. These are generally found to give somewhat higher loads due to the proximity of the rigid end walls of the box. Secondary tests were used initially to investigate the effects of different sizes of footing and to check on the repeatability of results. Latterly it has become the standard test procedure to use the secondary test positions to check the strength of the clay sample, by performing a load test with the footing directly on the clay (after removal of the fill) at one end and performing laboratory vane tests and taking samples of clay for triaxial testing and moisture content determination from the other.

The depth of clay affected by the tests is relatively small and it has proved possible to perform a series of tertiary tests using the bottom 200mm depth of the clay after removing the top 200mm and again allowing swelling under water to occur before placing the geogrid, if used, and the layer of aggregate. In fact the middle tertiary tests have given results which agree closely with the primary ones in directly comparable tests; the results of tertiary middle tests are therefore considered to be as reliable as those from the primary tests.

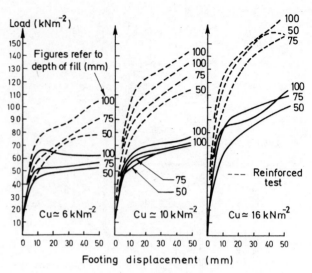

Fig 7. Load-displacement curves: summary plot

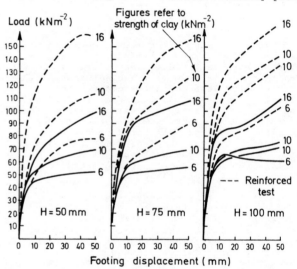

Fig 8. Load-displacement curves: summary plot

In addition to measurements of the load on and displacement of the footing, continuous recordings are made of the heave of the fill to either side of the footing using LVDT's. At intervals during the test, which lasts for only about 30 seconds (ensuring that the response of the clay subgrade is substantially undrained) photographs are taken through the side of the box. These are subsequently used to measure the movements of the surface of the fill, the surface of the clay subgrade and the markers embedded in the clay and attached to the grid, and to observe the positions of shear planes within the clay subgrade. Reference markers for these photographs are fixed to the side wall of the tank. Readings from all transducers are recorded by a computer-controlled system and stored on disc for subsequent analysis. Some use has also been made of video film to record tests to allow subsequent review of any part of a test.

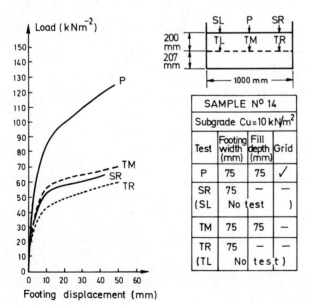

Fig 6. Load-displacement curves from one sample

For instance, to check whether the grid is being pulled from beneath the fill rather than straining locally under the load, light rods are tied to the ends of the grid and project vertically through the fill. The stage in a test at which these start to move may be observed on a video film; displacement mechanisms in the fill are also identified more clearly by repeated viewing.

RESULTS

The results presented here are for the main series of tests, details of which are given in Table 2; the numbers indicate the successive samples of clay prepared, and the letters P and TM denote the primary and tertiary middle test positions respectively. A typical series of plots, from tests on a single sample, of the load on the footing against its vertical displacement into the fill are shown in Fig.6. The curves for loading directly onto the subgrade (SR and TR) indicate that in this case the clay was somewhat weaker for the tertiary than for the primary/secondary level. Nevertheless, it can be seen that the layer of aggregate produces a relatively small improvement in performance (TM compared with TR) while inclusion of the grid reinforcement has a very marked effect (P compared with SR). A summary of all such load/displacement curves for the tests in the central position, with and without reinforcement, is presented in Fig. 7. The same data are replotted in Fig. 8 as sets of curves for constant fill thickness rather than constant subgrade strength.

In spite of minor variations in subgrade strength within individual samples, and some differences in fill grading and density between tests, a number of clear trends appear. In all cases the failure load, defined as the load at which displacement started to increase rapidly, is approximately equal to 5 c_u for footings bearing directly on the subgrade. An unreinforced layer of fill increases the bearing capacity, quite markedly for the softest subgrade, less so for the stronger ones. Although the bearing capacity increases with fill thickness for any one subgrade strength it does so less rapidly than might be expected. The response of these unreinforced systems tend to become more 'brittle' as the subgrade strength is reduced; on the stronger subgrades the load continues to increase after initial failure, while on the weak subgrade it remains constant or even reduces. In the field a failure in such a system is likely to be fairly serious, with the load punching right through the fill into the soft clay, and difficult to repair satisfactorily.

The systems incorporating a geogrid show in all cases a marked improvement in performance. Failure loads are typically about 40% higher than for unreinforced systems; however, failure loads are much less easy to identify for reinforced systems, as the loads continue to increase quite steadily with further displacement. The performance of such a system will generally be limited by excessive deflection rather than a catastrophic failure unless the

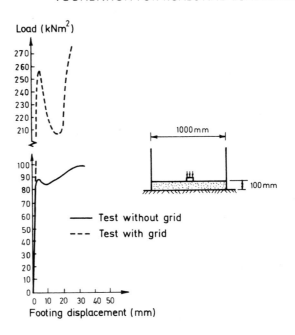

Fig 9.   Load-displacement curves with no clay

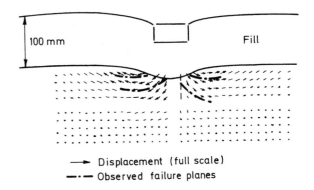

Fig 10.   Displacement vectors in clay: unreinforced system

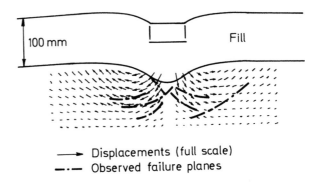

Fig 11.   Displacement vectors in clay: reinforced system

reinforcement breaks. In these tests the reinforcement always remained intact. Performance
clearly improves with subgrade strength, and
greater benefit is obtained from using thicker
fills than with unreinforced systems. The
initial stiffness of the system (the slope of
the load/displacement curve) increases with the
strength of the subgrade and also to some extent
with the thickness of the fill. For loads up
to about 50% of the failure load for unreinforced systems there is very little difference
in performance between the reinforced and
unreinforced systems; for higher loads the
stiffness of the unreinforced system reduces
quite rapidly as plastic flow begins to occur
in the subgrade clay.

Tests on samples 12 and 15 were nominally
identical, with a subgrade strength of 10 kN/m^2
and a fill thickness of 100mm. Curves for both
are plotted in Figs. 7 and 8 and give an
indication of the repeatability of individual
tests. Although there is some discrepancy
between the curves the repeatability is quite
good considering the difficulty of preparing
identical samples of soil, both subgrade and
fill, and the variability in the properties of
the miniature grid. In the tests using the
thickest fills there was some indication that
failure was occurring predominantly within the
fill, at least when no grid was present. Tests
were therefore conducted using 100mm of
aggregate placed directly on the base of the
test box, with no clay; the resulting curves
are shown in Fig. 9. With no reinforcement the
system response was initially very stiff, but
failure then occurred suddenly with a wedge of
fill being pushed out either side of the loaded
footing. When reinforcement was included the
performance of the system improved dramatically,
very high loads being reached before failure
occurred. Presumably the unreinforced system
failed by soil sliding outwards along the
fairly smooth base of the box; the curves for
the thicker fills on the strongest clay
subgrade are sufficiently similar to suggest
that the failure mechanism is similar, with
shearing occurring in the clay at or close to
the interface with the fill. When a strong
grid is present such lateral movement is
prevented; the upper curve in Fig. 9 is
probably an upper-bound curve for reinforced
fill on an infinitely stiff and strong subgrade.

Large quantities of data have been produced on
the displacements and strains within the
subgrade during successive stages of most
tests. These are found by digitizing the
positions of the markers in the clay visible
in the photographs taken during the test;
displacements are found from the initial and
final position of each marker and strains are
calculated from the distortions of quadrilateral
elements with nodes at four adjacent markers.
The main use of these measurements will be for
comparison with the results of finite element
analyses. They have also been useful, along
with observations of shear planes in the clay
and of the deformed shape of the surface of the
subgrade and of the fill, in clarifying the
different mechanisms of behaviour of the
reinforced and unreinforced systems. Typical

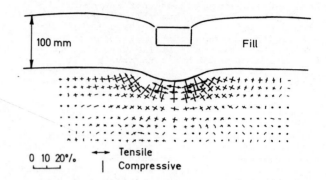

Fig 12. Strains in clay : unreinforced system

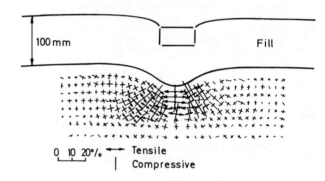

Fig 13. Strains in clay : reinforced system

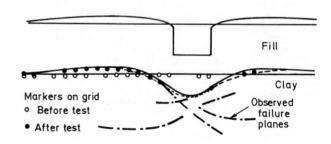

Fig 14. Displacements of markers on grid

results are shown in Figs. 10 - 13. When no
grid is used the vertical displacements in the
clay beneath the footing are quite small and
significant displacements are restricted to a
relatively limited body of clay (Fig. 10).
Failure planes within the clay, made visible by
the grease used to lubricate the sides of the
box, are also very shallow and clearly caused
by predominantly lateral movements in the clay
either side of the footing. Within the fill
the depth of aggregate beneath the footing is
much smaller than at the start of the test.
Little additional compaction has occurred and
the reduction in thickness is due to lateral
movement of aggregates from beneath the footing,

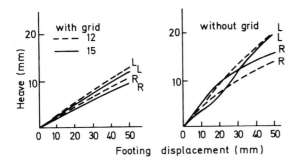

Fig 15.   Measurements of surface heave

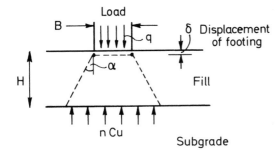

Fig 16.   Simple bearing-capacity analysis

resulting in wedge-type failures in the fill either side of the footing. By the end of the test these form clearly visible outcrops in the surface of the fill. Fig. 11 shows the results for the reinforced system with the same fill depth and subgrade strength, at the same displacement of the footing (but a much higher load). Inclusion of the grid has largely prevented loss of aggregate from immediately beneath the footing and no failure planes are observed in the fill to either side. Vertical displacements in the clay beneath the footing are greater than when no grid was present, while lateral displacements in the clay are slightly reduced. Shear planes in the clay penetrate deeper into the subgrade, and the body of deforming soil is significantly greater. The deformed shape of the subgrade/aggregate interface is quite different, as is the shape of the surface of the fill.

The different deformations within the subgrade show up even more clearly in the plots of strain in Figs. 12 and 13. When no grid is used the strains are restricted to a small zone beneath the footing in which large tensile strains are developed. When the grid is used the deformed zone is much larger; large tensile strains are again observed beneath the middle of the footing, but under the edges of the footing they are very much smaller than in the unreinforced system.

These deformation measurements were obtained from tests on sample 12. Tests on sample 15

were nominally identical and for the test with reinforcement an attempt was made to measure the deformations of the grid by attaching markers to the nodes of the grid against the side wall of the box. Unfortunately many of these became obscured by clay either during compaction of the fill or during the test; the initial and final positions of those that remained visible are shown in Fig. 14. Whereas with geotextiles there is a tendency for the fabric to pull out from under the fill to either side (Gourc et al 1983), the geogrid is securely anchored by the fill even at such a large displacement of the load. However, as has been observed at a late stage in many of the tests, particularly with the very soft subgrade, there is some tendency for the grid to be pulled down into the clay which is extruded through the apertures of the grid. Movements of the nodes of the grid are predominantly vertical, either up or down, in the area outside the footing; no information has as yet been obtained for the area immediately beneath the footing.

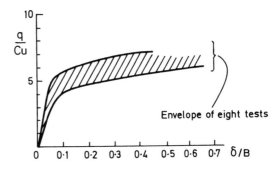

Fig 17.   Plot of q/Cu against δ/B for no fill

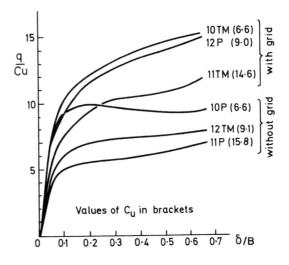

Fig 18.   Plot of q/Cu against δ/B for 100mm fill

135

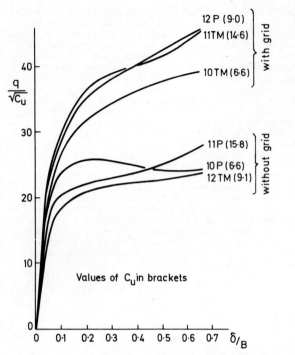

Fig 19. Plot of $q/\sqrt{C_u}$ against δ/B for 100mm fill

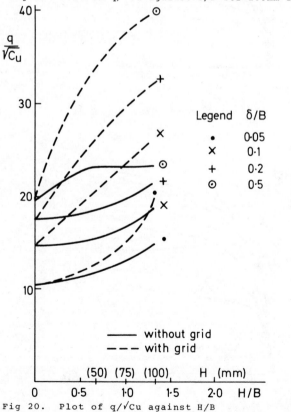

Fig 20. Plot of $q/\sqrt{C_u}$ against H/B

In all cases the volume of heaved clay is equal to the volume of displaced clay to within the possible accuracy of measurement; the clay is therefore deforming throughout at constant volume. The maximum heave at the surface of the fill is monitored throughout each test. Typical results, for tests 12 and 15, are shown in Fig. 15. Heave starts immediately and increases approximately linearly with the descent of the footing. Heave is almost twice as great in the unreinforced system as in the reinforced system, due to the more localised deformations of the whole system and the much greater distortion and hence dilation within the fill. In the field, with repeated loading, such dilation will leave the fill less dense and less able to cope with each successive load application. It is also observed that the effects of side friction on the heave of the fill are much more marked in the tests without reinforcement than in those with reinforcement. This probably happens because the heave in the reinforced system is largely a reflection of the deformations in the clay, which are little affected by side friction, while the heave in the unreinforced system is caused largely by deformations within the fill which are more influenced by side friction.

ANALYSIS

The bearing capacity of a layer of fill over-lying soft ground is often considered in terms of the simple model shown in Fig. 16. The angle of load spread α is typically taken to be $\tan^{-1} 0.5$; the coefficient n might be expected to vary with the vertical displacement of the footing δ. Neglecting the initial uniform surcharge of the fill and any forces along the boundaries of the trapezoidal block of fill beneath the footing, the equation of vertical equilibrium is:

$$qB = nCu \ (B+2H \ \tan\alpha) \qquad (1)$$

or

$$\frac{(q)}{(Cu)} = n \ \{1 + 2\frac{H}{B} \ \tan\alpha\} \qquad (2)$$

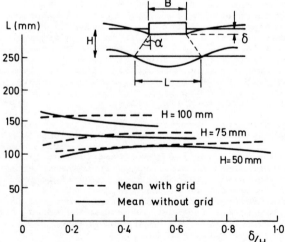

Fig 21. Spread of load through fill

This suggests that, for any particular value of fill thickness H, non-dimensional plots of (q/c_u) against (δ/B) should follow an unique curve, while for any particular value of δ/B (and hence n) there should be a linear relationship between q/c_u and H/B. The latter hypothesis was given some support by results from small-scale preliminary tests using a foam-rubber subgrade (Milligan 1982).

Non-dimensional plots of q/c_u against δ/B for all plots in which the loading was directly onto the subgrade (H=0) are shown in Fig. 17 and give quite close agreement between tests with the full range of c_u used. However similar plots for tests with the thickest fill (100mm) presented in Fig. 18 show a wide scatter for tests both with and without a grid. In each case there is a clear trend, with the curves for the weaker subgrades above those for the stronger subgrades. This suggests that the load-carrying capacity increases less rapidly with c_u than expected. Fig. 19 shows the same results as Fig. 18, only divided by $\sqrt{c_u}$ rather than c_u; the grouping of the curves is now much better though of course the agreement for the results with no layer of fill is then lost. The choice of $\sqrt{c_u}$ is arbitrary and may not be widely applicable to systems with different types of fill, loading, reinforcement etc. The main result of this analysis is probably to show that Equation (2) represents an over-simplification of the performance of a complex system.

Variation of q, again divided by $\sqrt{c_u}$ to minimise the scatter, against H/B is shown in Fig. 20. The superior performance of the reinforced systems is very apparent, the difference increasing with fill thickness and also with footing displacement. Such a plot appears to allow direct comparisons to be made between the performance of reinforced and unreinforced fills of different thicknesses. However, since $q/\sqrt{c_u}$ is not non-dimensional, direct extrapolation to field values of parameters should be done only with great caution. An analytical method is therefore needed which will reproduce the model test results in all major aspects of behaviour and allow extrapolation to field scale with greater confidence. Purely elastic analysis have been found to be quite inadequate to explain the difference in performance of reinforced and unreinforced systems (Milligan 1981). Various attempts have been made to allow for the elastic membrane action of a geotextile in combination with plastic yielding within the subgrade (Giroud & Noiray 1981, Sellmeijer et al 1982, Sowers et al 1982 for example). These generally treat the load-spread angle α as a constant, the same for both reinforced and unreinforced layers of aggregate. However, the experimental results reported above suggest that the greatest benefit from the use of a geogrid is in preventing the lateral spread of fill beneath the load by resisting the tensile strains at the base of the fill. This constraint on the aggregate layer might be expected to allow a much higher effective angle α to occur when a geogrid is being used; an analytical method based on this idea is developed by Giroud et al (1984).

The effective angle α has been determined from the experimental results, α being defined by the points at the fill/subgrade interface which have zero vertical displacement at any stage of a test (see Fig. 21). The appropriate points are probably more correctly the points of inflexion of the subgrade/fill interface, but these are always close to the zero-displacement points and in practice are much more difficult to define precisely. Because the deformations of the system are not always exactly symmetrical and the thickness of the fill varies during a test (particularly in an unreinforced test) it is convenient to define the load-spreading effect by a length L between the points of zero displacement, as shown in Fig. 21. It has been found that L remains practically constant or increases slightly, with α increasing, during the course of a test in which a geogrid is used. When no geogrid is used the value of L for a particular thickness of fill is initially much the same, but reduces during the course of a test, α remaining approximately constant, as shown in Fig. 21. The difference between reinforced and unreinforced systems is somewhat greater for the thicker fills. The back figured values of α at the start of a test vary from about 23° for the thicker fills down to about 19° for the thinner fills. The difference may reflect the greater degree of compaction obtainable in the thicker fills.

Multiplying L by the usual bearing capacity for a clay, $5.14 c_u$, gives total loads on the footing which are in general agreement with loads causing initial yield (rapidly increasing displacement) in tests in which a grid was used. This suggests that the membrane effect was negligible, as would be expected at the small displacements of the system at this initial yield stage. For the unreinforced systems, even when the measured values of α (or L) are used, such a simple bearing-capacity analysis greatly over-estimates the allowable loading on the footing; the lateral spreading of the fill beneath the footing must prevent development of the normal type of failure in the clay. Further analysis of the deformation measurements should allow this less safe mechanism to be defined more accurately.

Under normal working conditions the aim should be to maintain loads below the yield load. Even then some rutting would be expected to develop from repeated loading since the deformation will not be entirely recoverable. Limit analyses are not appropriate under such conditions, and work is in progress to model the complete behaviour, but particularly the pre-yield behaviour, using finite element techniques and realistic elasto-plastic soil models. At the same time the laboratory model testing will continue, using dual footings to model a complete vehicle axle and investigating the effects of repeated loading. A larger strong box for this testing is under construction at present. Some simple full scale testing is also envisaged if suitable opportunities arise.

CONCLUSIONS

By simplifying the problem of loading on a layer

of fill on soft ground it has proved possible to identify the essential differences in behaviour between systems with and without a geogrid at the base of the fill, by means of small-scale tests in the laboratory.

The significantly better performance of the systems incorporating a grid has been shown to be the result of the reinforcing action of the grid, which interlocks with the granular fill material and resists tensile strains which develop at the base of the fill. Membrane effects only become significant at large deformations. The bearing capacity of unreinforced fill is less than expected due to lateral spreading of the fill below the loaded area.

ACKNOWLEDGEMENTS

The experimental work reported in this paper was performed by J. P. Love at Oxford University under the supervision of Dr. G. W. E. Milligan. The work is supported by a co-operative award from S.E.R.C. and Netlon Limited. The authors are grateful for many useful comments and criticisms from members of the Steering Committee for the co-operative award, under the Chairmanship of Professor Sir Hugh Ford, and for the assistance of research students, technicians and staff of the Department of Engineering Science, Oxford University.

REFERENCES

Giroud, J. P. and Noiray, L. (1981). Geotextile reinforced unpaved road design. ASCE J. Geotechnical Div., vol 107, No. GT9.

Giroud, J. P., Ah-Line, C. and Bonaparte, R. (1984). Design of unpaved roads and traffiked areas with geogrids.

Symposium on Polymer Grid Reinforcement in Civil Engineering. I.C.E. London.

Gourc, J. P., Perrier, H. and Riondy, G. (1983). Unsurfaced roads on soft subgrade: mechanisms of geotextile reinforcement. Proc. 8th European Conf. on Soil Mech. and Found. Eng., (2), 495, Helsinki.

Jewell, R. A., Milligan, G. W. E., Sarsby, R. W., and Dubois, D. (1984). Interaction between soil and geogrids. Symposium on Polymer Grid Reinforcement in Civil Engineering. I.C.E. London.

Milligan, G. W. E. (1981). The use of mesh products to improve the performance of granular fill on soft ground. Report to Netlon Ltd., O.U.E.L. Report No. 1346/81. Dept. Eng. Science, Oxford University.

Milligan, G. W. E. (1982). Some scale model tests to investigate the use of reinforcement to improve the performance of fill on soft soil. Q. J. Eng. Geol. London, Vol. 15, 209-215.

Sellmeijer, J. B., Kenter, C. J. and Van der Berg, C. (1982). Calculation method for a fabric reinforced road. 2nd Int. Conf. on Geotextiles, Las Vegas.

Sowers, G. F., Collins, S. A. and Miller, D. G. (1982). Mechanisms of geotextile-aggregate support in low-cost roads. 2nd Int. Conf. on Geotextiles, Las Vegas.

Wroth, C. P., Randolph, M. F., Houslby, G. T., and Fahy, M. (1979). A review of the engineering properties of soils with particular reference to the shear modulus. C.U.E.D./D - soils TR 75, Dept. of Eng. University of Cambridge.

Stanley Airfield, Falkland Islands, 1982

The paper records the history of events relating to Stanley Airfield, starting with the invasion of the Falkland Islands on 2 April 1982 and leading to the landing of the first RAF Phantom in the following October. To achieve this the airfield had to be strengthened and extended by 2000 ft (610 m) over virgin areas of peaty soil. The paper describes the planning and design work, including the selection of geo-fabrics and grids, the post-surrender reconnaissance and, finally, the airfield construction. The practical aspects of laying the geogrid materials are covered

L. J. Kennedy, *Royal Engineers*

INTRODUCTION

The invasion of the Falkland Islands by the Argentine forces on 2nd April 1982 focussed immediate attention upon Stanley's small airport. The Headquarters of the Engineer-in-Chief (the professional head of the Royal Engineers) soon built up valuable engineer intelligence about the airfield which was provided by Falkland Islanders, resident in or visiting the UK, and others such as the consulting engineers for the original airport project. Messrs Rendel, Palmer and Tritton, who were able to provide their construction drawings and site survey data.

Preliminary studies considered what use the Argentine forces might be able to make of the airport, which provided the only means of conventional air access to the Falklands. It had been built between 1974 and 1976 on the Cape Pembroke peninsula near Stanley to accommodate a weekly feeder service by F27 aircraft of the Argentine military airline, operating from Comodoro Rivadavia in Patagonia, about 600 nm distant. The single runway was 4100 ft (1250 m) long and 150 ft (45 m) wide. It had a design LCN of 16 with a nominal construction of 32 mm of Marshall asphalt and 300 mm of crushed rock. The airport site was known to have a high water table and consist mainly of peat and sand with quartzite rock outcrops. The airport in its pre-invasion state is illustrated at Figure 1.

It was already clear in the early days of the occupation that the Argentines were flying Hercules C130 transport aircraft into Stanley and also lightweight ground attack aircraft such as the Pucara. The important consideration was whether they could develop the airfield to take fast jets such as their Mirage and Skyhawk fighters. The conclusion was that they would need to extend the runway and, from the extended centre-line profile provided by Rendel, Palmer and Tritton, there

appeared to be scope for extension without major earthworks. However, knowing the poor ground, it was thought that any extension work would take some months.

After a week or so the airfield problem was starting to be viewed from another direction. Plans were being laid for the re-occupation of the Falklands by our own forces and the use of the airfield by our own aircraft was considered. Evidence indicated that the runway LCN as built was greater than 16, probably nearer 30. Also an extension of 2000 ft (610 m) or more looked feasible. To build an extension at

Fig 1. Stanley Airport before April 1982, looking NW.

LCN 30, we examined our range of temporary, rapidly constructed expedients based upon aluminium planks and trackways. Our study intensified when the RAF decided that they needed instead to have a runway of at least LCN 45 capacity, as well as being 6000 ft (1800 m) long. They wanted the ability to operate very heavily laden transport aircraft and fast jets such as Phantom. They also required the enhanced runway in the shortest possible time after the re-taking of the

Polymer grid reinforcement. Thomas Telford Limited, London, 1985

airfield. Thus we arrived at a firm requirement for developing Stanley airfield.

INITIAL PLANNING AND DESIGN

Clearly, conventional civil engineering
options for airfield construction were not
possible, because of the time they would take
to execute. The only way forward seemed to be
by using the aluminium mat expedients already
considered. Our airfield mat system, called
PSA (prefabricated surfacing aluminium), had
been designed to carry Hercules aircraft with
low tyre pressures but it was not strong
enough for fast jets with high tyre pressures.
We had another expedient designated Class 60
Mat, which had been developed as tank trackway
and was strong enough. However, we had only
limited data on its performance under aircraft
loadings and we had never built and operated a
full runway system with anchorage and other
details established. We were therefore drawn
towards the American airfield mat known as AM2.
It had the major advantage that it had been
designed to carry fighter aircraft with high
tyre pressures. Also, it had been developed
into a complete airfield package, with
accessories for providing adjoining taxiways
and parking aprons, airfield lighting and
anchorages, and encouragingly it had given
proven service in Vietnam. With the strong
likelihood of poor ground conditions, the only
prudent course was to opt for AM2 Mat even
though this would entail a special purchase
from the US Government.

Apart from the extension work it was necessary
to strengthen the existing runway to LCN 45,
and we concluded that a simple over-slabbing
of the existing asphalt surface would suffice,
assuming of course that any war damage could
be adequately repaired.

The initial planning culminated in an intensive
weekend in Whitehall when a design team from a
unit called Commander Royal Engineers
(Airfields) occupied a top floor room in the
Old War Office building to make a complete
outline plan for the Air Staff for developing
the airfield. The planning work included the
runway layout, an enlarged parking apron with
extra access taxiways, fighter aircraft
dispersals, aircraft arrester gears and other
support facilities. In conjunction with the
staff of the Engineer-in-Chief's department,
the plan was phased to permit firstly, the
over-slabbing of the existing runway to allow
heavy transport aircraft to operate and then,
subsequently, the completion of the runway
extension and aircraft dispersals to allow
fast jets to operate.

Then while the plan was being considered at
high level, there was time to review the
planning estimates and look at the construction
specification in more detail. Without the
opportunity for a site investigation (not
possible under the circumstances!) we still had
to determine with as much accuracy as possible
the requirement for engineer plant, equipment
and materials, and it was at this time in early
May that the intention to use geo-fabrics was

examined and selections made.

We were fortunate to have bore hole logs along
the general line of the proposed runway
extension, provided for us again by Rendel,
Palmer and Tritton. An assessment of the
engineering characteristics of the soils showed
that the main soil types were fine sand and a
peaty sand, with the latter type predominant.
The extension area was flat, quite level and
poorly drained. The water table was very high
throughout and there were areas of standing
water in places. The peaty sand was forecast
to give very low bearing values with a likely
CBR of less than 3%.

The conclusion was that we should have a
ground membrane such as Netlon CE 121 laid
directly upon the subgrade to improve the
ground bearing qualities of the soil. It was
hoped to avoid cut and lay the geogrid
directly upon the undisturbed surface. It was
also thought worthwhile to lay a permeable
membrane such as Terram under the Netlon to
avoid the migration of the peaty subgrade into
the selected fill material and reduce any
pumping effect during compaction. In the
absence of further information we assumed
similar conditions elsewhere on the site and
made provision for geo-fabric and geogrid
materials for the apron extension, aircraft
dispersals and access taxiways. We further
assumed that we would import beach-won materials
to make up levels and provide an adequate base
for the AM2 mat.

Time was desperately short with the advance
party of the technical team of Royal Engineers
needing to catch QE2 leaving Southampton on
12 May 1982. This team was called Commander
Royal Engineers (Works) Falkland Islands (or
CRE (Works) FI) and the advance party contained
a team which would carry out the site survey
and site investigation work on the airfield
and finalise the design. It was necessary to
have this advance party travelling with the
second wave of invasion forces to be on hand
immediately Stanley was re-taken.

Meanwhile, those left behind carried on the
planning, leading to bids for equipment and
materials to be shipped south as soon as
possible. The Royal Engineer squadrons who
would form the construction force started
preparing for their deployment.

THE WAR AND THE AFTERMATH

In the South Atlantic the war took its course.
The airfield at Stanley received particular
attention because, with our Task Force
controlling the sea around the Falklands, it
provided an aerial lifeline to the occupying
forces. There was also the concern that the
Argentines would try to base their fast jets
at Stanley. Raids by Vulcan bombers and
carrier-based Harriers damaged the runway but
failed to close it completely.

Immediately Stanley was re-taken the airfield
site was reconnoitred. The runway was in an
adequate state to land lightly loaded Hercules

doing the short trip from the Argentine main-land but the hasty and poorly executed crater repair work of the Argentines would not be able to carry our own very heavily laden aircraft doing the long trip from Ascension Island. Emergency repairs were put in hand by the Royal Engineer troops who had been operating with the Task Force. We were fortunate to find Argentine supplies of AM2 mat that they had brought to the islands in a vain attempt to extend the runway - they had only laid about 50 ft (15 m) at one end. This matting provided us with excellent crater capping material. We also built a temporary Harrier base alongside the main runway using our PSA aluminium matting mentioned earlier to construct a short runway. This was to permit a continued air defence capability when we had to close the main runway for construc-tion work.

SITE INVESTIGATION

With the emergency work completed, attention was turned to the site reconnaissance for the airfield development project. The first task of course was to clear unexploded bombs and discarded Argentine munitions to permit access to all the site. In general, the desk study layout plan matched the ground quite well. Indeed there were few surprises, with the ground conditions as poor as had been expected. The weather conditions at this time were atrocious. In particular, the constant high winds and frequent blizzards made surveying very difficult. Deep beds of peat were found in some areas and where possible layout adjustments were made. Fortunately in the 2000 ft (610 m) western extension area there was consistently occurring peaty sand.

The water table was high and there was plenty of standing water in the runway extension area. This area was also very flat and level as expected. Figure 2 gives a good idea of the terrain. The main engineering problems were to fix the runway grade line to fit the required criteria while ensuring minimum earthworks, and to provide adequate drainage to the runway area without excessively deep ditches. After much juggling a good arrange-ment was found.

The major problem to which there was no ready solution was the lack of suitable material for the runway base. With hindsight, our original intention of using beach-won materials seems unrealistic. In the early days the idea of all the likely beaches being sown with mines had not occurred to us. There was plenty of dune sand for general fill but it seemed unlikely to provide an adequate base.

UK TRIALS

While we pondered the runway base problem in the Falklands, a trial at the Aeroplane and Armaments Establishment at Boscombe Down, which tested AM2 laid on ground of varying CBR with multiple passes of an aircraft at the maximum expected all-up-weight, concluded that we should have a minimum CBR of 5% under

the mat to avoid unacceptable rutting of the base and mat deformation. It was further deduced that if the sub-grade could not achieve that strength, a base layer of 150 mm of CBR 14% material was required; alternatively a layer of 300 mm of CBR 7% material would suffice. In Stanley, we tested the in-situ CBR of the peaty sand and found it to be in the range 1-3% near the surface. Also, we could not compact the proposed sand fill to achieve a firm surface because of its fine uniform nature. We needed, thus, a base layer to meet the bearing strength requirements of 7 or 14% CBR and there were no naturally occurring materials readily available.

Fig 2. A view of the runway extension site looking W.

A further trial was set up at Boscombe Down to try to find a suitable method of providing a strong enough base. The testing method was as for the first trial but the opportunity was taken to try out a variety of pavement specifi-cations under the AM2 mat. Four sections were prepared in the trial strip. The peaty sand subgrade for all sections was simulated by mixing sedge peat and pit sand. Its CBR was confirmed as 1.5% and a layer of Terram 1000 was laid on top. The first three sections had a further layer of Netlon CE 121 added before the fill was placed. The second section had an extra layer of Netlon CE 121 added at the mid depth of the fill. The third section had the top 150 mm of sand fill stabilised with 8% cement. The fourth section without the Netlon CE 121 was also stabilised. The pavement was topped with an impermeable membrane and the AM2 mat. The mat was trafficked by heavy aircraft and settlements measured.

The two unstabilised sections failed to provide adequate bearing though the second section with Netlon CE 121 at mid-depth (225 mm deep) was significantly stronger. The two stabilised sections showed no distress and there were no measurable differences to reflect on the omission of Netlon CE 121 at the subgrade level.

There were many useful lessons to come out of this trial. Cement stabilisation was a possible solution to the mat base requirement. In deep sand fill the Netlon CE 121 geo-grid seemed to have a stiffening effect and it was recommended for inclusion in the slopes of any deep sand fill embankments. The Netlon/Terram sandwich laid directly on the subgrade was recommended, especially in view of the poorer drainage conditions to be expected at Stanley. The problem was recognised of Netlon 'bow-waving' or opening at the joints when laid upon an uneven surface or when fill was hastily dumped without care.

FINAL DESIGN

The second Boscombe Down trial was followed by a trial of cement stabilisation on site at Stanley. Here, the stabilised layer did not gain any strength and would not compact in the top 50 mm or so. We concluded that the most likely cause of failure was the extremely poor weather with the temperatures hovering around freezing for much of the time. The other option for providing an acceptable foundation for the airfield mat was to manufacture crushed rock for the base layer. Equipment was already being provided to establish a quarry for concreting aggregates and road stone but the extra need for large quantities of crushed rock for the mat base was a daunting prospect. There was, though, no other choice.

The final design for the runway pavement, outside the extent of the existing runway, is illustrated in Figure 3. The subgrade was to be the undisturbed ground wherever possible. The addition of Tensar SS1 at the interface between the sand fill and the rock base was a later addition during construction when it was

found that stone was being lost in the sand during compaction. The basic requirement for the runway surface level was 1 inch in 12 feet (25 mm in 3.66 m). The rock layer was to be covered with an impermeable membrane called 'Trevira', to prevent water percolating through the mat joints and penetrating the base and weakening it. Finally the AM2 mat was to be laid on top. (The mat consists of interlocking panels measuring 12 feet by 2 ft by 1½ inch thick (3.66 m by 0.6 m by 38 mm), that are laid in a bonded pattern across the runway, the short dimension being parallel with the runway centre line. Each panel weighs 144 lb (65 kg)). The specification for deep fill embankments was as recommended from the UK and this is illustrated in Figure 4.

CONSTRUCTION

Even while the war continued, the stores and equipment for the airfield project were being gathered together for shipment, including 4000 tonnes of AM2 matting that was shipped from the United States to the UK prior to being incorporated with the rest of the stores. Certainly there was no question of our site reconnaissance having any effect on what we were being provided with to execute the work (it was an unusual situation to say the least!). Of course a small number of urgent items of equipment could be flown in on the 'air-bridge' via Ascension Island but, in general, we had to plan our task to suit what was already coming down to us by sea.

By mid July, just over a month after the surrender, the two shiploads of stores plus two fresh squadrons of Royal Engineers had arrived in Stanley. CRE (Works) staff were still working against the clock producing the working drawings and specification for the works. The construction force concentrated initially on getting the stores ashore and establishing a quarry. The stores unloading was a major task because Stanley possesses no deep water port facility. The stores ships were obliged to anchor in Port William outside Stanley Harbour and rafts (made up from 'Mexeflote' pontoon sections) and powered lighters operated to slipways constructed by the Royal Engineers. There were over 9000 tonnes of cargo for the airfield to be moved in this way including some very heavy items such as rock crushers (the primary crusher units weighed 45 tonnes). The unloading process took about three weeks.

The initial work on site covered drainage and the aircraft arrester gear foundations: these were massive concrete foundations founded at a depth of 2.2 m in running sand. Fortunately we had included well pointing kit in our equipment package; without it this task would have been impossible. The next stage was to overslab the existing runway with the AM2 mat. This involved more substantial repairs to the cratered areas and a re-shaping at each end of the runway to accommodate the vertical curves in the longitudinal profile joining into the proposed runway extensions. The AM2 mat was laid with a very large number of men working

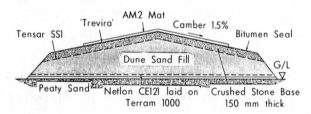

Fig 3.   Cross section through runway extension.

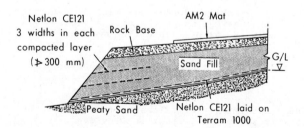

Fig 4.   Cross section through embankment.

3 hour shifts with 9 hours off. Handling the heavy mat panels in the high winds and at very low temperatures was an arduous task and 3 hours at a time was as much as could be managed. The runway closure was for 12 days in late August. Then the 4100 ft (1250 m) length of overslabbed runway was put back into use. During the closure, air defence cover was provided by the Harriers.

Attention then turned to the runway extension and other areas of new construction. By then it was September, spring was around the corner and the ground conditions became marginally better. Some levelling work was needed on the subgrade and a few bomb craters in this area needed filling. The digging of decent drainage ditches also caused a marked improvement in the ground conditions. The Terram and Netlon CE 121 went down very easily and proved an excellent platform for the subsequent filling work. The only difficulty with the Netlon was the tendency of it to wrinkle and gape at the overlap joints when fill was placed on top. Fortunately we had had plastic ties sent out to us on the airbridge and these cured the problem. The recommended spacing for the ties was 1.5 m but we found it better to reduce the spacing to about 0.6 m. Figure 5 shows the Terram and Netlon CE 121 in use.

The dune sand fill was placed if required or else the crushed rock base went down straight upon the Netlon. We found it essential to have at least 200 mm of loose material on top of the geogrid when doing rough grading. Otherwise the grid caught on blade corners and was ripped. The stone layer did then usually exceed the 150 mm requirement, though after compaction it was of course closer to that thickness. About half the runway extension length required sand fill.

In the strong winds and generally drying conditions in September it was impossible to retain sufficient moisture in the sand for optimum compaction. Its consistency soon resembled that of dry salt and it was impossible to stabilise the top surface. The Tensar layer that we introduced proved very useful as a separation layer between the sand

Fig 5. Terram and Netlon CE 121 on subgrade.

and the stone and it also of course provided more shear strength. The Tensar layer is shown in Figure 6.

Fig 6. Tensar SS1 used as separation layer.

The crushed stone production became the most critical activity of all. The decision forced upon us to found the new pavement areas upon a graded stone base led to a requirement for tens of thousands of tons of crushed rock. We had early problems with our rock drills. They were too lightweight and insufficient for the production rate required. Fortunately we were able to fly in more drills of a heavier variety. The critical equipment then became the rock crushers. It was as well that we had 2 sets of primary and secondary crushers, our suppliers had wisely provided at least two of everything where possible, and both sets were put to work. Still we had many difficulties. In part these were due to the very hard nature of the quartzite rock which caused crusher jaws in particular to wear very quickly. The other limitation was our own experience but we learnt fast and benefited from a number of technical advisory visits.

We started off with a specified grading for the base rock of a 40 mm graded layer topped by a 20 mm aggregate. In practice this was greatly modified. The secondary crusher could produce 40 mm stone but the penalty was excessive jaw wear. We settled in the end for a 100 mm down crusher run material topped with the 40 mm down fraction that we had to screen in a secondary operation. Further refinements were impossible if we were to meet the 1000 tons per day requirement. Our base strength requirement of CBR 14% was of course assured using crushed rock but we needed to achieve adequate compaction and a smooth grade to accommodate the AM2 mat. This we managed even with the large sizes of rock used.

In fact we achieved a very hard concrete-like surface upon which we rolled out the 'Trevira' material. We had selected this material from a sample sent to show the cladding material for the prefabricated aircraft shelters made by Rubb. We had been worried about the durability of the membrane sandwiched between the crushed rock and the metal mat which we knew would move under aircraft loadings. The very tough Trevira, though designed for building use, satisfied our requirement.

The AM2 mat laying was relatively straightforward and a less critical activity when only required to keep pace with the completion of the base layer. The second phase of the development work included the 2000 ft (610 m) runway extension, one aircraft dispersal area and a number of aircraft arrester gears and it was completed in time for the arrival of the first Phantom aircraft on 17 October 1982. The second phase of the work had taken 6 weeks and the overall task had taken about 2 months from the time the stores had been assembled on site. The time from the initial conception in London had been a little over 6 months. A view of the completed runway is shown at Figure 7.

Fig 7. Runway looking E with extension in foreground.

ROAD CONSTRUCTION

We also built many roads around the airfield site. Our early experience led us to use our geofabrics and geogrids for these too. Various combinations were tried until we arrived at our most economical and effective solution which is illustrated in Figure 8. We generally found 200 mm of crushed rock an adequate surfacing layer but later on, when putting in haul roads for the dump trucks carrying rock fill to other sites, we found it necessary to increase this layer to 300 mm. We found Netlon CE 121 perfectly satisfactory as a geogrid for roads but eventually opted for Tensar because it was lighter and cheaper and did the job just as well.

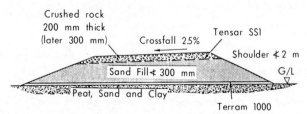

Crushed rock 200 mm thick (later 300 mm)   Crossfall 2.5%   Tensar SS1
Shoulder ≮2 m
Sand Fill ≮ 300 mm   G/L
Peat, Sand and Clay   Terram 1000

Fig 8.   Typical road section

CONCLUSIONS

It is not possible to put a precise value on the geofabrics and geogrids used in the development of Stanley airfield. In the emergency context, preliminary site trials were not possible. The design was based upon engineering commonsense (a subjective view!) and a prudent wish to use any means readily available to avoid pavement failure. We were not so constrained by financial limits but had an absolute need to guarantee results. It is difficult to assess the engineering risks taken and, at the time, there was little opportunity to dwell upon them. Also, the eventual base for the airfield mat was so strong that localised failure in the pavement was unlikely. It is worth then considering the longer term performance where inadequacies and more general failure might become apparent.

In fact the airfield has performed very well The maintenance problems have nearly all been related to the AM2 mat as a result of movements induced by aircraft. One failure of the base, shown by rutting along the Hercules wheel tracks in one section of the runway, was caused by the dislodgement of the Trevira membrane caused by the AM2 mat rocking across the runway crown and pulling the membrane by suction. This caused gaps for surface water to percolate through and led to localised weakening of the base; the base did not actually fail, as it was still necessary to attack it with picks to carry out repairs. The solution to the membrane problem was to place a layer of Netlon CE 121 between the mat and the membrane and this effectively broke the suction.

There has been some general settlement in areas of fill. The likely causes here are settlement in the sand fill which was compacted to a reduced standard when we could not maintain a high enough moisture content and, secondly, settlement of the subgrade following the drainage of the area.

We are left then with the conclusions that for the airfield pavement works the use of geogrids provided a sound base for the placing and compaction of the sand fill and crushed rock layers, provided a positive separation layer between the sand and the rock to save waste of the valuable crushed rock material, and provided a novel separation membrane under the AM2 mat itself. Also the geogrids most probably contributed to the overall strength and stability of the whole pavement system. As far as the road works are concerned we have more positive evidence that the geogrids were essential components in providing adequate roads without the excessive use of hard-won natural materials. We can only remain thankful that the failure limit for the runway itself was not demonstrated at Stanley in 1982.

Geogrid applications in the construction of oilfield drilling pads and access areas in muskeg regions of Northern Alberta

Gulf Canada Resources Incorporated uses Tensar polymer geogrids as a muskeg stabilizing agent for access road and drilling pad construction over muskeg areas in Northern Alberta. Drilling operations can last from 40 to 60 days and if the well goes into production the site may be in use indefinitely. Various techniques of subgrade preparation are discussed. Different design parameters including grid orientation, fill quality and supply are considered, to accommodate expected dynamic and static load conditions. The Tensar geogrid and fill system is evaluated in terms of overall settlement and reaction to static and dynamic loads over time. A comparison of geogrids to commonly used stabilizing agents and construction techniques is addressed.

J. B. Kroshus and B. E. Varcoe, *Gulf Canada Resources Inc.*

INTRODUCTION

During 1982-83, G.C.R.I. drilled 270 wells in Western Canada, approximately 8% of these were on muskeg. The Exploration and Development program for 1983-84 will increase the number of muskeg locations fourfold.

The purpose of this report is to illustrate the effectiveness of Tensar geogrids in providing stable soil conditions for short term drilling and production operations on muskeg locations.

DISCUSSION

Characteristics of Northern Alberta Muskeg

"Muskeg" is the term used to describe organic terrain containing a peat structure with an associated vegetation cover and related mineral sublayer (MacFarlane, 1969).

In Northern Alberta, the vegetation cover encountered is generally a Class A or Class B (trees from 1.5m - 4m). The subsoil varies from ablation tills to clay and is difficult to characterize.

The peat structures vary from fine fibrous to coarse fibrous with depths of 10cm to 10m.

Water content ranges from 50% to 760% and shear strengths vary from 2 kPa to 29 kPa.

DESCRIPTION OF LOADS

The major static loads applied to the drilling rig matting are the weight of the rotating equipment, mast, blocks, substructure, mud tanks and tubulars. The bearing stress below the matting* is relatively low (peak bearing stress = 30 kPa, average = 15 kPa). Our concern during drilling is not one of overall consolidation but rather that of differential settlement which would require jacking of the rig substructure or drillpipe storage racks.

*Assuming a 3600m drilling rig (well depth rarely exceeds 3500m in Northern Alberta muskeg regions).

The important static loads are: *

		WEIGHT (kN)	BEARING PRESSURE (kPa)
1.	Rotating equipment, mast, block, substructure and drillpipe (applied through matting)	3400	30
2.	Fully loaded pipe racks	980	150
3.	Water tank	620	100
4.	Cement tank	380	90
5.	Mud Tanks	710	15
6.	Overlying fill material (based on 1 m layer @ 20 kN/m^3)		20

The traffic patterns on the lease (See Figure 1) impart more critical shear and bearing stresses to the surface. The greatest dynamic load is the transport of the drawworks (approximately 450 kN) on a tandem-tandem tractor trailer truck.

Polymer grid reinforcement. Thomas Telford Limited, London, 1985

145

Typically, the dynamic loads are:

OPERATION	VEHICLE TYPE	AVERAGE LOAD (kN)	FREQUENCY (cycles)
Move in	Tandem Truck	220	25
Drilling	Vacuum Truck	220	160 - 240
	Cement Bulkers	400	3
	Cement Trucks	220	4
	Casing Trucks	400	15
	Pick-up Truck	10	200 - 300
	Miscellaneous	30	12 - 16
Move Out	Tandem Truck	220	25

Assuming a drilling duration of 40 -60 days.

FIGURE 1

RIG LAYOUT AND TRAFFIC PATTERNS

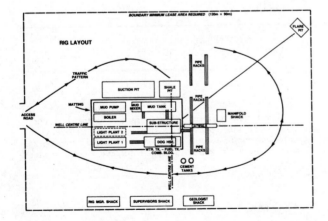

CASE HISTORIES

A.   Drilling pad construction using Tensar AR1

Location: Sturgeon Lake, Alberta
Date: August 1982

Site Details:

1. Lease Area = 9,300m^2
2. Condition of Lease = Drying after heavy rains
3. Depth of peat = 1.0m
   Water content = 350%
   Undrained shear strength = 5 kPa
4. Subsoil = 1.5m of sandy clay overlying rounded gravel
5. Maximum static bearing pressure = 150 kPa
6. Maximum dynamic load = 450 kN

The decision to use geotechnical reinforcement was motivated by a severe construction time constraint aggravated by wet weather. Tensar AR1 was chosen because of its competitive initial testing price and G.C.R.I.'s belief that Tensar was a better product than conventional geotextiles.

Construction:

The site had been levelled of brush and trees which were laid directly over the muskeg. The geogrid was unrolled perpendicular to the anticipated rig traffic with a grid overlap of 200mm along the length and 2m at the ends.

A low plastic fill material (40% sand, 60% clay) was applied as a 1m expanding wavefront. This "Rolling Surcharge" method of fill placement was used to reduce slack in the geogrid sheets and squeeze some of the water from the peat in an attempt to reduce future subsidence of the pad. Fill compaction was achieved using a sheep's foot compactor. The average height of fill at the end of construction was 880mm.

Observations:

With straight clay fill and no geotechnical reinforcement, the larger fill lumps punch through the muskeg root mat resulting in more material being used. With Tensar geogrids, there was a more uniform bearing pressure applied to the root mat and its integrity was maintained.

A calculation done by Pavement Management Systems Consultants predicted the maximum tensile force in the geogrid to be 12 kN/m, well below the biaxial tensile strength of Tensar AR1 (21 km).

The lease responded well to rain and rig traffic. No surface deformation was encountered and the average overall pad consolidation at completion of the drilling operation was found to be approximately 215mm.

Based on our construction and maintenance experience, geogrids provided quicker access to the location. Very few maintenance problems were encountered and pad subsidence was minimal. The pad was constructed in 5 days compared to 8 days for an adjacent lease (no geotechnical reinforcement and 1.25m of fill), thereby saving approximately 14% in construction costs.

CASE HISTORIES

B. A comparative study of Tensar SS1 and a heat-bonded geotextile for muskeg reinforcement during production operations.

Location: Pelican Lake, Alberta
Date: January 13, 1983

Site Details:

1. Lease Area = 10,800m²
2. Condition of Lease
   before construction = 1.0m of frost penetration
3. Depth of peat = 1 - 1.5m (deepening at the north end)
   *Water content = 50 - 200%
   *Undrained shear
   strength = 5 - 20 kPa
4. *Subsoil – 3 - 5m of low to medium plastic ablation till (penetration from 72 to 432% kPa)
   – 1.2m of sand
   – basal till (penetration readings 432= kPa underlies the sand and ablation till)
5. Maximum dynamic
   load = 250 kN (Tanker truck)

Construction:

Trees on the lease were levelled and the site divided into 3 north-south strips. The central strip was excavated to the mineral sublayer (removing an average of 1.3m of muskeg) and refilled with sandy clay. The geotextile was laid longitudinally on the muskeg west of the clay section and the geogrid was laid longitudinally to the east.

The entire lease was then backfilled with 1.0m of clay and sandy gravel. Compaction was very limited consisting mainly of D6 caterpillar traffic.

Observations:

Settlement data (Fig. 2) was obtained over the following year for each of the three lease strips. As expected the settlement of the clay strip (all muskeg removed) was less than the two strips where the highly compressible muskeg was not removed. Of the two muskeg strips, the performance of the Tensar geogrid strip was superior to that of the geotextile strip. Comparison of the two strips between April and August 1983, shows that on the south side of the site, settlement of the geotextile strip was 2½ times greater than settlement of the geogrid strip. On the north side of the lease where the depth of muskeg is greater, the settlement of the geotextile section was 38% greater.

From these results it can be concluded that total settlement can be minimized by removing the muskeg and replacing it with fill. This can only be achieved if the muskeg depth is shallow and replacement fill is economically available. At the Pelican Lake site, the muskeg varied from 1.0m - 1.5m, and the cost of fill was very low due to its proximity to the site. For this set of conditions, the muskeg removal and fill replacement alternative was 24% less expensive than the reinforced pad alternatives. Where the above cited conditions are not met, however, geogrid reinforcement has been shown to be cost effective and has achieved superior performance over the

*Taken from soils report describing the Pelican Lake Area.

geotextile reinforcement used in this study.

A 142m³ oil storage tank located on the geotextile/clay interface developed a severe list towards the geotextile section (after 4 months) and consequently was repositioned on the clay surface.

Visual inspection indicated the geotextile section was prone to rutting. (After 6 months heavy traffic avoided the area). Minimal rutting was encountered on the geogrid and clay sections.

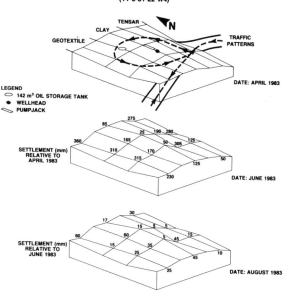

FIG. 2
RELATIVE SETTLEMENT OF
PELICAN LAKE DRILLING PAD
(11-5-81-22 W4)

CONCLUSIONS

1. Tensar geogrids have proven economical and effective in stabilizing muskeg for summer construction of drilling pads and access roads. They have proven effective in maintaining surface integrity of the pad during drilling operations. Fill cost and muskeg depth are the major factors in determining the economics of geogrid application. The observation period has not been long enough to quantify maintenance savings.

2. Without a major clay constituent, sandy fill (due to its low shear strength) is unacceptable as a traffic surface. Tensar geogrids (when placed at the muskeg/fill interface) have a very limited effect on providing surface stability (reducing surface rutting). Surface stability can be improved by placing a second layer of geogrid reinforcement at a shallower depth.

3. The "Rolling Surcharge" method of fill placement has been effective in reducing the future compressibility of muskeg by driving water away from the structure and preventing slack in the Tensar sheets.

RECOMMENDATIONS

1. When drilling pads and access roads are built on unfrozen muskeg, the bulk of settlement occurs during and shortly after construction, thereby allowing the surface of the fill to be properly graded prior to putting the pad or road into service. If these structures are built over frozen muskeg, the bulk of the settlement occurs shortly after the subsequent spring thaw. Although Tensar geogrids are effective in reducing the differential nature of spring thaw settlement, the presence of reinforcing will not prevent total settlement from occurring. However, where geogrids are used to reduce the thickness of fill required, settlements are reduced because the total weight of fill is reduced.

2. Winter leases on muskeg should be constructed with a minimal amount of fill to take advantage of support provided by frost penetration. If the well goes into production, a layer of geogrid and more fill should be added after the first major thaw to stiffen the surface and compensate for the thaw consolidation.

3. Care must be taken not to puncture the root mat when preparing a muskeg location for construction. Levelling the trees on muskeg provides the most stable base for Tensar geogrid application.

4. A market potential for Tensar geogrids of several million dollars exists for the 1983-84 drilling season. In order to utilize the product more effectively, G.C.R.I. suggests that design curves be established to optimize the use of this reinforcing material in muskeg with a variety of fill materials.

REFERENCES

1. MacFarlane I. (1969), Muskeg Engineering Handbook. 1-130, University of Toronto Press. Toronto, Ontario.

Evaluation of geogrids for construction of roadways over muskeg

P. M. Jarrett, *Royal Military College of Canada*

Large scale plane strain loading tests have been made on gravel fills compacted on a peat subgrade. The fills were tested with and without Tensar geogrid reinforcement. Detailed measurements were made, of the deformed shape of the geogrid, of tensile loads in the geogrid and of geogrid movements under load. The results are being used to develop simple design methods for geogrid reinforcement of expedient roads and to assist development of comprehension of the mechanics of behaviour in geogrid reinforced soils.

INTRODUCTION

The organic soils and peats, which form the surficial layers of muskeg or peatlands, represent a formidable obstacle to the road-builder and to those wishing to establish temporary or permanent earth structures on such deposits. The basic problems in such construction lie with the extremely compressible nature of virtually all peats and organic soils and in cases of well decomposed organic soils with their low shear resistance.

In many situations, roads and railways have been constructed over deep muskegs using the tensile resistance of the natural surface mat as a primary supporting mechanism. Such surface mats are formed from interwoven roots and other plant remains. The use of corduroy construction, in which a continuous bed of logs is formed beneath the road, is also common in such circumstances. It is, therefore, not surprising that geogrids represent a potentially valuable aid as a tensile reinforcement in earthworks constructed over peats. Geogrids possess a more uniform and predictable tensile resistance than a natural surface mat and would be easier to use in construction than a full corduroy roadbed.

The tests reported in this paper are an attempt to quantify the benefits of Tensar geogrids in such construction and to allow a better understanding of the mechanisms involved in their use as a tensile reinforcement in soils.

TESTING APPARATUS

Large scale plane strain loading tests have been made in the laboratory on a series of compacted gravel fills, constructed on a 1 m deep peat subgrade. The tests were carried out in a test pit which is 3.7 m long by 2.4 m wide and 2 m deep. Loading was applied to the compacted gravel surface through a 203 mm wide beam spanning the full width of the test pit. The load was generated using a computer-controlled MTS hydraulic actuator. The test pit is shown during operation in Plate 1.

Plate 1
Test Pit Arrangement

The peat subgrade was formed using bales of air dried, finely fibrous, horticultural sphagnum peat with a low degree of decomposition. Water and peat were mixed in the test pit to a depth of about 1.3 m after which excess water was drawn out of the mixture using a series of drain pipes on the floor of the pit.

Polymer grid reinforcement. Thomas Telford Limited, London, 1985

149

This downward drainage consolidated the peat to a thickness of 1.0 m with the following typical average properties:

Moisture content – 850%

Vane shear resistance – 4 kN/m^2

After each test, the peat was loosened by digging, water was then added to bring the volume back to the 1.3 m depth and compressed air was bubbled through the system to complete dispersion of the peat-water mixture. After the dispersion, the peat was reconsolidated using downward drainage to provide the subgrade for the next test.

Gravel fills 305 mm thick were compacted on the peat subgrade in two lifts using a Wacker vibrating plate compactor. The nominally 20 mm gravel was formed from crushed limestone with the following gradation:

    100 % passed 20 mm
     50 % passed 10 mm
     22 % passed #4 Sieve
      9 % passed #10 Sieve
      4 % passed #20 Sieve
      1 % Fines

Compacted dry densities were in the order of 1920 kg/m^3 at a moisture content of 1.8%. Fills were compacted with and without geogrid reinforcement. In the reinforced cases the Tensar Geogrid Type SS2 was placed at the peat to gravel interface.

The following monitoring procedures were used when appropriate:

1. The load applied and the displacement of the actuator piston were monitored through built-in transducers and the computer controlled data acquisition system.

2. Cross checking of the beam displacement with the movement of the actuator was carried out using dial gauges reading directly onto the beam.

3. Vertical movements at the peat-gravel interface were measured at various distances from the beam centerline. This was achieved primarily using settlement plates at the interface with attached rods passing through the gravel so that a series of level readings could be taken as displacement occurred.

4. Horizontal movements of points on the geogrid were observed. To do this, small plywood boxes were inserted into the gravel exposing a section of geogrid. Travelling microscopes were then used to sight on a fiducial mark scratched on a geogrid junction point.

5. Load cells were manufactured which could be inserted as a joint in a 305 mm wide strip of geogrid. The cells transmitted the tensile load between sections of the geogrid strip through 8 Aluminum fingers. Each finger had 2 electrical resistance strain gauges attached to it. The strain gauge readings were calibrated against tensile load. Three such cells were inserted in one strip of Tensar. They measured tensile loads directly beneath the loading beam and at distances of 400 mm and 800 mm from the beam centerline. The instrumented strip was placed along the longitudinal centerline of the pit and as plane strain conditions prevailed this should not have influenced test behaviour.

TESTING PROCEDURE

In order to simplify analysis, an incremental static loading procedure was adopted. Each load level was applied to the beam until the rate of displacement became less than 0.02 mm/min. At this time, the load was increased unless it occurred at a time when the test operator was not present. In this case, the increment would be applied as soon as he returned and made the necessary manual readings of the settlement plates and travelling microscopes. Loads were increased until at least 200 mm of beam displacement had occurred. At this point, a series of 5 unload-reload cycles were performed. The load on the beam was reduced to 2.5 kN for 100 minutes and then increased to its previous maximum for 100 minutes. This represented one cycle. At the end of the cycling the beam was completely unloaded, removed and the rut formed by the test, filled by the addition of compacted gravel. The beam was then replaced and reloaded in increments to the previous maximum, after which a further series of unload-load cycles was instituted. Following these cycles, incremental loading was continued to higher levels. At the maximum load level further unload-load cycles were also carried out. The complete beam load against beam displacement results are given for a Tensar reinforced test in Figure 1. This test was repeated twice with virtually identical results.

The procedure for testing the unreinforced fill was similar in the initial phase. However, the load increment which took the beam beyond 200 mm of displacement had to be terminated, because the rate of displacement did not fall to the required value. As the system was showing signs of distress with a punching failure in progress, the cyclic loading was not performed. The rut was, however, filled and the beam incrementally reloaded. After the load was increased and again showed signs of failure, a series of unload-load cycles were carried out. The beam load-beam displacement results from this test are given in Figure 2, where they may be compared with an abbreviated form of the reinforced test.

In addition, beam load tests were performed directly on the surface of the peat as a measure of its compressibility characteristics. The rate of displacement criteria was the same in these as in the other tests. Displacements

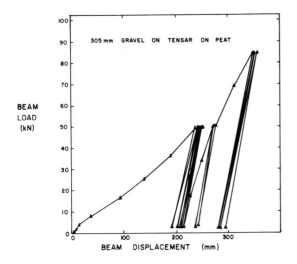

Figure 1
Beam loading results for reinforced fills

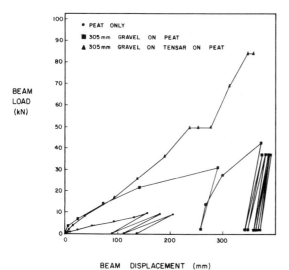

Figure 2
Beam loading results for all tests

could not be taken so far in these tests as the beam and foot of the actuator were disappearing beneath the water table into the peat. The results for these tests are also given in Figure 2.

DISCUSSION OF BEAM LOAD–BEAM DISPLACEMENT RESULTS

The unreinforced fill appeared to be in a state of distress under the application of the 31 kN load level, as its rate of displacement did not appear to be decreasing to the required level. On the other hand, the geogrid reinforced fill took over 50 percent more load without undue distress and also withstood the load-cycling procedure.

The decision to fill the ruts after 200 mm of displacement was arrived at in relation to the probable maximum rut depth allowable in a temporary haul road. It may be seen that after filling the ruts loads were increased to 85 kN and this produced only 120 mm of new displacement. This form of result gives credence to the procedure of producing a rutted fill during construction and then filling the ruts prior to use.

At the start of the tests it may be seen that no difference between the unreinforced and reinforced cases was measured until close to 100 mm of displacement had occurred. This is a common finding in such studies and indicates that the reinforcement is acting in this situation primarily as a stretched membrane. It is possible, therefore, to use developments of the types of analysis suggested by Giroud and Noiray (1981) and Sowers, Collins and Miller (1982) to assess the tensile loads carried by the geogrid. The only major difference between the peat problem and the clay problems dealt with by the quoted authors is that peat represents a drained situation, where constancy of volume cannot be assumed. However, using the results from the three forms of tests shown in Figure 2 and the procedures outlined by Jarrett (1983), estimations may be made of the relative contributions of the peat, of the gravel and of the geogrid to the overall bearing capacity. From these results and the deformed shape of the membrane calculations of the tensile forces in the geogrid can be made.

TENSILE LOAD MEASUREMENTS

The load cells in the Tensar strip were used in one of the two reinforced tests and functioned well. The results obtained are summarized in Figure 3 for Cell 1, which lay directly beneath the centerline of the beam and in Figure 4 for Cells 2 and 3, which lay respectively 400 mm and 800 mm from the centerline. The loads have been presented in terms of kilonewtons per meter for convenience of reference.

Considering Figure 3, tensile forces of approximately 10.8 kN/m were generated under a beam load of 48 kN. During load cycling some

151

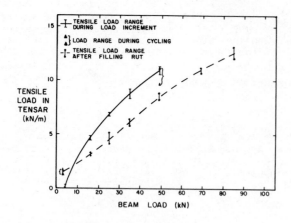

Figure 3
Tensile loads measured in Cell 1 on beam centerline

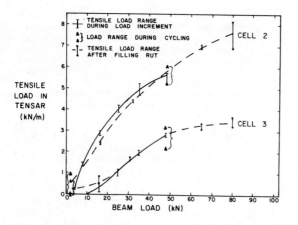

Figure 4
Tensile loads measured in Cells 2 and 3

and coincided with the inflexion point in the curvature of the geogrid from concave beneath the beam to convex towards the outside. At this point the effect of the geogrid changes from decreasing the load on the subgrade beneath the beam to increasing it as the load is spread laterally. The difference between the tensile force before and after filling the rut was much smaller at this load cell. Little extra gravel was added this far from the centerline and so the tensile behaviour remained the same.

At Cell 3 little vertical movement was measured and so the tensile force was being applied virtually horizontally. This means that this cell is in an anchorage zone. The facts that the tensile force appeared to reach a maximum and that the travelling microscope readings showed the Tensar to be slipping towards the centerline indicate a limitation in the testing arrangement. In essence, the geogrid is not reaching its full tensile capacity because of an insufficient anchorage length.

This may have practical implications as the geogrid does extend 1.85 m laterally from the centerline of the load. This may well be a greater distance than is usually provided outside the wheel paths on a single track haul road.

Despite the absolute limitations of the anchorage, valuable information is available from these results on the mechanism of geogrid anchorage. This point can be more fully developed when the tensile loads measured in Cell 3 are considered in conjunction with the lateral movements of the geogrid.

LATERAL GEOGRID MOVEMENTS

Using travelling microscopes, horizontal movements of two points on the geogrid reinforcement were observed. These points were initially 1.59 m and 1.08 m from the beam centerline. The results for the initial incremental loads are presented in Figure 5. The basic movements recorded were towards the load. The first point of note is that no movement occurred prior to 8 kN of applied beam load. The second point is that little differential movement occurred between the two observed points until 25 kN of applied beam load. At this stage one expects that the geogrid will be developing tension in this outer region. This is consistent with the results given by Cell 3 and shown in Figure 4. This stage also coincides with the beginning of the separation of the reinforced from the unreinforced cases in terms of beam displacement, Figure 2. Thus one may consider that the full mobilization of all aspects of the geogrid reinforcing action are underway at this point.

The difference in movements observed between the two geogrid points can be reduced to terms of strain by considering them over a gauge length of 0.51 m. Thus at a beam load of 48 kN a differential movement of 0.16 cm would represent a strain of 0.3 percent. If

tensile force remained locked into the system in the unloaded phase and very consistent values of tension were measured over all 5 cycles. After the rut was filled application of the same beam load produced smaller tensions in the geogrid. This is presumed to be due to the better load spreading caused by the thicker gravel layer. In fact, the maximum tension reached only increased to 12.5 kN/m under application of the higher load levels. This, it is felt, was due partly to limitations in the lateral anchorage of the geogrid that may be better explained after study of Figure 4.

Cell 2 lay 400 mm from the centerline

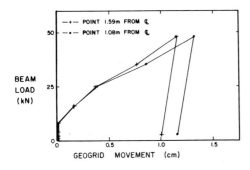

Figure 5
Lateral movements of geogrid

indeed this is a true strain in the geogrid it appears to remain "locked in" as the differential movement remained the same during the unloading cycle. This would be consistent with qualitative observations that the outer anchorage zones appear to move as a block.

Unfortunately tensile load cell 3 lay 800 mm from the centerline and the first movement gauge 1080 mm. As such they were not coincident. If one takes the liberty of assuming that they were then a combination of the two results produces an in-situ "pull-in" test or interface shear resistance test. A reasonably consistent curve of tensile load against displacement is obtained by making this assumption. If one considers the maximum tension recorded in Cell 3, 3.5 kN/m, to be pulling a block of gravel 0.305 m deep and 1.05 m long. Then this represents an average interface shear resistance of 3.33 kN/m^2 under a normal stress of 5.74 kN/m^2. The average movement of the geogrid at this stage of the test was 1.8 cm representing a mobilized shear strain of only 1.7 percent. The value of shearing resistance ties in reasonably with the vane shear resistance measured for the peat but the mobilized strain is very small compared to the usual strains necessary to produce failures in peats. It is possible therefore that greater tensions could be mobilized with greater deformations.

CONCLUSIONS

1. Tensar geogrids have been shown to have a significant effect on the bearing capacity of compacted fills over a peat subgrade.

2. Significant tensile forces were measured in the geogrid and these developed despite the very compressible nature of the peat.

3. The full spectrum of results obtained open the way for the checking and development of analytical methods and the development of a better understanding of the mechanics of geogrid reinforcement.

ACKNOWLEDGEMENTS

The author wishes to thank the Tensar Corporation for its direct sponsorship of this work and also the Department of National Defence of Canada for their continuing support.

Special thanks go to Dr. R. Douglas, now of the University of New Brunswick, who designed and built the load cells, to Mr. J. Bell who performed the testing, to Miss Martina Lahaie who typed the manuscript and Mr. J. DiPietrantonio who prepared the figures.

REFERENCES

Giroud, J.P. and Noiray, L. (1981). Geotextile-reinforced unpaved road design. PASCE, Vol. 107, No. GT9, p. 1233-1254.

Jarrett, P.M. (1983). Reinforcement of roads on organic terrain. Proc. 7th Pan American Conference on Soil Mechanics and Foundation Engineering, p. 329-342.

Sowers, G.F., Collins, S.A. and Miller, D.G. (1982). Mechanism of geotextile-aggregate support in low-cost roads. Second International Conference on Geotextiles, Vol. 2, p. 335-340.

Foundation for roads and loaded areas: report on discussion

R. J. Bridle, *Mitchell Cotts plc*

Dr P M Noss, SINTEF, Norway made an invited contribution which gave information about a field test of a road constructed in Norway during the fall of 1983.

The test section is 180m long and is based on clay. The clay is highly susceptible to changes in moisture content and it suffers a significant loss of bearing capacity during the spring/fall period.

The pavement structure consists of a wearing course of 35mm of asphalt concrete and a granular base where the thickness varies between 800mm and 400mm. Two different materials have been used as a base material - a crushed stone where a maximum size of 120mm is used, and a well-graded gravel in the second part. The test section is sub-divided into smaller sections, and on all sections a thin fabric is used as a split between the subgrade and the granular materials.

On some of the sections Tensar Geogrid SS2 has been placed on the top of the subgrade in order to study how it will affect the bearing capacity of the road. A steel mesh has been put in another of the test sections.

Sections without any reinforcement, and the section reinforced with the steel mesh, have been modelled in the inter-bed programme at the University of Illinois, and the stress model has been used to describe the elastic properties of the base and the subgrade material. Where the steel mesh has been used, computations show that we should expect a reduction of the elastic deflection beneath a 50kN single-wheeler of about 15%-35%, depending on the thickness of the crushed stone layer of the bearing capacity of the subgrade during the spring/fall. Stress-dependent moduli have to be used in these computations.

It is believed that the measured deflection and the curvature of the sections which are reinforced with Tensar Geogrid will lie between the sections with no reinforcement and the sections with steel reinforcement.

During the spring/fall period this year, elastic deflection measurements will be made at the test section. A falling-weight deflectometer simulating the moving loading post of a 50kN single-wheel load will be used in this study. By installing strain gauges in the pavement structure the mean elastic and plastic strain in the base layer can be measured, and that should give us some additional information about how compaction can affect the bearing capacity in the sections where the SS2 Geogrids are used.

The design charts presented by Dr Giroud and shown at Fig 11 of Paper 4.1 attracted a good deal of attention. Other papers were discussed in the light of the evidence they gave to support or challenge the use of the design charts for everyday engineering use.

The arguments challenging the charts cited the limited evidence available for determining, quantitively, the mechanisms through which reinforcing meshes, either elastic or visco elastic, contribute to improved performance. Pavements represent a very expensive part of the infrastructure and public authorities are risk averse. The pragmatic basis of the curves may fit observed results but extrapolation required caution. This is because their basis is not so clearly related to the engineering mechanisms used by the engineering community that they generate confidence in use. In particular the method's reliance on elastic behaviour, when demonstrably it is inelastic, caused some delegates to express doubt about its use.

The argument in support of the design method was that it does not rely directly on elastic theory but rather employs elastic theory to establish direct comparison between the reinforced and unreinforced cases. This reduced any inaccuracies attributable to the use of the elastic theory. Since the charts rely on a number of variables, calibration against actual cases takes some account of non-linear behaviour through their interaction. The curves can be improved by further consideration of non-linear behaviour. They can also be improved by using them to forecast cases outside the existing experimental observations and making comparisons, that is attempting to radiate the curves. Furthermore, for most of the life of a road, induced strains would be modest. Now since existing policy in developed countries is to intervene at reasonably low levels of permanent deformation, to offset increasing vehicle operating costs, elastic analysis used on a comparative basis is probably adequate.

It was generally agreed that the curves should help engineers. Engineers can produce target designs to compare with conventional proposals on a basis of equal life. If the engineer felt the risk involved in using the reinforced design is greater than a traditional design, he can interpret it conservatively and yet achieve some of the benefits in direct cost and increased life.

There was a need expressed for the development of a design method which deals explicitly with the strength parameters of the materials involved, the variation in pavement loadings, the variation in the strength and non-linearity in the stress strain curve of the materials used.

One important variable is the elastic stiffness of granular materials. Their stiffness varies with the stress levels to which a particular element of a granular pavement base is subjected. This had been demonstrated many times in extensive testing at the University of Nottingham. The plot of elastic stiffness against initial effective stress, generally that due to over-

154

burden pressure, has a very sharply curved form showing a high stiffness with high effective stress. Furthermore, good confinement will also yield high values of stiffness. However, the effect of deterioration with time, due to high repeated value of shear stress, is to reduce the stiffness to relatively low values.

If this phenomena is ignored, then the results of any design equations based upon that shortcoming will be misleading.

It is evident therefore, that increasing confinement will be an advantage because it will increase the stiffness of the base. Contributors to the discussion speculated on how this may be done. Two examples were given. Firstly, it was suggested that pre-stretching will increase confinement, or at least make its achievement more secure. Secondly, deflecting the reinforcing layer to sag below the wheel loads and hog at other places was thought by contributors to be beneficial.

Both techniques aim at enhancing the membrane effect. Pre-stretching is also aimed at increasing bond, and thereby the confinement, while profiling, will provide vertical components of the stress in the membrane, to increase the spread of load across more of the pavement.

It was doubted whether practical and economic techniques to achieve these aims can be devised. However, pre-stretching below asphalt layers is reported in papers in other sessions. Transverse deflection can be achieved by shaping a first layer of stone, returning and pulling the reinforcement horizontally over it and then pouring the second layer to stretch the membrane into the profile of the first. The description shows that the increased number of activities meant that profiling was unlikely to prove economic.

The discussion on membrane tensioning showed that more work is needed to identify the mechanism of confinement and bond. Slippage between the reinforcement and base at the higher levels of strain, or later stages of loading will have an adverse effect on performance.

Delegates were prompted to speculate on the mechanisms by which improved performance was achieved. The aim of further research should be to better identify and quantify such mechanisms, see how to enhance their beneficial effect and how to better represent them in design equations.

The discussion confirmed the basic mechanisms at work. In general an aggregate layer will deteriorate with time, due to contamination of the base by the subgrade fines moving upwards and the base aggregate moving downwards. In addition the aggregate in the base will be displaced laterally. This deterioration over time is modelled in the design method in paper 4.1 by applying different stiffness moduli for the effect of contamination and for the effect of base cracking.

The contamination can be prevented by a geotextile acting as a separating layer. However, since geotextiles are generally highly extensible, that is they induce only a little load for a large strain, it was argued that they did not contribute to confinement. Figure 9 of paper 4.1 gives the improvement due to separation as no more than 10%. In comparison containment of tensile strains by the interaction of a strong, inextensible grid with the base is attributed with an improvement of 70% even without separation. Together the improvement is enhanced to 100%.

It was acknowledged by contributors that the presence of reinforcing mechanisms is more important than separation. Nevertheless the thickness of the base and position of reinforcing layer are evidently significant variables in determining the degree of improvement due to reinforcement. It therefore surprised contributors to see that the choice of factors in Figure 9 takes no account of the ratios of material strengths of the various constituents or the thickness of the base.

Fig 9 also infers that greater benefits are achieved with time and after many repeated loadings. The inference derives from the fact that greater benefits are shown for the contribution of reinforcement to contaminated bases, and bases will contaminate more with time and loading. However, at higher strains and many repeated loads, the assumption of continuing quality of confinement might be doubted since the effect of migration of fines and repetition of load might affect bond. In the papers the expression of improvement in performance is taken to be an increase in the angle α, the angle of spread from the load through the base. That is, it is assumed that the pressure on the subgrade is more widely distributed, just as though the base is thickened. The adoption of such a concept stems directly from static considerations. In reality the problem is one of repeated loads and the aim of design is to maintain some restraint on the accumulation of permanent deformation.

However, the majority of discussion concentrated on the contribution each of the three mechanisms makes to the angle α rather than their contribution, under repeated loading, to restraining the accumulation of permanent deformation.

Many considered that the tensioned membrane effect makes no significant contribution because, in static terms, it is evident that the membrane load needs to be deflected to provide a vertical contribution to load carrying. The Chairman noted that if the membrane is loaded, it absorbs energy and thereby must help to reduce cumulative deflection.

It was agreed that the effect of confinement in strengthening the load carrying capacity of the base, particularly at loads which will otherwise induce large tensile cracks, is a significant mechanism in improving load carrying capacity. In addition, subgrade materials are damaged by movement at the interface and Geogrids helped restrain such movement.

The design method adopted in paper 4.1 does take account of these advantages by a reduction of strength in the unreinforced case. It was also observed that the findings recorded in paper 4.2 show an over prediction of bearing capacity for the unreinforced case if such softening is not considered.

There was a measure of agreement between authors of papers 4.1 and 4.2 about the equivalent angle α and its improvement due to the mechanisms described. For small strains, induced under static loading, it is generally expected that the effect of reinforcement spreading load to the subgrade would be small. However, for larger strains, it seems that the reinforcement makes a much greater contribution to load spreading. As time passes or more axle passes are experienced the angle α reduces in both the reinforced and unreinforced cases due to increased contamination.

In summary, the Chairman concluded that more work is needed to establish how the mechanisms worked in

DISCUSSION

improving performance. He suggested that improved performance for the road engineer is not the change in the angle α but rather that the residual deflection under each load pass is reduced. He believed that if the approach to the mechanisms is to establish their contribution to dynamic rather than static performance, engineers will increase their application of reinforcing techniques. Engineers are inherently conservative. Large economic losses result from poor prediction of performance in the design of civil works and premature failure of road pavements has already induced a high degree of political sensitivity.

Increase in bearing capacity in the case of foundations, and increase in the life of pavements, are not radically different problems. However, they are dissimilar in the nature of the loading and the criteria in respect of limit states. They therefore require separate design methods. Some departure from the traditional methods for designing roads based on load spreading seem warranted. In addition, road design needs to allow explicity for the great variations in strength due to materials and workmanship in use in construction.

The key question is "What percentage of failures before design life is complete are clients prepared to accept?"

It is evidently very small and therefore high percentile values of material strengths need to be adopted and used in designing pavements.

The Chairman thanked all authors and contributors for a thought provoking Session which had contributed significantly to the general understanding of the behaviour of pavements with Geogrids.

Participants: Dr Bonaparte

Dr Giroud

Dr Milligan

Mr Wood

Professor Brown

Mr Bridle

Mr Knutson

Foundation for roads and loaded areas: written discussion contribution

B. J. Robinson, Cheshire County Council

I would like to comment on the use of Tensar SS1 geogrid on the A51 Tarvin SW by-pass. A length of carriageway 315 m in shallow cutting comprised of silty, sandy clay had a very low CBR. A decision was made to strengthen the subgrade using two layers of SS1 with 600 mm thick 75 mm down crusher-run limestone. Over a section of 120 m the subgrade was very soft and wet and a layer of Terram 1000 fabric was placed below the bottom layer of geogrid to prevent the ingress of the wet fines into the rock layer.

The crusher-run material delivered to site was 100 mm down which proved to be too coarse and resulted in the geogrid mesh being cut in places. However, the overall result was successful and plate bearing tests were done on the top of the rock layer and also on top of the overlying sub-base which was 150 mm thick type 1 to the DTp specification. Table 1 shows the equivalent CBR results which were obtained. The recommended minimum CBR value on the sub-base is 30 and this value was easily achieved with the exception of one result in the very soft area.

Table 1. Equivalent CBR values

Chainage	CBR on rock layer	CBR on sub-base
770 E/B	40	70
850 W/B	44	>75
910 W/B	30	50*
923 E/B	12	33*
938 CL	36	48*

* Area includes use of geotextiles

Polymer grid reinforcement. Thomas Telford Limited, London, 1985

157

Tensar reinforcement of asphalt: laboratory studies

The potential for using polymer grid reinforcement in pavements is discussed with particular reference to applications in asphalt. A laboratory study still in progress, is investigating the properties of reinforced asphalt using realistic simulative experiments. These include wheel tracking tests on slabs to investigate permanent deformation development and special beam tests to study the problem of reflection cracking in asphalt overlays. In addition, the mechanical properties of Tensar AR1 polypropylene grid have been determined together with its response to asphalt paving temperatures. The preliminary results presented in this paper show that this product has considerable potential for use in asphalt pavements.

S. F. Brown, B. V. Brodrick and D. A. B. Hughes, *University of Nottingham*

INTRODUCTION

In most countries, pavement design methods and techniques for specifying asphalt mixtures are largely empirical. Encouraging progress is being made in the application of theoretical concepts and the use of mechanical properties of paving materials towards improving design practice. Economic advantages will be apparent as present research knowledge is implemented. There are, however, a number of problems in asphalt pavements for which solutions using conventional materials are unlikely to be completely successful, even with the use of new design techniques. These problems include the development of rutting caused by permanent deformation in heavy vehicle wheel tracks under high temperature conditions and the propagation of cracks through asphalt surfacings (overlays) placed on existing cracked pavements. These latter are known as "reflection cracks".

MECHANICAL PROPERTIES OF THE POLYMER GRID

The material supplied for use in this project was Tensar AR1 polypropylene grid which has the geometric properties shown in Fig. 1. This is a biaxially orientated product with larger dimensions in the "primary" direction and hence, greater strength and stiffness in this direction too. All the tests were carried out on the same batch of material.

Stiffness of Tensar AR1

For pavement applications, the grid would be subjected to repeated applications of low strains (or loads). The test arrangement developed to measure the load/strain relationship under these conditions involved a 0.5m wide x 0.8m long test specimen held by special clamps in a servo-hydraulic testing machine. The test involved applications of increasing amplitudes of controlled deformation at 1 Hz using a sinusoidal waveform and measurement of

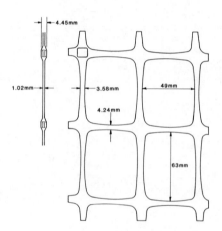

Fig. 1 Tensar AR1 Polypropylene Grid

the corresponding load amplitude. Cyclic strains up to 0.8% were applied.

In view of the importance of assessing the influence of exposure to the elevated temperatures experienced in asphalt paving, a number of grid specimens were heated in various ways and retested after cooling. Fig. 2 shows the load displacement curves for a typical specimen at one amplitude traced from the X-Y plotter used to monitor results. A reduction in stiffness is apparent as a result of the exposure to elevated temperature (in this case hot asphalt placed at 165°C). The load displacement curves exhibit slight hysteresis, but the response is essentially linear. The tests were carried out at 20°C.

Fig. 3 shows the results from a typical set of tests on another specimen and has been developed from data such as that in Fig. 2. Each point represents the peak load and strain during increasing and decreasing increments.

Polymer grid reinforcement. Thomas Telford Limited, London, 1985

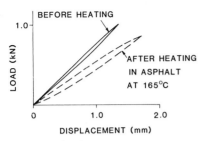

Fig. 2 Load-displacement response at 1 Hz and 20°C.

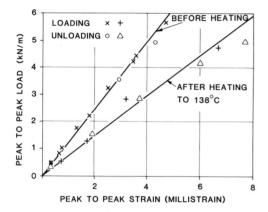

Fig. 3 Load-strain relationships for Tensar AR1

In all tests the load was applied in the primary directions.

From the slopes of the lines, such as those in Fig. 3, the values of Young's modulus and of elastic stiffness were calculated. The former was calculated on the basis of stress determined using the minimum cross-sectional area of the ribs (see Fig. 1). This area was 3.65mm² per rib and the average Young's modulus from nine tests on unheated grids was 14.7 GPa. Since the specification of cross-sectional area is somewhat arbitrary, it was considered more useful to describe elastic behaviour in terms of elastic stiffness, being the load per unit width divided by the strain. The mean value from all nine tests for this was 1.12 MN/m.

The reduction in stiffness after heating and cooling depended on the exposure temperature and the medium in which it was applied. Fig. 4 summarises all the test results to date including the use of a variety of containing media offering different degrees of restraint to the grid which had a tendency to shrink on heating. For typical asphalt paving temperatures of 140°C, stiffness is about 60% of that for new material.

The tests were all carried out on relatively small pieces of grid and the restraint may not be representative of that which could

develop in a pavement. A significant force (typically 0.8 KN/m) builds up in the grid if it is rigidly restrained while being heated. Developments in the production process and placement techniques on site will change the detailed characteristics, so additional testing will be required when these matters have been pursued further.

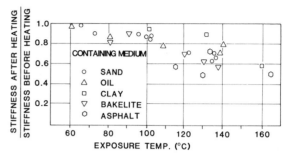

Fig. 4 Stiffness reduction on heating

Fatigue Characteristics of Tensar AR1

Four cyclic load fatigue tests were carried out in the servo-hydraulic machine on test specimens 360mm long x 250mm wide at 20°C. A mean strain of between 8.7 and 13% was applied and a cyclic strain of ± 0.25% was superimposed using a sinusoidal waveform at 13 Hz. The load was monitored and a typical result is shown in Fig. 5. This indicates that the cyclic load remained unchanged over the 370,000 cycles involved. This implies no decrease in elastic stiffness and, hence, no suggestion of fatigue failure. The mean load, however, relaxed in a manner characteristic of visco elastic materials. The same pattern was apparent in all four tests and this stress relaxation behaviour had earlier been noted in static tests at lower mean strain levels.

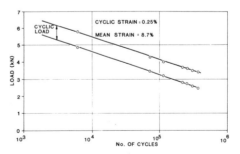

Fig. 5 Fatigue test results on Tensar AR1

THE ASPHALTIC MATERIALS

The mixtures used in this investigation were all made to current British Standard Specifications (British Standard Institution, 1973, a and b) and included hot rolled asphalt (HRA) and dense bitumen macadam (DBM) wearing course

and hot rolled asphalt basecourse. The
aggregate grading curves are shown in Fig. 6.
In U.K. practice, the wearing course materials
would be used in approximately 40mm thick
layers at the road surface while the basecourse
mixture would be used in layers of about 60mm
thickness below. Both materials could feature
in either overlays or new construction.

The maximum aggregate sizes varied between 10
and 20mm (see Fig. 6) and this factor should
be considered in relation to the grid aperture
size of 63 x 49mm (see Fig. 1).

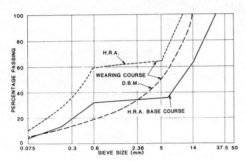

Fig. 6 Asphalt grading curves

Bitumen does not stick to polypropylene so
continuity between the grid and the asphalt
relies entirely on interlock. Furthermore,
to avoid creating a plane of weakness (low
resistance to shear stress) by inclusion of
the grid, there must be sufficient continuity
of asphalt through the grid apertures. The
proportions of Tensar AR1 were considered
adequate for aggregate sizes up to 20mm but
the aperture size is a parameter which could
be varied and is the subject of current
research.

Table 1 gives details of the binder content
for each mix and the grade of bitumen used
characterised by its penetration (British
Standards Institution, 1974). The average
air void contents are also indicated as a
measure of the state of compaction and this is
discussed in more detail in a subsequent
section of the paper.

The grading curves of Fig. 6 show that the HRA's
have gap gradings while the DBM has a continuous
grading. Resistance to permanent deformation
is inherently better for the DBM type of mix,
which mobilises higher aggregate interparticle
friction than the HRA. The latter, however,
because of its higher binder content and
harder bitumen, has the better tensile strength.

Various slabs and beams of these mixtures were
prepared, both reinforced and unreinforced.
Details are presented in the appropriate section
of the paper.

Table 1 - Details of Asphalt Mixtures

Mix Type	Bitumen Content (% by Mass)	Bitumen Penetration Grade	Air Void Content (%)
HRA wearing Course	7.9	50	5
DBM wearing Course	5.0	100	9
HRA Base Course	5.7	50	2

ELASTIC STIFFNESS OF REINFORCED ASPHALT

The relationship between uniaxial stress and
strain in bituminous material is termed
"stiffness" after Van der Poel (1954). Under
conditions of high strain rate such as occur
in pavements subjected to moving wheel loads,
asphalt behaves in an essentially elastic
manner, although small residual strains do
develop and the accumulation of these lead to
surface rutting.

The stresses transmitted to the lower layers
of a road and those set up in the asphalt
layer are strongly dependant on the elastic
stiffness of asphaltic material. This parameter is therefore most important for design
computations.

Elastic stiffness is a function of temperature
and loading time. For a rolled asphalt
wearing course mixture at 20°C and typical
vehicle speed of 80 km/hr, the elastic stiffness would be about 4 GPa. The DBM would be
similar while the HRA basecourse would be a
little stiffer. Under similar conditions, the
Young's modulus of Tensar AR1 was found to be
about 15 GPa which could reduce to, say, 9 GPa
as a result of its exposure to elevated temperatures during paving. Hence, the order of
magnitude of the stiffness is similar. In a
unit cross-sectional area of asphalt, the
percentage of reinforcement ribs is likely to
be very low. In view of these factors it
seems unlikely that the presence of the grid
will increase the effective stiffness of the
composite. Exceptions could occur at high
temperatures when asphalt stiffness will be
low and the stiffness modular ratio would be
larger. It seems unlikely, though, that major
improvements to stiffness under low strain
conditions will be realised. The real benefits
will come from situations where large strains
can develop (permanent deformation or opening
of cracks) when the grid should be effective.
A few experiments were carried out to check
the low strain stiffness of reinforced HRA
wearing course. A pair of rectangular test
specimens 100 x 89mm in cross-section
and 250mm long were cut from a reinforced and
an unreinforced slab. In the reinforced case,
two ribs of Tensar AR1 were located along the
centre of the specimen. Steel Loading plates

were glued onto each end.

Cyclic, axial load tension-compression tests were performed with LVDT's attached over gauge length's on opposite faces of each specimen to monitor the small deformations accurately. Stresses between 600 and 1200 kPa were applied at a frequency of 16 Hz which corresponds to about 100 km/hr traffic speed. The test temperature was 20°C. The results are detailed in Table 2, readings have been taken after as few cycles as possible.

Table 2 - Results of axial load elastic stiffness tests on HRA wearing Course

Specimen Details	Axial Stress (kPa)	Axial Strain (microstrain)	Elastic Stiffness (GPa)
Reinforced	697	133	5.2
Void Content	921	200	4.6
= 9.1%	1135	300	3.8
Unreinforced	652	130	5.0
Void Content	888	183	4.9
= 7.8%	1124	267	4.2

It will be noted that the levels of stiffness were almost identical on both specimens. The state of compaction in the reinforced one was poorer than in the other, which is one of the consequences of including the grid. Hence, there may be some compensating effects resulting in the same stiffness, but no improvement was obtained.

A further pair of tests were conducted using the configuration shown in Fig. 7 which is a beam on an elastic support. Cyclic loads were applied through the loading pad at 5 Hz and a temperature of 20°C. The horizontal strain at the bottom of the beam was determined from the LVDT shown in Fig. 7. The results for reinforced and unreinforced beams are shown in Fig. 8 from which it will be noted that there is no significant difference between the two. The strains involved in this case were larger than in the axial load tests.

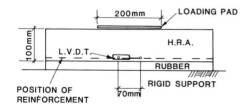

Fig. 7 Asphalt beam on elastic support

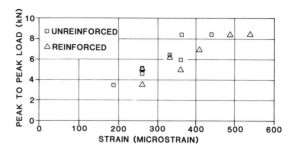

Fig. 8 Stiffness from beam tests

RESISTANCE TO PERMANENT DEFORMATION

A series of wheel tracking tests was carried out on 1.2 x 0.34m slabs of all three mix types and, in each case, an idential pair was manufactured with one containing the reinforcing grid. Table 3 presents details of all the slabs showing the method and level of compaction, the thickness and the position of the grid.

The slabs were cast in steel formwork and compacted using a hand operated vibrating roller. For tests E to H, the slabs were compacted in two stages with the grid being placed upon the first layer after rolling. This was considered to be most representative of likely site practice. In the other cases, the grid was placed on an uncompacted first layer with rolling being effected after placement of the second.

A study of the void contents in Table 3 shows that the presence of the grid did inhibit compaction slightly. In some of the HRA wearing course slabs, the void contents in the top and bottom parts of the slab were determined and these results are shown in Table 4. Clearly, compaction at both levels provides more uniform densities in the asphalt.

The slabs were tested in the Nottingham Pavement Test Facility (Brown and Brodrick, 1981). Some staging was constructed to provide a constant support for all the test slabs with a resilience representative of the site situation.

All tests were carried out at 30°C. For all but tests I and J a wheel contact pressure of 415 kPa and speed of 8 km/hr were used. The width of the contact area was about 150mm. A profilometer was used to measure the transverse surface profile at intervals during the tests which generally were continued to 50,000 wheel passes.

For tests I and J, the contact pressure below the tyre reduced to 250 and the lateral position of the wheel was varied to produce a realistic transverse distribution of passes.

Table 3 - Details of Slabs after compaction

Test	Slab No.	Material	Construction Method	Slab Thickness (mm)	Tensar depth divided by layer thickness'	Air Void Content (%)
A	1 T1	Hot Rolled Asphalt	Compaction on complete slab	89 86	0.71	7.8 9.1
B	2 T2			85 86	0.77	4.7 6.1
C	3 T3			93 93	0.54	5.1 7.4
D	4 T4			77 78	0.47	6.9 4.6
E	5 T5		Compaction at both levels	104 104	0.49	2.4 2.8
F	6 16			105 103	0.63	3.0 3.4
G	7 T7	Dense Bitumen Macadam		87 87	0.40	8.7 9.7
H	8 T8			89 89	0.69	8.9 8.7
I	9 T9	H R A Base Course	Compaction on complete slab	107 108	0.53	1.9 2.5
J	10 T10			110 108	0.5	2.0 2.5

In order to illustrate the results, data from tests A, E, G and I have been reproduced in Figures 9 and 10. This includes the HRA wearing course with both types of construction (A and E), the DBM (Test G) and the HRA base course (Test I).

Fig. 9 shows the build up of permanent deformation during each test and Fig. 10 illustrates the final transverse surface profiles. It is clear that substantial reductions in permanent deformation are apparent in all tests with that in the DBM particularly marked. The lower levels of deformation in Test I resulted from the reduced applied stress on a more resistant mix. The tests on DBM (G and H) only went to about 5000 passes as this material was poorly compacted (see Table 3) and early failure developed. The lateral restraining effect of the reinforcement was, however, particularly apparent in these tests as illustrated by the data in Table 5. These were obtained from horizontal measurements between markers on the slab surface 200mm apart across the wheel track.

The reductions in rut depth for the various groups of tests are summarised in Table 6.

Omitting G and H, where the test conditions proved rather extreme, it is apparent that better performance was achieved when the grid was placed prior to compaction of the asphalt above and below it in one operation. Placing the grid at a well defined interface probably results in poorer interlock.

RESISTANCE TO REFLECTION CRACKING

The experimental arrangement illustrated in Plate 1 and Fig. 11 was used to apply cyclic loads to beams cast over a preformed crack representative of an overlay above a concrete pavement joint, considered to be the most severe source of reflection cracking.

The beams were 525mm x 150mm x 100mm deep and a layer of Tensar AR1 was placed in various positions as detailed on the schedule of tests in Table 7. The beam width was such that three primary ribs provided the reinforcement. All beams were made of HRA wearing course material and compacted to the same void content (10%).

The 10mm wide discontinuity below the beam was provided by a pair of high quality plywood

Table 4 - Influence of Reinforcement and Construction
Method on Compaction

Construction Method	Type of Slab	Air Void Content (%)	
		Top of Slab	Bottom of Slab
Compaction of complete slab	Unreinforced	7.3	4.9
	Reinforced	7.2	9.8
Compaction at both levels	Unreinforced	2.3	2.4
	Reinforced	3.3	2.3

Table 5 - Transverse Surface Deformation for Tests G and H
between 1000 and 30,000 Passes

Test	Transverse Deformation (mm)	
	Unreinforced	Reinforced
G	47	5.7
H	29	5.0

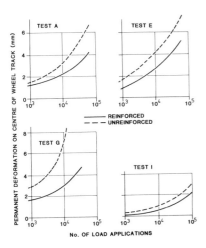

Fig. 9 Build up of permanent deformation

sheets. A resilient support in the form of a
piece of rubber was located below the plywood.
(see Fig. 11).

Load was applied through a rubber based
loading platen 200mm wide placed at the beam
centre. A loading frequency of 5 Hz was used
and the load was cycled between 1.3 and 8.3
KN in all tests. This gave a peak contact
stress below the loading pad of 275 kPa.

An LVDT was located to measure horizontal
displacement across the plywood gap (Fig. 11).
Under test, the gap acted as a crack inducer
and the progress of crack propagation on

Table 6 - Summary of Rut Depth Data

Material	Construction Method	Average rut depth		Ratio Y/X
		Unreinf.	Reinforced	
		(X)	(Y)	
HRA wearing course	Single compaction	10.2	4.8	0.47
HRA wearing course	Compaction at both levels	18.4	14.8	0.80
DBM wearing course	Compaction at both levels	16.1	6.8	0.42
HRA Base course	Single compaction	5.8	2.6	0.45

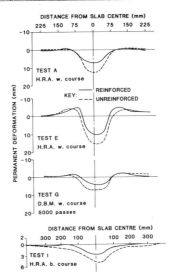

Fig. 10 Transverse profiles after testing

either side of each beam was monitored visually.
This was facilitated by painting the beams
white. The average height of the crack
tip above the bottom of the beam was recorded
at intervals during the test. Initially the
cyclic deformation across the gap was approximately 0.05mm giving an average local strain of
0.5% which is very high for asphalt.

Fig. 12 summarises the results of these experiments showing mean crack lengths from the
replicated tests. When the grid was placed at
the bottom of the beam no cracking was observed.
In similar unreinforced beams after corresponding numbers of load cycles, the cracks had
propagated above mid-depth and the beams could
be regarded as having failed. When the grid
was placed at intermediate depths, although
there was crack development and growth, it was
controlled. In addition, the cracks were of
hairline type indicating the effectiveness of
the grid in holding the asphalt together.
(This point is well illustrated in Plate 2).
The results for the intermediate grid positions
were apparently inconsistent as better perfor-

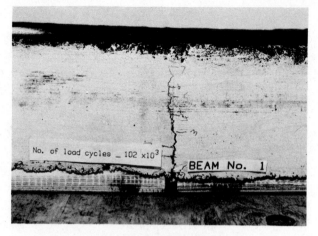

(a) Unreinforced

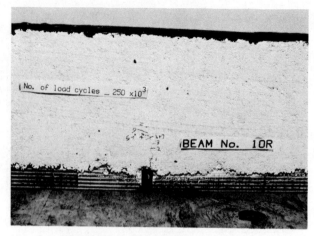

(b) Reinforcement at mid-depth

Plate 2. Typical beams after testing

nance was observed with the grid at mid-depth than at 3/4 depth. Further testing should resolve this point.

DISCUSSION AND CONCLUSIONS

The experimental results reported herein have come from a set of preliminary tests designed to assess the potential for using polymer grid reinforcement in asphalt with particular reference to overlays. The tests have been carried out with some care and the results are considered to provide reliable guides for continuing research.

Only one grid type and size has been used to date and although this appears successful, evidence from early site trials suggests that a larger aperture could provide improved continuity of asphalt without decreasing the effectiveness of the reinforcement. It will be important for future laboratory research to be closely linked with site trials because of the influence which the techniques to be used for grid placement in asphalt will have.

This paper has dealt only briefly with the influences of asphalt paving temperature and associated grid shrinkage since development in manufacturing and laying techniques may make current results inapplicable. It has, however, been established that temperatures in the region of 135°C can be used without damaging the variety of grid currently produced (Tensar AR1. Kennepohl (1984) has described early work on the installation of this material.

On the basis of the work described in this paper, the following conclusions can be drawn:

1. Tensar AR1 can be installed, without damage, in asphalt mixes at temperatures appropriate for adequate compaction.

2. Tensar AR1 has an elastic stiffness at low strains which is of the same order but slightly higher magnitude than a typical asphalt mix at 20°C and a loading time representative of traffic moving at 80 to 100 km/hr.

3. Under cyclic loading representative of severe conditions in a pavement, Tensar AR1 showed no indication of fatigue failure but did exhibit mean load relaxation.

4. The presence of Tensar AR1 in hot rolled asphalt did not increase its elastic stiffness.

5. Inclusion of Tensar AR1 in three different asphalt mixes increased the resistance to rutting substantially.

6. Best rutting resistance was obtained when the grid was compacted in an asphalt sandwich rather than by placing it on a precompacted surface and then covering with further asphalt.

7. The presence of Tensar AR1 impeded compaction of the asphalt slightly but this did not prevent the improved performance noted in 5.

8. A layer of Tensar AR1 placed immediately over a discontinuity, representative of a concrete road joint, prevents the development of a reflection crack in the asphalt overlay. Inclusion of the grid within the overlay thickness inhibits the crack growth and limits its width.

ACKNOWLEDGEMENTS

The research described in this paper has been made possible by two co-operative research awards from the Science and Engineering Council in conjunction with Netlon Ltd. The Authors are grateful to Dr. F.B. Mercer and his colleagues at Netlon and Professor Sir Hugh Ford, Chairman of the SERC/Netlon research steering committee together with its members for providing a stimulating environment in which to pursue their studies. The support and encouragement of Professor P.S. Pell, Head of the Civil Engineering Department at Nottingham is also gratefully acknowledged

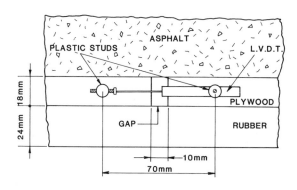

Fig. 11 Details of crack initiation

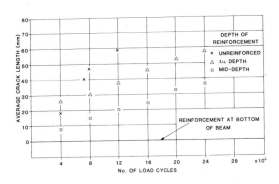

Fig. 12 Crack growth

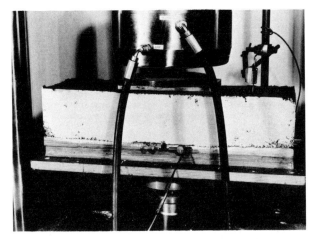

Plate 1. Apparatus for reflection crack
          testing

Table 7 - Details of Beam Tests

Beam No.	Total No. of Load Cycles (thousands)	Position of Reinforcement
1	102	Unreinforced
2	140	Unreinforced
3	206	Unreinforced
4R	103	Bottom
5R	167	Bottom
6R	218	Bottom
7R	241	3/4 depth
8R	279	3/4 depth
9R	245	mid-depth
10R	250	mid-depth

together with the advice of Mr. K.E. Cooper.

REFERENCES

British Standards Institution (1973, a).
"Specification for rolled asphalt (hot process)
for roads and other paved areas". B.S. 594.

British Standards Institution (1973, b).
"Specification for coated macadam for roads
and other paved areas". B.S. 4987.

British Standards Institution (1974).
"Method for determination of penetration of
bituminous materials". B.S. 4691.

Kennepohl, G. (1984). "Construction of Tensar
reinforced asphalt pavements". Proc. Symp. on
Polymer Grid Reinforcement in Civil Eng.,
Inst. of Civil Eng.

Van der Poel, C. (1954). "A general system
describing the visco-elastic properties of
bitumen and its relation to routine test data".
Journ. App. Chem. Vol. 4, pp.221-236

Brown, S.F., Brodrick, B.V. "Instrumentation
for the Nottingham Pavement Test Facility",
Transportation Research Record 810, 1981,
pp. 71-79.

Structural behaviour of Tensar reinforced pavements and some field applications

R. Haas, *University of Waterloo*

The concept of reinforcing flexible pavements has recently become a viable option with the availability of Tensar geogrids. Depending upon the traffic, environmental, subgrade and pavement material conditions, reinforcement may be considered for the surface, base or subgrade layers.

A comprehensive program of analytical and full-scale "model" experiments has been conducted to evaluate the structural behavior of Tensar reinforced asphalt pavements. In addition, a field program of using Tensar geogrids in pavement spot improvements and trench cuts has been carried out.

The results of the work show that: (a) reinforced pavement sections can carry up to three times the number of loads for equal thicknesses, or substantial thickness reductions are possible if the criterion is equal number of loads carried, (b) reinforcement has a relatively more significant effect under weak subgrade conditions, (c) elastic layer theory can be used to describe the behavior of reinforced pavements, and (d) the use of Tensar reinforcement in pavement spot improvements and trench cuts was very effective.

INTRODUCTION

If pavement reinforcement could reduce the thickness of layers and/or extend pavement life in a cost and performance effective way, it would certainly be a viable alternative to conventional designs.

The potential of a new polymer geogrid, Tensar, in achieving this objective was recognized in 1979 shortly after it was introduced, and a comprehensive research program was initiated in 1980. The objectives of the program were to evaluate the behavior and effectiveness of Tensar in paved road structures. An integrated set of laboratory and field experiments were planned and carried out. The laboratory experiments were felt to be vital not only for assessing behavior and establishing design parameters but also as a basis for intelligent planning of the field trials.

The laboratory experiments were carried out as "model" tests to simulate full-scale pavements and full-scale dynamic loads, but under controlled conditions.

It is the purpose of this paper to describe the laboratory experiments and results, the analyses conducted and some of the field applications.

Experimental and Analytical Program Objectives

Reinforcement can occupy several alternative locations in a road structure, ranging from the subgrade to the subbase and base layers to the asphalt layer. It's relative effectiveness depends on the strength of the layers, particularly that of the subgrade, their thicknesses and the number of loads to be carried. Thus, each individual design situation dictates where the reinforcement should be placed, and its relative effectiveness.

We carried out an initial analytical investigation, using elastic layer theory, for a wide range of paved road designs.(1) The results indicated that the Tensar geogrid offerred particular potential if used in thicker asphalt layers where high traffic volumes occur. Consequently, the experimental program concentrated on this application, as described in the following sections, but current investigations are covering the full range of road structures.

The experimental program was designed to include varying thicknesses of reinforced and unreinforced asphalt on subgrades of varying strength. Its main goal was to thoroughly investigate, under a variety of controlled conditions, the mechanical behavior and load-carrying capabilities of Tensar reinforced flexible pavements and to compare with unreinforced (control) sections.

As well, the results of the program were to be used to verify and/or modify the elastic layer theory and develop initial design procedures for reinforced asphalt pavements.

THE EXPERIMENTAL INVESTIGATION

Test Facility

The pavement sections were constructed in a test pit at Royal Military College (RMC) in Kingston, Ontario. Dimensions of the pit are 4 m by 2.4 m wide by 2 m deep.

Dynamic loads, to simulate dual truck tires, were applied through electro-hydraulic actuators and a circular plate.

Controlled Variables and Sequence of the Experiments

The experimental program was divided into five series of tests, called loops. Each loop involved one set-up of the test pit, in which half of the pit was reinforced and the other half was left as a control section. For each loop, the asphalt thickness and the subgrade condition (either dry or saturated) were the controlled variables. Between 4 and 9 tests were performed for each loop, with each test representing a different location within the test pit. The five loops are described in Table 1.

The design of these test loops involved a logical sequence to assess certain parameters related to reinforcement evaluation. For example, the first loop was designed to compare the behavior and performance of the reinforced section with an unreinforced section of same thickness (150 mm) on a weak subgrade. Permanent deformation and vertical deflections were monitored throughout the test until complete failure occurred on the unreinforced section.

In loop 2, strain carriers were installed at the bottom of the asphalt layer to compare tensile strain, in the critical zone, between reinforced and unreinforced sections of the same thickness (165 mm) on a stronger subgrade. Results of these two loops were of major

Polymer grid reinforcement. Thomas Telford Limited, London, 1985

importance since they compare the reinforced sections with unreinforced sections under identical geometric, loading and environmental conditions.

TABLE 1

TEST LOOPS AND CONTROLLED VARIABLES

Loop No.	Test No.	Asphalt Thickness	Subgrade Condition	Description
(1)	1	150 mm	Dry	Control
	2	150 mm	Dry	Reinforced
	3	150 mm	Saturated	Reinforced
	4	150 mm	Saturated	Control
(2)	1	165 mm	Dry	Control
	2	165 mm	Dry	Reinforced
	3	165 mm	Dry	Reinforced
	4	165 mm	Dry	Control
(3)	1	250 mm	Saturated	Control
	2	150 mm	Saturated	Reinforced
	3	150 mm	Saturated	Reinforced
	4	200 mm	Saturated	Control
	5	250 mm	Saturated	Control
	6	150 mm	Saturated	Reinforced
(4)	1	200 mm	Dry	Reinforced
	2	200 mm	Dry	Reinforced
	3	250 mm	Dry	Control
	4	250 mm	Dry	Control
	5	250 mm	Saturated	Control
	6	250 mm	Saturated	Control
	7	200 mm	Saturated	Reinforced
	8	200 mm	Saturated	Reinforced
	9	200 mm	Saturated	Reinforced
(5)	1	115 mm	Dry	Control
	2	115 mm	Dry	Control
	3	115 mm	Dry	Reinforced
	4	115 mm	Dry	Reinforced

Upon achievement of the first objective (basic comparisons between reinforced and unreinforced), the second objective was to find the equivalent thickness of the reinforced layer. Loops 3 and 4 were designed for this purpose. In loop 3, two unreinforced sections (200 mm and 250 mm) were tested against a thinner reinforced section of 150 mm on weak subgrade. Results of this loop, subsequently described, showed that a value of (50 - 100 mm) equivalent thickness may represent the reinforcement effect.

Based on this finding, loop 4 tests were performed with an unreinforced section of 250 mm and a reinforced section of 200 mm to confirm the minimum saving value (50 mm).

The last loop was designed to compare the vertical stresses on the subgrade (strong subgrade) for reinforced and unreinforced asphalt sections of the same thickness (115 mm).

Load Applications

Loads were applied through a 300 mm (12 in.) diameter rigid circular plate placed on the pavement surface. The loading pulse was sinusoidal, with an amplitude or peak of 40 kN for each cycle, and a frequency of

10 Hz. The loading program was designed to represent typical traffic loadings on pavements under operating conditions. The cyclic loading was carried out until certain defined criteria for failure, as subsequently described, were reached.

After certain numbers of selected cycles, dynamic loading was discontinued and static loading sequence [5-10 static cycles] applied as a time lengthened, step-wise approximation of one cycle of loading. In addition to obtaining static load response per se, this static loading sequence was necessary for monitoring the array of displacement gauges, strain gauges and strain carriers in each section.

Test Materials and Construction

The subgrade for each loop consisted of a 1.2 m depth of medium to coarse sand, compacted at an optimum moisture content of 11.5 percent using a Plate tamper. Moisture content and compactive effort were carefully controlled using a Troxler nuclear densitometer. For the "weak" subgrade tests it was flooded to full saturation from below, to the sand-asphalt interface.

The asphalt used was a local Ministry of Transportation and Communications of Ontario grade HL4 hot mix. A 25 mm lift (125°C) was first placed on the subgrade for all tests. The mesh was then placed on half the pit, the other half being left unreinforced, as the control section. Next, reinforced and unreinforced halves were covered with one additional 50 mm of asphalt and compacted using four passes of a plate tamper. Additional uniform 25 mm to 75 mm lifts were then placed and compacted.

The Tensar grid used was an ARI type with about 50 mm by 50 mm openings. For loops 1, 2, 3 and 5, strain gauges were bonded to the top and the bottom of the ribs at locations covering a wide area under the loading plate in order to monitor strains.

Instrumentation

The general arrangement of instrumentation used to monitor the pavement sections during testing is shown in Figure 1. Data access was through a PDP 11/34 computer at pre-programmed intervals (2).

Foil-type (120 ohm) strain gauges were used to record the magnitude and distribution of elastic and plastic tensile strains generated in the reinforcement elements as a result of the loading.

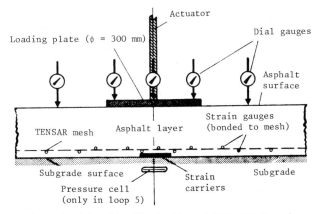

Figure 1 Schematic Illustration of Test Set Up and Instrumentation

Mastic "strain carriers", consisting of two (120 ohm) strain gauges embedded in a 150 mm square by 12 mm thick mastic plate, were used for each test set-up of the last four loops. These strain carriers were placed directly under the centerline of the loading plate and at the subgrade-asphalt interface.

Dial gauges were located on the rigid loading plate and at radial distances. They were read during static load cycles to determine the elastic and plastic surface deflection profile after various numbers of load cycles.

For loop 5, a circular plate pressure cell was embedded in the subgrade directly below the centre of loading for each test set-up. Depth of burial was about 50 mm below the sand/asphalt interface.

Failure Criteria

In order to objectively compare the performance of the reinforced and control sections, certain failure criteria were established. Failure was said to have occurred if:

1. A permanent deformation of 30 mm was measured;
2. The development of extensive cracks, or
3. A steady increase in the measured values of stress, surface deflection and/or horizontal strain at the interface.

It is noteworthy that the failure of the mesh did not have to be considered in the criteria. The reason is that the strains on the mesh on all loops did not exceed 30% of its yield strain of 15%.

RESULTS OF THE EXPERIMENTS

A large amount of information was collected. Some of it is summarized in Ref. (3, 4), while Ref. (5) contains a comprehensive documentation. It is only possible herein to present some typical results illustrating the absolute and relative behavior of the reinforced sections. There are many fundamental considerations and implications of this behavior, which are covered in Ref. (5).

Loop 1 was designed to compare reinforced and unreinforced (control) sections at a constant asphalt layer thickness of 150 mm under both strong (dry) subgrade and weak (saturated) subgrade conditions. The results (5) showed that the unreinforced section on a strong subgrade started to fail after 200,000 cycles of the 80 kN load, in terms of increasing static deflection and fatigue cracking. Both sections were carried to 500,000 cycles, with the reinforced showing no signs of failure. As well, the "angle of curvature" (5), a measure of the load spreading capabilities of the reinforced layer, was significantly less for that layer. The foregoing results were confirmed for the saturated or weak subgrade condition, although the absolute number of loads carried were substantailly lower.

Loop 2, involving a strong subgrade condition, utilized extensive instrumentation (see Fig. 1). Some of the results are shown in Figures 2 to 5. Figure 2 shows that maximum elastic deflection (under static loading) after various numbers of load cycles is not substantially higher for reinforced as compared to unreinforced. However, the angle of curvature (Fig. 3) is about 50% less for the reinforced section, as is the elastic tensile strain at the bottom of the layer (Fig.4), both of which have major implications for permanent deformation and fatigue behavior. Figure 5 shows that permanent surface deformation is substantially less for the reinforced section. At the 20 mm failure criterion shown, the unreinforced at best carried only 110,000 loads while

the reinforced section carried 320,000 loads, a threefold increase.

Loop 3 was designed to investigate the equivalent thickness effect of reinforcement, under weak subgrade conditions. Figure 6 illustrates some of the results, in terms of permanent deformation. The 150 mm reinforced section carried about 80,000 loads compared to only 34,000 loads for the 200 mm unreinforced and 92,000 loads for the 250 mm unreinforced. In other words, 150 mm of reinforced nearly compares to 250 mm of unreinforced. When the much longer total loading time is taken into account for the reinforced section (because of having to read the instrumentation, under static load conditions, after various numbers of cycles), it actually performed as well or better than the unreinforced section, as shown by Abdelhalim (5).

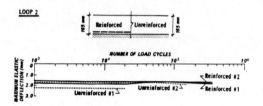

Figure 2   Maximum Elastic Deflection

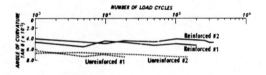

Figure 3   Angle of Curvature

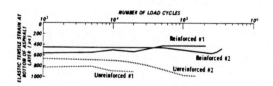

Figure 4   Elastic Tensile Strain at Bottom of Asphalt Layer

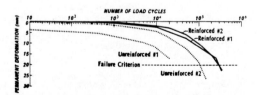

Figure 5   Permanent Deformation at Surface

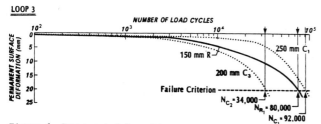

Figure 6   Permanent deformation

As well, the unreinforced sections were severely cracked at the end of the test while the reinforced section surface was still quite sound.

Loop 4 was designed to confirm the previous results. The major conclusions were that thickness savings for reinforced sections can range from 50 to 100 mm and that the reinforcement would have a relatively greater effect for weaker subgrade and lower stability mix conditions. Another important conclusion was that maximum elastic surface deflection, a major criterion of behavior used in the pavement field, was inadequate for evaluating reinforced pavements. A much better criterion is angle of curvature.

Loop 5 involved pressure cells in the subgrade under each test section, using a thin asphalt layer (115 mm) and a strong subgrade. Figure 7 shows that the vertical compressive stress was 20 to 40% higher for the unreinforced section. Also, the angle of curvature was about 20% higher by the end of the test. Both these results represent an important observation as to the load spreading ability of the reinforcement.

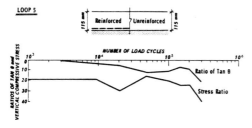

Figure 7   Angle of Curvature and Vertical Compressive Stress Ratios for Reinforced and Unreinforced Sections - Dry Subgrade

APPLICATION OF ELASTIC LAYER THEORY

The experimental program also provided the opportunity to examine the validity of elastic layer theory under simulated field conditions. Furthermore, it also provided enough information about asphalt pavement responses (elastic deflections, horizontal tensile strains

and vertical stresses) under a wide range of variables.

A multi-layer elastic model, BISAR (6) was used to predict surface deflections, horizontal strains and vertical stresses. Table 2 shows that the predicted values under the maximum load (40 kN) were quite comparable to measured values, for the unreinforced sections.

For reinforced sections, the application of the theory is more difficult because of assumptions which must be made on interface conditions, effective thickness of the reinforcement, etc. An example, using a minimum thickness of 2.5 mm for the reinforcement, an effective modulus of 10 GPa (1,500,000 psi), and a Poisson's ratio of 0.35 plus moduli of 690 MPa (100,000 psi) for the asphalt and 34.5 MPa (5,000 psi) for the subgrade, is given in Figure 8. It shows the vertical compressive strain reduction on the subgrade ranging from a maximum of about 40% for a 400 mm asphalt layer to about 10% or less when a thicker (300 mm or greater) layer is used. This compares with the measured reductions in vertical compressive stress on the subgrade (see Figure 7).

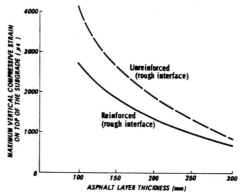

Figure 8   Comparison of Maximum Vertical Compressive Strain at Top of Subgrade for Varying Asphalt Layer Thickness - (rough interface)

SOME FIELD APPLICATIONS

The use of Tensar geogrid in field applications is being carried out in three areas: (a) regular paving operation, where the Tensar is in the asphalt layer, (b) installation of Tensar at some other point in the raod structure; i.e., at the subgrade-base interface, and (c) pavement spot improvements and trench cut repairs where the placement is by hand. Construction technology is being developed for the first area. The second area has seen some field trials, primarily where soft subgrade soils such as peat are involved. The third area was seen as an opportunity to obtain some early field installations and observations of behavior.

TABLE 2. COMPARISON OF MEASURED AND PREDICTED DEFLECTIONS AND STRAINS UNDER MAXIMUM LOAD (40kN)

Thickness: mm	Asphalt modulus: MPa	Sub-grade modulus: MPa	Measured Deflection: mm	Tensile strain: $\mu\varepsilon$	Predicted Deflection: mm	Tensile strain: $\mu\varepsilon$	% error in deflection	in strain
115	2997	33.3	1.57	605	1.54	589	2.0	3.0
165	1399	23.3	1.81	680	1.80	715	1.0	5.0
200	965	17.6	1.98	805	2.05	765	4.0	5.0
250	1233	15.2	1.69	440	1.70	452	1.0	3.0

Consequently, a program of such pavement spot improvements and trench cut repairs was undertaken in 1983, with an approximately 20 installations being completed. These have been more fully reported elsewhere (7, 8) and are only briefly summarized as follows. Basically, the spot improvements involved removal of the deteriorated asphalt, and any underlying material if required, followed by preparation and levelling of the affected area, hand placement of the Tensar (after cutting to size), hand placement of the asphalt and finally compaction with conventional rollers. Trench cuts involved similar procedures, with more than one layer of Tensar used where required by depth and backfill soil conditions.

Figures 9 and 10 show a typical pavement spot improvement area with Tensar. In first photo, the deteriorated surface has been removed and the Tensar is being trimmed to size. The second photo shows how the hot mix is hand placed over the grid. This was followed by conventional rolling.

Figure 9: Typical pavement spot improvement area, with Tensar being trimmed to size.

Figure 10: Hand spreading of the asphalt hot mix over the Tensar.

Because the spot improvements and trench cuts involve relatively softer areas, reinforcement can be a particularly useful approach to gaining more uniformity in support conditions. This was verified by a limited amount of dynamic deflection testing (using a Dynaflect device) which showed reductions usually in the order of 30% in surface deflection when Tensar was used.

These field applications were felt to be quite successful. Field crews found it easy to work with the material. The only caution is that the asphalt mix should not be over about 140°C· otherwise the Tensar will distort. Observations of performance will continue in 1984 and beyond.

CONCLUSIONS

The research described in this paper has shown that flexible pavements can be effectively reinforced with the polymer geogrid, Tensar. This involves asphalt thickness savings ranging from 50 mm to 100 mm, or the ability to carry two to three times more traffic loads for equal thicknesses.

The material can also be very effectively used in pavement spot improvements and trench cut repairs.

ACKNOWLEDGEMENTS

This paper is based largely on the work of Dr. A.O. Abdelhalim. His contributions are gratefully acknowledged, as are those of many others such as Professors Jarrett and Bathhurst of Royal Military College, Mr. William Phang of the Ontario Ministry of Transportation and Communications, Mr. Jamie Walls and Ms. Louise Steel of the PMS Group and many colleagues at Gulf Canada Ltd. and the Tensar Corporation.

REFERENCES

1. Abdelhalim, A.O. and Ralph Haas, "Flexible Pavement Reinforcement: Assessment of Available Materials and Models and a Research Plan", Interim Rept. for Project No. 21130, Ontario Joint Transportation and Communications Research Program, March 31, 1981.

2. Bathhurst, R.J., J. Walls and P.M. Jarrett, "Summary of Large-Scale Model Testing of Tensar-Reinforced Asphalt Pavement", Interim Rept. for Project No. 21130, Ontario Joint Transportation and Communications Research Program, March, 1982.

3. Abdelhalim, Omar Abdelhalim, Ralph Haas and William A. Phang, "Geogrid Reinforcement of Asphalt Pavements and Verification of Elastic Theory", Paper presented to Transp. Res. Bd., Wash., D.C., Jan., 1983.

4. Abdelhalim Omar Abdelhalim, Ralph Haas, Jamie Walls, Richard Bathhurst and William A. Phang, "A New Method for Effective Reinforcement of Asphalt Pavements", Paper presented to Roads and Transp. Assoc. of Canada, Halifax, Sept., 1982.

5. Abdelhalim Omar Abdelhalim, "Geogrid Reinforcement of Asphalt Pavements", PhD Thesis, University of Waterloo, 1983.

6. DeJong, D.L., Peutz, M.G.F. and Korswagen, "Computer Program BISAR, "Layered Systems Under Normal and Tangential Surface Loads", External Report Shell - Laboratrium, Amsterdam, January, 1973.

7. Haas, Ralph, "Progress Report on Research and Development Activities Related to Tensar Geogrids", Prepared for SERC/NETLON Steering Committee, July 6, 1983.

8. Steele, Louise, "Development Project for Using Tensar Geogrids in Pavement Spot Repairs and Trench Cuts", Report prepared by the PMS Group for the Tensar Corp., Nov., 1983.

Construction of Tensar reinforced asphalt pavements

G. J. A. Kennepohl, *The Tensar Corporation,* and N. I. Kamel, *Gulf Canada Ltd*

The viability of reinforcing asphalt pavement structures with Tensar geogrid has been established by studies carried out in the U.K. and Canada. Substantial material cost savings and/or performance advantages can be realized for the reinforcement of asphalt concrete in new construction, overlay applications, granular bases and subgrades.

In case of pavement spot repair and unbound granular base stabilization the geogrid can be hand laid. For the reinforcement of asphalt concrete in new construction and overlay applications the installation method must be automated and compatible with conventional paving procedures. Correct positioning and proper installation of the Tensar grid is crucial to performance.

Tensar laydown equipment for tensioning the geogrid before placement of the paving mix has been developed and field tested. An alternate installation method is under development which uses a set of discs to roll the geogrid into the hot mix immediately after the paver. This paper also discusses the recycling of Tensar reinforced pavements and future developments.

BACKGROUND

There has been no lack of attempts to reinforce asphaltic concrete in layered pavement systems to assure longevity of satisfactory pavement performance. Asphaltic concrete exhibits an ultimate tensile strength which is usually less than the ultimate compressive strength by an order of magnitude and, therefore, reinforcement is expected to yield several important advantages, such as:

- increased tensile strength
- greater resistance to cracking
- longer fatigue life
- coherence of pavement after cracking
- improved shearing resistance due to lateral restraint
- reduced cost from material savings

Previous attempts to improve the tensile strength of bituminous concrete included the incorporation of natural and synthetic fibres and fabrics - in either random or aligned orientation - as well as steel welded wire mesh.

Most field trials with fibres, starting from the addition of cotton fibre in the 1930's (1) to the use of current pavement fabrics (2), have experienced modest or unqualified success owing mostly to low tensile strength and high creep under load of the fibre.

Welded wire or expanded steel mesh, normally associated with reinforcement of Portland cement pavements, have been extensively field tested on air field runways and highway pavements in 1950's and 1960's. Following the construction of test sections in the U.S.A. (California, Illinois, Massachussetts, Michigan, Minnesota, New Jersey, New York, Pennsylvania, Texas, Wisconsin) and Canada, (Ontario), the use of steel mesh in asphalt concrete was discontinued, in spite reports claiming significant reflection cracking reduction.

While factors such as corrosion and cost of steel had an influence, the lack of success is primarily attributed to construction problems. According to Davis (3) the steel mesh had a tendency to curl at the edges, and irregularities in

Fig 1   Paving over geogrid which has been laid down under tension ahead of the paving train.

the mesh produced bulges and caused pavement failure. Transverse cracks at the mesh splices were caused by expansion of the wire under the hot mix, some creep of the mesh occurred in front of the paver allowing significant bulges to develop and the mesh was often springy during installation and compaction. In summary, experience with wire mesh paving has shown that the mesh must be kept absolutely flat on the underlying road, otherwise bulging and springiness under rolling develops and later causes break-up of the pavement.

With the advent of Tensar geogrid, a new candidate for pavement reinforcement is now available with a more suitable set of material properties. The objective of this paper is to describe in subsequent sections geogrid property requirements for performance as well as for construction, types of reinforced pavement structures and the development of geogrid installation procedures (Figure 1).

Polymer grid reinforcement. Thomas Telford Limited, London, 1985

171

GEOGRID PROPERTY REQUIREMENTS

A comparison of suggested, property requirements with the properties of various reinforcement materials presented in Table I shows that Tensar geogrid ARI meets the desired criteria best. The key features are high tensile strength, high modulus, and a compatible coefficient of expansion. Specific tensile strength is defined as tensile strength/specific gravity.

TABLE I - CHARACTERISTIC OF REINFORCEMENT MATERIALS FOR ASPHALT PAVEMENTS

Property	Desired Range	Steel Mesh	Glass Fibre	Polypropylene	Tensar Geogrid
Spec. Tensile Strength (MPa)	>5	45	500	38	225
Spec. Elastic Modulus (GPa)	>4	26	29	1.5	12
Coefficient of expansion (X10⁻⁴)/°C	5-7	0.12	0.2	1	-

The geogrid ARI has been specifically designed for asphalt pavement reinforcement. The biaxial open mesh structure provides for bond and interlock with the matrix. Alternate structure designs can be made easily if required for optimization of performance. The geogrid is manufactured from polypropylene polymer for high thermal stability in contact with hot paving mix.

In addition, a heat soaking process is applied to reduce thermal shrinkage to a maximum of 3% at 140°C. Excessive shrinkage during placement of the hot mix can result in movements of the mat and build-in cracks.

The extent of shrinkage during installation and cooling of the mat is not only a function of the initial mix temperature, but also the cooling rate of the mat and the degree of restraint due to aggregate interlock, friction, and weight. Figure 2 illustrates the synergistic effects on shrinkage from a pilot test simulating various temperature, mix types and mat thickness conditions.

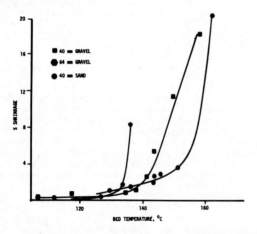

Fig 2   Temperature and type of paving mix mat on ARI geogrid shrinkage.

The effective shrinkage under field conditions was studied in several off-road, full-scale experiments. Shrinkage and associated edge cracking were not observed in test sections with heat stabilized ARI geogrid for specified temperature and laying procedure.

Shrinkage force measurements reported by Brown (4) show that on cooling some tension is retained. Therefore, under controlled conditions the geogrid can possess a useful pre-tensioning force in hot laid asphalt pavements.

TYPES OF REINFORCED PAVEMENT STRUCTURES

There are several options or types of structures for reinforcing layered, flexible pavement systems which will determine the applicable construction procedure. Depending on the design criteria, environmental subgrade and paving material condition, the geogrid may be placed in the surface, base or subgrade layers.

The most common geogrid applications can be represented by the following three types of reinforced pavement structures:

1. Asphalt concrete in new construction
2. Asphalt concrete overlays
3. Granular base or subgrade

In full-depth pavements, the geogrid is placed at the bottom of the asphalt concrete slab for largest reduction in tensile strain as reported by Abdelhalim (5). If such pavement is placed on unbound base or subgrade, the geogrid may be installed at the interface assuring adequate interlock with the hot mix. However, in case of a very hard surface, such as an old pavement, the geogrid may have to be sandwiched between the first and second lifts to provide sufficient interlock. For the reduction of permanent deformation and rutting, the geogrid is placed near the centre of the slab. The postulated mechanism as reported by Brown (4) for this behaviour of Tensar geogrid is the reduction of plastic flow of the asphalt mix away from the stressed area.

For overlayed pavements the preferred construction procedure is the placement of the geogrid between a levelling course and the surface lift. This method of construction will give a better interlock with the asphalt concrete and also reduce possibility of debonding. If the geogrid is placed directly upon the aged asphalt surface or alternately upon a Portland concrete surface, a tack coat should be applied. Further field trials are needed to investigate long term performance. It may be necessary to improve the bond with the old pavement by increasing the contact area using a geogrid with a larger mesh opening. Further information is also required for a proper strength and thickness design of the overlay to avoid excessive shearing which could result in delamination.

The reinforcement of unbound base or subgrade does not require special installation techniques. The geogrid can be laid by hand as in established geotechnical applications, such as unpaved roads.

In pavement spot repairs the reinforcement with geogrids is somewhat easier to apply because the construction procedure, including the placement of the paving mix, is done manually. Therefore, it is appropriate and feasible to install the Tensar geogrid also manually.

INSTALLATION METHODS

Pavement Spot Repair

Geogrids have been successfully used in pavement maintenance for repairing of spot distresses on asphalt and Portland cement pavements and in utility trench cuts. While the utility trench cuts apply to roads and streets, the repair of spot distresses extends to highways, parking lots, industrial yards, etc. Installations of Tensar geogrid at more than 20 job sites indicated the following major benefits:

o improved load bearing characteristics
o greatly reduced differential settlement
o increased pavement life from reduced deflection of up to 30%

The installation of the geogrid involves no changes in construction procedure. It can be done with conventional equipment and materials. The geogrid is simply cut with tin snips to the required size and hand laid. Based on experience from various types of field applications an installation manual has been developed for the repair of a variety of localized pavement distress conditions including potholes, ruts, alligator cracking and base distortions. There are three types of repairs where it is advantageous to include geogrids:

1. Surface (Skin) Patch

   Hand placed overlay (patching mix) is spread directly on the distressed area of the pavement. A piece of AR1 approximately cut to the size of the patch area is placed at the bottom for reinforcement.

2. Replacement (Deep) Patch

   The cracked pavement and in some cases also the base is removed as deep as is necessary to reach firm support. In repairing SS1 or SS2 (for DTN >100) is placed in the granular material and AR1 in the asphalt mix for reinforcement of the patch.

3. Trench (Utility) Cuts

   When pavement and subgrade is removed to install or repair a utility, Tensar geogrid can be used effectively in the backfill and patch as shown in Figure 3.

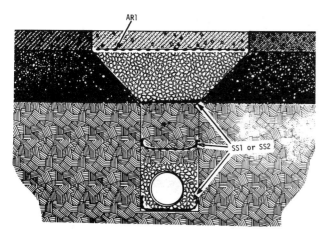

Fig 3   Schematic of a trench cut repair with Tensar geogrid reinforcement.

Placement of Tensar SS1 on the bottom of the excavation with the edges curled up on the sides of the trench will provide a firm work platform, support for the pipe or conduit, and reduce contamination of granular backfill (e.g. silt or clay). Geogrid placed over the pipe or conduit will protect the utility from damage by large aggregates and reduce the stress on the pipe. Where appropriate, the backfill may be reinforced with additional layers of geogrid. This will reduce differential settlement, lateral pressure against pavement base structure and associated damage during compaction.

The pavement spot repair with geogrid is now being introduced to the paving industry.

The use of geogrids in spot repair applications will, no doubt, help to acquaint the highway and paving engineer with the potential uses and benefits of geogrids in pavement structures, new construction and rehabilitation projects alike. However, for full-scale paving projects manual installation of geogrid is not economical and specialized geogrid laydown equipment must be used.

Tensioning Method

Tensar laydown equipment has been developed for tensioning of the geogrid flat against the base before placement of the paving mix by the finisher. This technique is applied to paving with geogrid in new construction or overlay applications which involve the installation of large volumes of geogrid at regular paving speeds. To-date, six full-scale field trials have been carried out in the United Kingdom and Canada (Table 2), as well as several off-road test sections.

TABLE 2 - FIELD TRIALS

LOCATION	DATE	TYPE OF CONSTRUCTION
Highway #5 Ontario, Canada	November/81	overlay on asphalt
Canvey Island U.K.	Spring/82	reconstruction over PCC (hand layed)
Carriageway A 631 Nottinghamshire U.K.	Fall/82	overlay on PCC
Motorway M 621 West Yorkshire U.K.	Fall/82	reconstruction/resurfacing (under base course)
Highway #6 Saskatchewan Canada	May/83	cold mix surfacing
Sudbury, Ontario Canada	November/83	overlay on asphalt (instrumented)

The tensioning of the geogrid proved to be quite a critical feature of the installation. Similar to the experience reported for paving with steel mesh (3), it was found that the geogrid must be held absolutely flat against the underlying base in order to attain proper compaction. The applied tension force must not only flatten the geogrid, but also overcome the shearing forces acting on the geogrid by the wheels or crawlers of the finisher. The required tension is in the order of 2.5-3.0 KN/m depending on such variables as the type of surface, paving mix fluidity and thickness, paving train wheels (or crawlers) and weight, including size and load on dump truck.

the high tensile strength polymer geogrid is more
than steel, it can be handled easier than steel mesh.
kness in the geogrid that might develop by the action
ving equipment or minor relaxation of the geogrid can
d up by the continuous tensioning equipment which
moves immediately ahead of the paving train.

Fig 4   Tensar laydown equipment takes a test run
deomonstrating the tensioning method.

Fig 5   Paving over geogrid which is being held
in tension by the laydown equipment
ahead of the paving train.

The first prototype tensioning equipment, shown in Figure 4 was
designed and constructed by K.C. Welding Corporation,
Bartlesville, Oklahoma. It consists of a 4 metre wide steel
frame with a dispensing hanger for a roll of Tensar geogrid and
two rubber-sleeved steel drums. These drums function as
nipping rollers and release the geogrid as the equipment moves
ahead. The lower roller is piloted and maintained at a preset
tension by a disc brake at each end.

The laydown equipment can be used as an attachment to a
front-end loader or tractor with a 3 point hitch. For shipment
from jobsite to jobsite it is mounted on a small skid (1.2 x
3.7 m) and transported on a truck or carrier.

Several test sections have been constructed successfully using
the tensioning method (Figure 5). However, further experience
is still required to develop and perfect the operation and
construction procedures for full-scale applications. The major
advantages of this method include:

   o minimum interference with normal paving operations

   o low installation cost, i.e., 2 unskilled laborers, plus
     about $5,000 for the laydown equipment

   o in-place pretensioning of reinforcement mesh and
     elimination of shrinkage in longitudinal direction.

When the geogrid is placed at the interface of the base and the
asphalt concrete mat the interlock with the mix may be weak,
especially where the base is solid and hard. Also, on hard
surfaces some damage may be incurred to the tensioned geogrid
by heavy paving equipment. It is mainly for these reasons that
an alternate installation, the roll-in method is being studied.

Roll-in Method

Rolling the Tensar geogrid into the hot mix immediately after
the paver and prior to compaction has been investigated as an
alternate method to incorporate the geogrid into asphalt
pavement. The task to push the mesh into the mat appeared
initially quite difficult because the mix is not fluid like
Portland concrete mix and any load exerted onto the mesh is
distributed in a snow-shoe like manner.

Initial experiments with a first generation roll-in equipment
consisting of a set of steel discs and a 1 metre wide vibrator
screed have been encouraging. A second 4 metre wide
prototype is shown in Figure 6. It features a holder for a roll of
geogrid, a set of steel discs and immediately following an
adjustable vibrator screed with weights. The vibrators are
connected by hose to and driven by the hydraulic system of the
finisher.

Fig 6   Prototype geogrid roll—in equipment
pulled behind a paver.

The weight of the roll-in equipment and the resulting drag on the paver proved to be a limiting factor. Further experiments and development of this equipment is still required. Major advantages expected from the roll-in method include:

o improved interlock of the geogrid with the mix
o no extra vehicle needed for installation
o no damage to geogrid by paving equipment or overtensioning

RECYCLING OF TENSAR REINFORCED PAVEMENTS

As the recycling of old and worn asphalt pavements becomes more and more the accepted procedure for rehabilitation and reconstruction of pavements in North America, it is important to establish (a) whether the reinforced asphalt pavement can be recycled and (b) if the cost of recycling is increased because of the reinforcement. In a cost benefit analysis of geogrid a significant demerit would be assessed if the geogrid reinforced pavement had no recycling value.

The first field trial of rotomilling a Tensar reinforced pavement took place in October 1983 at the Nelson Quarry, Burlington, Ontario (Figure 7). Several test sections constructed there previously with Tensar AR1 were milled successfully to a depth of 5 cm with a Goetz "Scaraplane" rotomill. At normal milling speeds the Tensar geogrid was cut into pieces of about 3 or 4 mesh units. The milling drum became wrapped in pieces of Tensar geogrid, but did not appear to affect the operation.

Fig 7   Rotomilling of Tensar reinforced pavements for recycling.

About ten tons of the rotomilled asphalt pavement has been stockpiled at the quarry for a future plant-scale recycle trial. Before recycling it will be necessary to remove by screening the geogrid pieces from the rotomilled product. Separate laboratory experiments indicate that geogrid pieces smaller than 2 cm do not affect the quality of the asphalt paving mix. While the evaluation is not yet complete, it has already been established that Tensar reinforced pavements can be recycled without any significant increase in operating cost.

CONCLUSIONS

Significant progress has been made to develop the required construction technology for asphalt pavement reinforcement after the viability, cost and performance benefits have been established over the past 3 years. The main conclusions of this development as well as points for future activities are summarized here.

1. Tensar geogrid has distinct advantages over other reinforcement materials with regard to required construction and performance properties.

2. The installation and construction procedure varies with the type of reinforced pavement structure.

3. Manual installation of geogrids can be used for pavement spot repair and reinforcement of unbound granular base and subgrade.

4. Automated laydown equipment and specific installation technology has been developed for the reinforcement of asphalt concrete.

5. Reinforced asphalt pavements can be recycled.

6. Field data must be obtained to support and confirm laboratory results and to develop application design procedures.

REFERENCES

(1) Busching, H.W.; Elliott, E.N. and Reyneveld, N.G, "A-State-of-the-Art Survey of Reinforced Asphalt Paving, Proc. of the Association of Asphalt Paving Technologist," Vol. 39 (1970), pp. 766-798.

(2) Federal Highway Administration, U.S. Department of Transportation, "Report on Performance of Fabrics in Asphalt Overlays," (Sept. 1982).

(3) Davis, N.M., "A Field Study of Methods of Preventing Reflection Cracks in Bituminous Resurfacing of Concrete Pavements," Report No. 12. University of Toronto, Toronto, Ontario, (1960), 16 pp.

(4) Brown, S.F.; Hughes, D.A.B.; Broderick, B.V., "Grid Reinforcement for Asphalt Pavements," Science and Engineering Research Council Meeting Presentation, Strathclyde University, U.K., November 5, 1983.

(5) Abdelhalim, A.O.; Haas, R.C.G. and Phang, W.A., "Geogrid Reinforcement of Asphalt Pavements and Verification of Elastic Theory," Transportation Research Board, Washington, D.C., January, 1983.
also
Abdelhalim, A.O., "Geogrid Reinforcement of Asphalt Pavements," Ph. D. Thesis, University of Waterloo, Ontario, 1983.

Construction of a reinforced asphalt mix overlay at Canvey Island, Essex

This paper describes the design, specification and construction of the first application in April 1981, and a subsequent application in 1982, of a reinforced asphalt or bituminous material overlay using Netlon geogrids.

The reinforced asphalt mix overlay was designed to strengthen a severely damaged concrete road pavement over a very weak saturated esuatine silt and clay subgrade. Polymer grid reinforcement being specified to increase tensile strength of the thick asphalt mix overlay and deter propogation of reflective cracking from the underlying cracked concrete layer.

G. R. Pooley, *Mobil Oil Company Ltd*

CONSTRUCTION OF A REINFORCED ASPHALT MIX OVERLAY AT CANVEY ISLAND, ESSEX

INTRODUCTION

Canvey Island is a primarily residential area but has located on the island extensive gas and petroleum product storage facilities. The site of the application of the geogrids was the B1014 road known as Somnes Avenue. This road is one of only two roads connecting the island with the mainland, and consequently, besides providing a major route for every day traffic, would also play a vital role in the event of a large scale emergency.

The road which is of two lane construction was originally built in 1972 of reinforced concrete, 150 mm thick, on 175 mm of ash sub-base, constructed over considerable depths of eustuarine silt and clay with a very high water table. By 1978, after carrying less than one million standard axles, (msa), the concrete had suffered extensive damage over its 1.5 km length requiring constant maintenance. The necessary maintenance repair cost was estimted at £576,000 in 1981.

Various strengthening and reconstruction options were considered, but following the design recommendations of the author a thick asphalt mix overlay incorporating Netlon geogrids was selected for the first phase of restrengthening in 1981. With the success of this phase a second phase was undertaken in 1982, at a total cost for the two phases of only approximately £220,000.

PAVEMENT AND ASPHALT MIX DESIGN

Pavement Design Options

Soil tests on the very weak saturated subgrade soil indicated a California Bearing Ratio, (CBR), generally lower than 2% and in many cases less than 1%. For the pavement design the following design criteria were stated by the highway authority.

> Design life, 20 years
> Commercial vehicles in each direction in 1981, 433
> Annual traffic growth rate, 3%.

Initially various reconstruction alternatives involving the removal of the existing reinforced concrete and ashes pavement were considered. These may be summarised as follows.

Pavement Type	Layer Thicknesses
1. Flexible bituminous on granular sub-base	205 mm bituminous 650 mm sub-base
2. Reinforced concrete slab on granular sub-base	250 mm concrete 410 mm sub-base
3. Full depth asphalt directly on to subgrade	420 mm bituminous

In addition overlays of flexible polymer modified bituminous materials and a 175 mm thick concrete overlay to the existing damaged concrete pavement were considered. However, the uncertain performance of such overlays, and the cost, inconvenience and impracticability of completly reconstructing the damaged concrete pavement led to the selection of a thick asphalt mix overlay incorporating Netlon geogrids.

As part of the construction process it was decided to break up what was remaining of the existing concrete slabs to form a 'flexible sub-base', and to prevent any stress concentrations, thus reducing the possibility of reflective cracking through the overlay.

The thickness of the asphalt mix overlay was selected following pavement analyses using the Nottingham University analytical computer

Polymer grid reinforcement. Thomas Telford Limited, London, 1985

design programmes. Any improvement in tensile strength due to the incorporation of the geogrids was not included in the pavement analysis. The overlay design consists of a 210 mm thick basecourse layer of gravel gap graded rolled asphalt with 65% coarse aggregate content, topped with a 40 mm thick rolled asphalt wearing course as the surfacing, giving a total overlay thickness of 250 mm of asphalt mix.

Asphalt Mix Design

To achieve both high fatigue resistance and low permanent deformation the rolled asphalt mixes were desiged using the total mix Marshall method. Based on the pavement design analyses calculated values for voids, binder content, and quotient and stability were set to achieve the required flexible but deformation resistant performance. The gap grading being selected in preference to a continous aggregate grading because of its greater fatigue resistance. The mix design criteria together with the actual material properties achived are summarised below.

	Wearing Course Surfacing 30% coarse aggregate content 14 mm nominal size, 50 pen bitumen	
Mix Property	Design Criteria	Actual Properties
Bitumen Content %	7.0-8.2	7.0
Mix Voids, %	3-5	3.5
Minimum VMA, %	14.5	19.3
Min Stab., KN	3.0	7.9
Min Quotient KN/mm	1.0	1.3
Lab Density g/cc	-	2.316

	Basecourse 65% coarse aggregate content 20 mm nominal size, 50 pen bitumen	
Mix Property	Design Criteria	Actual Properties
Bitumen Content %	4.9-5.8	5.0
Mix Voids, %	5-10	5.8
Minimum VMA, %	14.0	16.4
Min Stab., KN	4.5	5.8
Min Quotient KN/mm	1.25	2.1
Lab Density, g/cc	-	2.324

Geogrid Reinforcement

To increase the tensile strength of the overlay and deter propogation of reflective cracks from the concrete layer, Netlon geogrids made from polypropylene were included as horizontal reinforcement near the base of the overlay.

Tensile strength tests were undertaken on single lengths of Tensar AR1 geogrid using a Houndsfield Tensometer at a loading rate of 3mm/min. These tests being of single lengths of Tensar only, took no account of the important grid effect when in service. However, they did confirm the considerable potential tensile strength of the geogrid as an asphalt reinforcement. The results obtained are summarised below.

Failure mode - splitting across the mesh joint or node.
Load at failure - 0.99 to 1.09 KN
Approx stress at failure - 150 to 200 MPa
Strain at failure - 16 to 24%

PAVEMENT CONSTRUCTION

The construction of phase I commenced in April 1981 and was planned so that only one lane of the road was closed at any time. Construction was undertaken in the following stages.

1. Break existing concrete.

2. Lay first layer of asphalt mix.

3. Place geogrid.

4. Lay second lay of asphalt mix.

5. Install new road edge kerbs.

6. Complete laying of asphalt mix.

During the construction of phase I a number of practical problems were expereined and overcome, and the experience gained was used to develop the procedure for phase II. The most important aspects of the experience gained are discussed in the following sections.

Breaking of Existing Concrete Pavement

The design intention was to break the concrete slabs into a system of interconnected flexible blocks of approximately 300 to 400 mm in size. By this means it was hoped to achieve a firm foundation for the reinforced asphalt overlay without the problems of stress concentrations and subsequent reflective cracking through the asphalt layers. Fracture of the concrete was to be achieved by use of a dead weight drop hammer falling onto the concrete slabs in a regular pattern. Upon completion of the fracturing of the slabs a thin layer of small size graded stone was to be spread over the concrete and rolled as a blinding and regulating layer prior to laying the asphalt mix overlay. However, experience showed that this procedure was unsuitable because of the following problems.

1. A spade pointed tip was provided with the drop hammer instead of the club foot tip anticipated. This pointed tip tended to penetrate the concrete slab rather than cause cracking of the surrounding area.

2. A regular pattern of hammer drops was proposed, but the concrete proved to be of variable strength. In some areas the hammer tip fully penetrated the slab and in others several drops were required to break the concrete.

3. Some areas of the concrete slab proved to be doubly reinforced with steel reinforcement also in the top of the slab. Where the slab surface was badly broken by the hammer tip some of the steel reinforcing had to be cut away.

To overcome these problems and produce the fractured pattern required the following procedure was recommended.

A club foot hammer tip was used to weaken the concrete slabs in a suitable pattern determined by the concrete strength and the operators experience. By this method the slab was only weakened and the required complete fracturing of the slab into interconnected blocks was achieved by subsequently rolling the slab with a heavy 20 Te dead weight roller. This caused the weakened slab to bend and crack without extensive shattering of the surface. It was found using this method that no stone blinding and regulating layer was necessary and a firm but flexible foundation was achieved.

Any particularly weak areas of concrete that did shatter extensively or had suffered previous damage making them unsuitable for overlay strengthening were removed and replaced with approximately 150 mm of compacted asphalt mix base. To reduce penetration of this asphalt base into the very weak subgrade a layer of Netlon CE 131 geogrid was placed on the subgrade prior to laying the asphalt mix. The alternative to the use of this full depth asphalt replcement would have required excavation and placing of 650 mm of crushed rock material as required by conventional procedures.

Achieving the Geogrid - Pavement Bond

The importance of obtaining a good bond between the asphalt pavement and the geogrid, to achieve both pavement strength and the required tensile reinforcement, was recognised throughout. In the original design concept it was proposed to achieve this by penetration of the asphalt mix through the geogrid by rolling the geogrid into the surface of the first layer of asphalt mix. This first layer was to be hand laid with no compaction so that it would allow penetration of the geogrid. However, again problems occurred and amendments to the proposed procedure were necessary.

1. At the start of Phase I the specified Tensar AR1 with its rectangular mesh size of 50 mm x 65 mm was not available, and it was necessary to use Netlon CE 131 with its circular mesh size of only 27.5 mm. The nominal size of the asphalt mix was 20 mm and insufficient mix penetration was achieved causing poor bond at the geogrid layer with the Netlon CE 131.

2. The original intention was to lay a loose layer of asphalt mix 40 mm thick, either by hand or without use of the paver screed tamper and compaction. Then place the goegrid on the surface of this loose layer and roll it into the asphalt mix. This would then be followed by a second layer of asphalt mix 60 mm thick laid by paver and compacted normally. Both methods of laying the first loose layer were attempted but due to inadequate asphalt mix thickness, distortion and movement of the geogrid during rolling into the surface, little satisfactory penetration of the geogrid was achieved. A method of 'sandwiching' the geogrid between two layers of asphalt mix was subsequently developed.

To achieve a satisfactory bond using the Tensar AR1 geogrid reinforcing the following procedure was adopted on the basis of these experiences.

A 50 mm thick layer of asphalt mix was laid and compacted using a conventional paver and roller. Onto the surface of this layer the geogrid was unrolled and a 50 mm thick layer of asphalt mix was hand spread over the geogrid. The sandwich of asphalt mix and geogrid was then completely compacted to give the required pavement strength and bond. Satisfactory bond being proven by the cores cut from the completed pavement. Following the sandwich layers the remaining layers of 150 mm of asphalt mix were placed by conventional paver methods.

Placing the Geogrid

Because of the design intention to roll the geogrid into the surface of a loose layer of asphalt mix and the subsequently developed sandwiching procedure, it was felt unnecessary to fix the goegrid to the asphalt mix layer in any way. However, various methods of nailing the geogrid to a compacted layer of asphalt mix were attempted but due to geogrid distortion these were not very successful and were abandoned as unnecessary. However, a number of problems were experienced mainly because of the distortion of the gogrid, both during manufacture and when placing on site.

1. Phase I of the project commenced before the manufacturing plant producing Tensar AR1 was in production, and the first rolls of Tensar subsequently received were manufactured before the commissioning of the new plant was complete. As a consequence some of the geogrids used were distorted not only in their mesh pattern

and size but in the roll shape itself. This caused two problems directly related to the distortion.

    a.   the geogrids did not lay flat on the surface of the asphalt mix and in some cases penetrated the top of the second ashalt mix layer.

    b.   because of the distortion of the rolls, ensuring continuous overlapping of adjacent rolls of geogrids was difficult, and the overlap width was not constant.

Later non-distorted production runs of the Tensar solved both these problems.

2. During the compaction of the second layer of asphalt mix it was found necessary to roll the asphalt mix in one direction only, and particularly not to change direction of the roller on the asphalt layers until they had stabilised. If this was not done, rolling in two directions caused the geogrid to stretch in opposite directions with severe distortion and pushing of the geogrid through the surface of the asphalt layers occuring. Again this problem was particularly apparent if the geogrid was already distorted during manufacture.

3. The original rolls of Tensar AR1 were manufactured with a thick edge strip approximately 30 mm in width. To make it easier to place the goegrid flat it was necessary to remove this edge strip before placing the geogrid on the asphalt mix.

To place the geogrid on the first asphalt layer and hold it flat, with the appropriate overlaps of 150 mm between adjacent rolls, the following procedure was adopted which does not require any method of fixing or tensioning the geogrid other than hand placing of the second asphalt layer. This method also allows the geogrid free movement and avoids any localised tensioning of the geogrid.

Place by paver and compact asphalt mix layer 50 mm thick. Unroll geogrid onto surface of first asphalt mix layer, closely followed by the second asphalt mix layer which is discharged from the rear of the delivery vehicle and spread by hand. Heaps of asphalt mix being used to hold the geogrid flat as it is unrolled further, but not restraining, thus avoiding any distortion as the second asphalt mix layer, 50 mm thick, is spread over the geogrid. Roll the sandwhich of asphalt mix and geogrid, avoiding change of direction of the roller on the fresh asphalt mix until the sandwich has stabilised.

Pavement Compaction

To ensure satisfactory compaction of the asphalt mix layers was achieved the contractor was required to undertake a compaction trial area, 9.0 m x 3.5 m, including the proposed method of sandwiching the geogrid. Compaction achieved being measured by the density of cores cut from the trial area. Percentage requirements of not less than 95% of the Laboratory Design Mix Density (LDMD), were required in the trial areas, and from these trials not less than 97% of the Job Standard Mix Density was required for the actual pavement overlay. Results from the cores showed that no problems of achieving satisfactory compaction were experienced with either the use of Netlon CE 131 or Tensar AR1 geogrids. The following core density results were achieved for the basecourse asphalt mix.

Laboratory Deisgn Mix Density, (LDMD) - 2.3240

Trial Area Density - 2.2550, (97% of LDMD)

Overlay Densities - 2.2213 to 2.3146, (96 to 99% of LDMD)

In order to avoid damaging the polypropolene polymer used to manufacture the geogrids the maximum temperature of the asphalt mix was restricted to 130°C when in contact with the geogrid. The core density results quoted demonstrate that this restriction caused no problems in achieving satisfactory asphalt mix compaction. Temperatures of generally 135°C are considered a suitable maximum based on this experience. With the development of new geogrid polymers specially treated to remain heat stable this maximum temperature may be increased if considered necessary to achieve satisfactory compaction with less workable asphalt mixes.

PAVEMENT PERFORMANCE

During construction of the overlay it was found necessary to open one section to traffic before the basecourse was completed to the specified thickness of 210 mm. Only 100 mm of asphalt mix incorporating the geogrid had been placed when the section was opened to traffic for two days, resulting in some extensive damage, visible as cracking and crazing of the surface of the asphalt mix. It was decided to leave this section in place and continue construction of the overlay as a comparative section for considering subsequent performance of the overlay. To date this 'weaker' section of overlay is performing totally satisfactorily.

A further small section of overlay in part of phase I was also constructed without benefit of the geogrid reinforcement. Again to date this section is performing as well as adjacent sections, but this may be a consequence of the greater foundation strength in the area of this unreinforced section.

To assess the performance of the overlay design, pavement deflection readings using a deflectograph vehicle were undertaken in October 1981. Average deflections of 46.5 10^{-2} mm at 14°C were recorded. These indicate a life expectancy of 5 to 8 million standard axles, (msa), compared to the original design requrirement of 4 msa.

Having now carried up to 0.9 msa all sections of the overlay are performing totally satisfactorily. Performance monitoring is being undertaken which confirms the suitability of the overlay design and the construction methods adopted.

CONCLUSIONS

A thick asphalt mix overlay incorporating Netlon geogrids to strengthen a severely damaged concrete pavement on a very weak subgrade has performed satisfactorily since 1981. A total of approximately 1,500 m^2 of Netlon CE 131 and 10, 500 m^2 of Tensar AR1 being employed to reinforce the overlay which was completed in two phases, the second phase being undertaken in 1982.

Deflectograph measurements taken on the completed pavement overlay confirm that the predicted life of the strengthened pavement exceeds the design life by between 20 to 100%.

A method of sandwiching the geogrid between two asphalt mix layers, with the second layer laid by hand has been developed and used successfully. Holding the geogrid in place with heaps of asphalt mix allows the geogrid to take up its required placement without distortion or the need for elaborate fixing or pre-tensioning techniques.

Because of its larger aperture size Tensar AR1 is to be preferred to Netlon CE 131 to ensure that satisfactory bond with the asphalt mix layers is achieved. No problems in achieving satisfactory compaction of the asphalt mix with conventional plant were experienced even though a maximum asphalt mix temperature in contact with the geogrid of 130°C was specified.

Cost of supplying and placing the geogrid was approximately £1.20 per sq. m. The additional cost, including hand laying as necessary, as a proportion of the pavement cost depends upon the pavement thickness but is probably between 5 and 15%. This additional cost will obviously reduce as more experience is gained, larger applications are undertaken, and mechanical placing methods are developed. However, even with these additional costs the use of geogrids offers considerable financial advantages as demonstrated by the project described in this paper where the total restrengthening overlay cost was 60% lower than the alternative estimate for maintenance repairs alone.

In addition geogrids offer the possibility of technical solutions where previously no solutions existed. The use of geogrids, used sensible as an option within a pavement and asphalt mix design system that can take account of the beneficial effects of these geogrids, offers considerable potential to the design engineer in the future.

Pavement reinforcement: report on discussion

T. L. H. Oliver, *Netlon Ltd*

The discussion session was preceded by a contribution from the floor by Dr Jamnejad of Sunderland Polytechnic Civil Engineering Department. The department has recently initiated a research programme to investigate the behaviour of three-dimensional grids for road base construction. The grids comprise 200mm deep vertical walls welded together intermittently. The shape of the cells is a function of the tension applied prior to filling with sand. The grid may be expanded up to 48 times the folded thickness.

Using a 1.5 cubic metre test pit, plate bearing tests have been conducted on layers of reinforced and unreinforced sand. Some plate bearing testing of block paving, with and without reinforcement of the sand road base has also been carried out. Pressure pads were installed beneath the sand road base to measure subgrade stress.

Preliminary test results have shown that subgrade stress is reduced for the reinforced case due to improved stress distribution through the reinforced road base. Vertical deflection of the pavement was also reduced.

It was noted that in paper 5.1, under the heading "Mechanical Properties of the Polymer Grid" the authors state that exposure to elevated temperatures results in a reduction in stiffness of the grid. Current UK guidelines permit asphalt to be delivered at temperatures between 125 degrees Centigrade and 190 degrees Centigrade. Concern was expressed by the session chairman that as a surfacing contractor he would be forced to work within a reduced range of laying temperature when using polymer grid reinforcement. This may make the achievement of adequate compaction and surface finish difficult especially during winter on wind chilled sites.

Prof. Brown agreed that laying temperatures would need to be restricted to below 160 degrees Centigrade adding that for most mixes workability is adequate at this temperature. He also noted that laying temperatures have been restricted to 150 degrees Centigrade for sulphur asphalt in order to reduce pollution from fumes, and this had been implemented quite adequately. The 190 degrees Centigrade referred to is the maximum <u>delivery</u> temperature in the UK. The normal maximum <u>compaction</u>

temperature in the UK is 145 degrees Centigrade, so this would be the critical temperature.

Reference was also made to paper 5.4 which concluded that adequate compaction was achieved with a maximum laying temperature of 135 degrees Centigrade.

During his presentation Prof. Brown had mentioned that pre-stressing of the reinforcement could be advantageous though it may be impractical. Commenting on this, one delegate suggested that a significant pre-stress may already be occurring during installation due to the shrinkage effect acting on the restrained grid. It was agreed that this would occur in the field, and indeed had been reported in paper 5.3.

When asked where the reinforcement would be located in new construction consisting of base course plus 40mm of wearing course, Prof. Brown stated that he would not comtemplate putting reinforcement shallower than 40mm. Ideally the grid would be sandwiched within the base course, or conceivably placed between base course and wearing course.

In paper 5.2, the experimental results from Loop 2 indicated a significant reduction in the angle of curvature (Fig 3).

The question was aked whether this improvement was truly the effect of the reinforcement during loading, or whether it was possible that the reinforcement serves to improve compaction of the asphalt, and the resulting reduction in air voids content is the reason for the apparent increase in stiffness. In answer, reference was made to paper 5.1, Table 3, which shows that the presence of a grid actually impedes compaction of the asphalt. Despite the lower densities achieved under the same compactive effort, the reinforced sections still gave better performance.

In paper 5.3 Table 1 compares the specific tensile strengths of various materials. One delegate commented that this made the polymer materials appear far stronger than steel and this was misleading. What was important he argued, was that in order to reinforce anything a material needed to be extremely stiff, and in this aspect Tensar did not have the best performance.

Dr Kennepohl explained that the table had been included to enable some cost/benefit comparisons to be made. He apologised if the information had proved misleading. With regard to the second point, reference was made to paper 5.1 where the elastic stiffness of reinforced asphalt is discussed. The authors of that paper explained that as there was no significant difference between the elastic properties of the reinforced layers and the unreinforced layers the analysis would remain the same. The reinforcing effect comes into play when assessing the life of the pavement. The reinforcement will limit cracking and reduce rutting. This is how the performance of the reinforcement should be assessed and not purely on the initial elastic analysis.

In response to a question concerning recycling Dr Kennepohl reported that a reinforced pavement had been taken up with a road milling machine. Laboratory testing of the material has indicated that it is suitable for re-use. 10 tonnes of the material have been stored for recycling trials due to commence in Spring 1984.

While accepting that the subject of the Symposium was Polymer Grid Reinforcement, one delegate commented that

Polymer grid reinforcement. Thomas Telford Limited, London, 1985

DISCUSSION

in reality the subject was Tensar. He expressed concern that some of the papers lacked objectivity, and that no comparisons were made with other criteria or methods of operating. In this connection the author of paper 5.4 was questioned as to what basis had led to the incorporation of a grid in 1981, since the pavement design used did not appear to take reinforcement into effect.

Mr Pooley explained that at the time of design not only was the effect of reinforcement unable to be modelled, but the performance of a cracked concrete base was also unknown. An assumption was made that when broken up it would perform similarly to a granular base. Because of concern with the accuracy of this assumption, some additional insurance was sought. What was known of Tensar at the time indicated that it was stronger than other fabrics available and it was selected for that reason.

In response to the question of objectivity attention was drawn to reference 1 of paper 5.2. Prof. Haas pointed out that Reference 1 consists of a rigorous comparison of all available reinforcement materials and methods. He explained that throughout the research programme, comparisons had been made with other materials such as geotextiles and that at an installation due to commence shortly in Houston, Texas, comparisons are to be made with steel mesh, which is still used in some parts of the USA.

One delegate suggested that much of the discussion had tended to be based on the incorrect assumption that the properties of asphalt pavements are predictable with a high degree of accuracy. In practice pavements vary to a great degree. Much of the experimental work carried out at the UK Transport and Road Research Laboratories has been aimed at reducing the variability likely to be experienced on site. Even if the mean pavement performance remains unchanged, any increase in the predictability of pavements would result in immediate cost benefits by enabling the insurance against failure to be reduced. With this in mind the delegate asked whether grid reinforcement will improve predictability of performance. Variability is most likely to be reduced by improving control at the mixing plant and on site.

Participants:

Mr Burman, Chairman

Professor Brown

Professor Haas

Dr Kennepohl

Mr Pooley

Dr Jamnejad

Dr Hoare

Mr Bridle

Professor Bell

Mr Smith

Design and construction methods

Reinforced soil structures can be very economical when compared with conventional structures. To achieve the optimum benefits close attention must be given to the idealisation of the structural concept. Only when this has been undertaken should the proposed structure be analysed.

The economical construction of a reinforced 'soil structure requires a recognised technique and the use of proven details; without these all or many of the benefits may not be realised.

C. J. F. P. Jones, *West Yorkshire Metropolitan County Council*

1. CONCEPTIONAL DESIGN

The development of modern soil reinforcing techniques has been rapid, even so the benefits to be gained from their use have been demonstrated not only in the financial savings achieved but also in their ability to produce novel solutions to construction problems.

Due to the extensive lead times involved in civil engineering schemes it is probable that the first consideration of soil reinforcing systems will be as an alternative to a conventional solution. The disadvantages of substitution can be considerable; contractors inexperienced in the technique may tender high, short lead times for material delivery can cause logistical difficulties and the lack of knowledge relating to specific subsoil conditions may create design problems. The fact that reinforced soil can frequently provide financial benefits when used as a late alternative to a conventional design suggests that greater benefits could be obtained if the use of soil strengthening systems were considered at the conceptual design stage of any scheme.

Full benefit of soil strengthening systems can only be obtained if the designer is aware of the advantages and limitations of the technique and has access to the necessary analytical, testing and estimating procedures required for design. An essential requirement is a comprehensive soil survey which must be planned with the understanding that soil reinforcing techniques could form part of the design solution. In particular if finite element techniques are to be used in the analysis the soil survey may need to be supplemented to provide information relating to the initial stresses in the subsoil.

1.1 Retaining Walls

When viewed at the conceptual design stage reinforced soil walls present few problems, although their cost effectiveness may suggest

vertical and horizontal alignments which could not be contemplated with conventional structures. Design and analytical techniques for conventional reinforced soil walls have now been established, although the complex mechanisms involved are not properly understood.

1.2 Bridge Abutments

Medium or small span bridges, if constructed using abutments, will have a significant proportion of the total cost invested in the substructure. Split costs of decks and abutments on these bridges have indicated substructure costs rising above fifty per cent of overal cost. Since reinforced soil has been shown to produce economies in abutment costs, significant reductions in total bridge costs are possible. The use of reinforced soil abutments cannot be accomplished without some change to the deck design, the span of which will almost certainly be increased at a cost dependent upon span and skew, Fig 1. In addition the possibility of differential settlement of reinforced soil abutments raises concern in articulation of the deck. However, this problem can be resolved with the use of a low torsion deck.

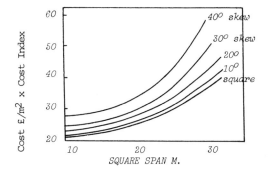

Fig 1. Deck Costs Relative to Span and Skew

Polymer grid reinforcement. Thomas Telford Limited, London, 1985

On weak foundations bridge abutments are frequently supported on piles and the same solution can be adopted with reinforced soil abutments. However, an alternative idealisation may be possible based upon a different treatment of the overall stability analysis. In many analytical circumstances the abutment, piles and the adjacent embankment will each be considered separately and the effects on one element superimposed upon the others. If the abutment is subjected to vertical loads from the deck and the piles to lateral loads from the embankment, a frequent conclusion of this analysis is that the abutment will move away from the embankment and the piles should be raked forward. A global assessment of the behaviour of the abutment, piles and embankment, in which the ground movement caused by the placing of the embankment is also considered, may show an alternative behaviour. Abutment rotation may be towards the embankment; introducing an increase in moment in the abutment stress due to increased lateral pressures. Instead of providing compressive reinforcement in the form of conventional piles, a concept of tensile strain arc reinforcement may be used and the abutment reduces to a reinforced earth structure forming one end of the embankment, Fig 2.

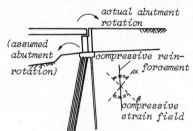

(a) *Conventional idealisation in abutment construction on weak soil.*

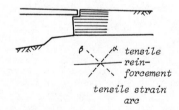

(b) *Alternative idealisation*

Fig 2.

Practice indicates that the pile lengths associated with the idealisation of Fig 2(a) may be substantial, >30m on occasions, whereas the reinforcement length needed with the condition of Fig 2(b) may be substantially less. An additional benefit of the reinforced soil solution is that this idealisation effectively eliminates the problem of differential settlement between the abutment and the embankment. The design concept in these circumstances is similar to the procedure adopted in mining areas where piled foundations cannot be used due to

the problem of differential subsoil strain caused by a moving subsidence wave. One solution is to provide a substantial bearing pad, up to seven metres thick, of compacted granular material under the abutment and to accept any residual differential settlement. The use of a thinner reinforced soil foundation formed as an integral part of the reinforced approach embankment is a practical alternative which has the advantage of minimising the incident of differential settlement which frequently occurs behind conventional abutments. The same concept can be used under central piers of a two span structure, although a degree of sophistication may be required in the analysis to permit the settlements of the abutments and the pier to be of the same order.

2. ANALYSIS

Walls and abutment structures are normally constructed using horizontal reinforcement and take the form illustrated in Fig 3. The vertical spacing of the reinforcement may remain constant throughout the depth, but the density may vary. Analysis covers two stability conditions:

(i) External stability.
(ii) Internal stability.

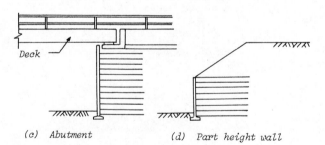

(a) *Wall elements*   (b) *Stepped wall*

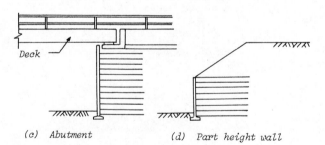

(c) *Abutment*   (d) *Part height wall*

Fig 3.   Typical forms of walls and abutments

2.1 External Analysis

External analysis covers the basic stability of the reinforced soil structure as a unit, covering; sliding, tilt/bearing failure, and slip within the surrounding subsoil or slips passing through the reinforced structure. Failure mechanisms of this nature are represented by (c), (d) and (e) in Fig 4. In addition, stresses imposed upon the reinforced earth structure due to particular external conditions such as the creep of the subsoil have to be considered. Fig 4(f).

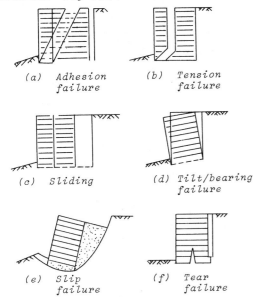

(a) *Adhesion failure*

(b) *Tension failure*

(c) *Sliding*

(d) *Tilt/bearing failure*

(e) *Slip failure*

(f) *Tear failure*

Fig 4. Failure Mechanisms of Reinforced Soil Walls

External Stability: The stability of the structure against forward sliding, cracking, tilting and overall stability of the supporting foundation and the adjacent retained fill may be checked as follows:

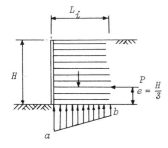

(a) *Tilting*

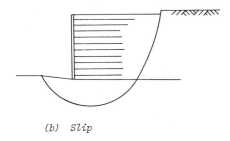

(b) *Slip*

Fig 5.

Forward sliding:

Sliding forces $= \frac{1}{2} K_a \gamma H^2$

Resisting force $= \mu_f \gamma H L_i$

Factor of Safety $= \left[\dfrac{2 \mu_f L_i}{K_a H} \right]$ (1)

where μ_f is the coefficient of friction at the base of the wall.

Overturning:

Overturning moment with respect to the toe $= P_e = \frac{1}{2} K_a \gamma H^2 \times \frac{H}{3} = \left[\frac{1}{6} K_a \gamma H^3 \right]$ (2)

Resisting moment with respect to the toe $= \dfrac{W \cdot L_i}{2} = \dfrac{\gamma H L_i^2}{2}$ (3)

∴ Factor of safety against overturning $= \left[\dfrac{3 L_i^2}{K_a H^2} \right]$

Tilting: For structures situated on good subsoils a trapezoidal pressure distribution beneath the structure may be assumed. An allowable bearing pressure in the subsoil may be taken as half the ultimate bearing capacity, provided any resulting settlements can be tolerated by the wall and any superimposed structure.

From Fig 5(a) $W = \frac{L_i}{2}(a+b)$; $P_e = (a-b)\dfrac{L_i^2}{12}$ (4)

from which $a = \frac{1}{L_i}\left[W + 6 \dfrac{P_e}{L_i} \right]$ (5)

$b = \frac{1}{L_i}\left[W - 6 \dfrac{P_e}{L_i} \right]$ (6)

where $P_e = \frac{1}{2} K_a \gamma H^2$; $e = \frac{H}{3}$ (7)

$a = \gamma H \left[1 + K_a \left(\frac{H}{L_i} \right)^2 \right]$

$b = \gamma H \left[1 - K_a \left(\frac{H}{L_i} \right)^2 \right]$

If the computed value of (a) in equation seven exceeds the allowable, the stability of the base with respect to the ultimate bearing capacity may be improved by widening the base width of the structure to $(L_1 + L_2)$, Fig 3(b).

With a poor subsoil widening the base width may not be sufficient to satisfy the ultimate bearing capacity criteria for stability with respect to tilting. In this case support to the base may be provided by external means, Fig 6. Alternatively the overall stability of the structure and the surrounding soil may be considered on a global basis using continuum systems.

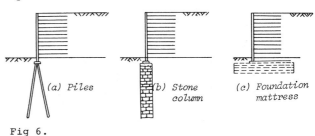

(a) *Piles*

(b) *Stone column*

(c) *Foundation mattress*

Fig 6.

Rotational Slip: All potential slip surfaces should be investigated, including those passing through the structure, Fig 5(b). Where slip planes already exist, residual soil strength parameters should be adopted. The factor of safety for reinforced soil structures against rotational slip is the same as for conventional retaining structures.

2.2. Internal Stability

The internal stability is concerned with the estimation of the number, size, strength, spacing and length of the reinforcing elements needed to ensure stability of the whole structure, together with the pressures exerted on the facing, Fig 4(a,b).

Numerous analyses to check for internal stability have been developed, most fall into the following categories:

(a) Those in which local stability is considered for the soil near a single element of reinforcement, and

(b) those in which the overall stability of blocks or wedges of soil is considered.

The most widely used analytical procedures are semi empirical systems which can be considered to represent the serviceability and ultimate limit state.

2.2.1 Coherent Gravity Hypothesis

The coherent gravity hypothesis relates to a reinforced soil structure constructed with a factor of safety and in a state of safe equilibrium. The design stresses relate to actual working stresses not to failure conditions. Thus the coherent gravity hypothesis relates to the serviceability limit state.

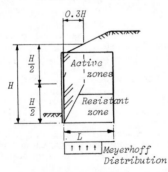

Fig 7.   Coherent Gravity Hypothesis

The coherent gravity hypothesis assumes:

. The reinforced mass has two zones, the active zone and the resisting zone, Fig 7.

. The state of stress in the fill, between the reinforcements is determined from measurements in actual structures constructed using well graded cohesionless fill.

. An apparent coefficient of adherence (μ^*) between the soil and reinforcement is derived from an empirical expression developed from pullout tests.

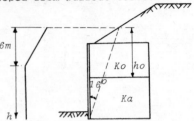

Fig 8.   Variation of apparent friction co-efficient μ^* with depth. (After Schlosser & Segresstin 1979)

For a structure using strip reinforcement, the maximum tension (Tmax) per element at depth h:

$$\text{Tmax} = k\sigma_v \frac{\Delta H}{N} \qquad (8)$$

where $\sigma_v = \gamma h$
 ΔH = zone of action of the reinforcement
 N = number of reinforcements per area considered

$$K = k_0\left(1 - \frac{h}{h_0}\right) + K_a \frac{h}{h_0} \; ; \quad h \leqslant (h_0 = 6m) \qquad (9)$$

$$k = k_a \qquad ; \quad h > (h_0 = 6m) \qquad (10)$$

Similarly, the maximum adhesion force (Tad) per element of reinforcement, assuming a well graded, cohesionless fill, B=reinforcement breadth, L_r= length of reinforcement within resistance zone;

$$\text{Tad} = 2B \int_{L-Lr} \mu^* \sigma_v \, dL \qquad (11)$$

where $\qquad (12)$

$$\mu^* = \mu_0\left(1 - \frac{h}{h_0}\right) + \frac{h}{h_0}\tan\phi' \; ; \; h \leqslant (h_0 = 6m)$$
$$\mu^* = \tan\phi' \qquad\qquad ; \; h > (h_0 = 6m) \qquad (13)$$

μ_0 for rough reinforcement* is defined empirically as: $\mu_0 = 1.2 + \log C_u$

where C_u = coefficient of uniformity $\frac{D_{60}}{D_{10}}$ $\qquad (14)$

$\mu_0 = 0.4$, for smooth reinforcement $\qquad (15)$

2.2.2   Tie-Back Hypothesis

The tie-back hypothesis is based upon the following design requirements for vertically faced structures:

. The design criteria is simple and safe.
. The design life of the structure is 120 years.
. The design procedure is consistent with the use of a wide range of potential fill materials, including frictional and cohesive-frictional soil.

Internal stability considerations include:

. The stability of individual elements, Fig 9.
. Resistance to sliding of upper portions of the structure.

* geogrid may be defined as rough in this analytical model.

The stability of wedges in the reinforced fill, Fig 10. (Note: Centrifuge studies indicate that at failure a wedge failure mechanism can develop. The tie-back hypothesis therefore relates to the ultimate limit state rather than the serviceability limit of the coherent gravity hypothesis.)

The following factors which influence stability are included in the design:

. The capacity to transfer shear between the reinforcing elements of the fill.
. The tensile capacity of the reinforcing elements.
. The capacity of the fill to support compression.

The state of the stress within the reinforced fill is assumed to be (Ka). The at rest (Ko) condition measured in some structures, Fig 7, is assumed to be a temporary condition produced by compaction during construction. The active state of stress is assumed to develop during the working life of the structure.

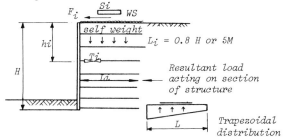

Fig 9.   Tie-Back Analysis, Local Stability

Local Stability: The maximum tensile force Tmax is obtained from the summation of the appropriate forces acting in each reinforcement as follows:

$$T_{max} = T_{hi} + T_{wi} + T_{si} + T_{fi} + T_{mi} \quad (16)$$

where T_{hi} = reinforcement tension due to fill above the reinforcement layer,

T_{wi} = reinforcement tension due to uniform surcharge,

T_{si} = reinforcement tension due to a concentrated load,

T_{fi} = reinforcement tension due to horizontal shear stress applied to the structure,

T_{mi} = reinforcement tension due to bending moment caused by external loading action on the structure.

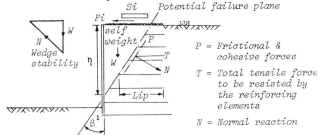

Fig 10. Tie-Back Analysis - Wedge Stability

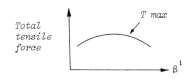

Fig 11.   Angle of Failure Plane

A Coulomb wedge failure is assumed to be possible within the reinforced soil structure. For design, wedges of reinforced fill are assumed to behave as rigid bodies of any size of shape and all potential failure planes are investigated (i.e. η and β may both vary, Figs 10 and 11.

2.2.3 Tie-Back Analysis - Geogrids

Although both the tie-back - wedge analysis and the coherent gravity analysis may be used with grid reinforcement, simplified analyses based upon the superior adhesion capacity or pullout performance of grids have been developed. Analysis is based upon an assumption of a Coulomb failure mechanism, Fig 12.

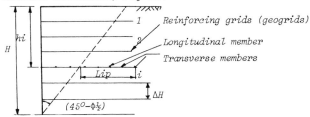

Fig 12.

Local tensile stress: For a uniform vertical distribution of horizontal grids the force exerted (Ti) on geogrid (i) at depth hi:

$$Ti = Ka \gamma hi \, \Delta H \quad (17)$$

Pullout resistance: The total pullout resistance (Ft) is a combination of the frictional resistance (FF) presented by the grid, plus the anchor resistance (FR) of the grid.

$$F_t = F_F + F_R \quad (18)$$

The frictional resistance (FF) per unit length of longitudinal wire diameter d:

$$F_F = \mu . \pi d \, \sigma_v \quad (19)$$

The Anchor resistance per transverse member based upon the Terzaghi-Buisman bearing capacity expression is defined as:-

$$\frac{F_R}{N_W} = d \, c' N_c + \frac{\sigma_v}{2} d^2 N_\gamma + \sigma_v d \, N_q \quad (20)$$

where N_W number of transverse members outside the Coulomb failure wedge and,

$$N_c \,;\, N_\gamma \,;\, N_q = \text{Terzaghi bearing capacity} \quad (21)$$
$$\text{factors}$$

JONES

Since d is small, for a cohesionless fill

$$\frac{F_R}{N_w} = \sigma_v d N_q \qquad (22)$$

Total pullout resistance per unit width

$$F_t = (\sigma_v \pi d L \mu M) + (\sigma_v d N_q N) \qquad (23)$$

$$= \sigma_v d (\pi L \mu M + N . N_q)$$

where M = number of longitudinal members in grid per unit width

N = number of transverse elements outside the Coulomb wedge

μ = coefficient of friction between longitudinal members and the fill.

3. CONSTRUCTION

Construction of reinforced soil structures must be of a form determined by the theory and in keeping with the design and analysis assumed. The theoretical form of the structure might be quite different to an economical prototype, and attention should be paid to the method of construction throughout the design process.

Speed of construction is usually essential to achieve economy, and this may often be achieved by the simplicity of the construction.

3.1 Construction Methods

Constructional techniques compatible with the use of soil as a constructional material are required. The use of soil, deposited in layers to form the structure, results in settlements within the soil mass caused by gravitational forces. These settlements result in the reinforcing elements positioned on discrete planes moving together as the layers of soil separating the planes of reinforcement are compressed. Construction techniques capable of accommodating this internal compaction within the fill are required. Failure to accommodate the differential movements may result in loss of serviceability or worse.

The three constructional techniques which can accommodate differential vertical settlements within the soil mass are shown in Fig 13. Except for some special circumstances, every reinforced soil structure constructed above ground uses one or other or a combination of these forms of construction.

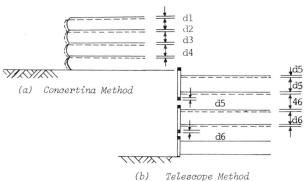

(a) Concertina Method

(b) Telescope Method

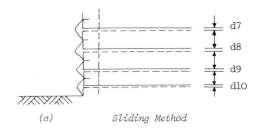

(c) Sliding Method

Fig 13. Three Methods used for Constructing Reinforced Soil Structures

Concertina Method: The constructional arrangement of the concertina method developed by Vidal (1963) is shown in Fig 13(a). Differential settlement within the mass is achieved by the front or face of the structure concertinaring. Some of the largest modern reinforced earth structures have been built using this approach, and it is the form of construction frequently used with fabrics and geogrid reinforcing materials in both embankments and cuttings. Since the facing must be capable of deforming, a flexible hoop shaped unit made from steel or aluminium is normally used with strip reinforcement. Geogrids usually provide their own facing. The method is often used with temporary structures.

Telescope Method: In the telescope method of construction the settlement within the soil mass is achieved by the facing panels closing up an equivalent amount to the internal settlement. This is made possible by supporting the facing panels by the reinforcing elements and leaving a discrete horizontal gap between each facing panel (i.e. the facing panels hang from the reinforcing elements).(Vidal 1978). The closure between panels will vary from structure to structure depending upon the geometry, quality of fill material, size of the facing panels and the degree of compaction achieved during construction. Typical movements, reported by Finlay (1978), show vertical closures of 5-15mm for facing panels 1.5m high.

Sliding Method: In the sliding method of construction settlement within the soil mass is accommodated by having the reinforcement embedded within the soil slide down the facing whilst still remaining connected to the facing. The facing may be made up of discrete elements or may be single full height units.

Labour and Plant: Labour and Plant requirements for the construction of reinforced soil structures are minimal, and no specialist equipment or skills are required. Erection of a normal vertically faced structure of 500-1000m exposed area is usually undertaken by a small construction team of 3-4 men deployed to cover the main construction elements, namely, erecting the face, placing and compacting the fill and placing and fixing the reinforcement.

A comparison in labour requirements for different forms of retaining walls has been given by Leece (1979), see Table 1.

Table 1.

Type of Wall	Labour Content Manhours/m²
.Reinforced earth - without traffic barrier	4.1
.Reinforced earth - with traffic barrier	4.7
.Mass concrete	11.2
.Reinforced concrete	11.5
.Crib walling	13.3

The plant requirements during construction normally include aids to the placing and compaction of soil, and some form of small crane or lifting device, although the latter is not required when a non-structural facing is used. Where a method specification is employed, as with DTp Memorandum BE 3/78, the compaction plant used within 2m of the facing normally consists of the following forms:

(a) Vibro tampers,
(b) Vibrating plate compactors with a mass <100kg,
(c) Vibrating rollers with a mass/metre width <1300kg and a mass <1000kg.

Rate of Construction: Construction of reinforced soil structures is normally rapid. Construction rates for vertically faced structures of 40-200m² per day may be expected and usually the speed of erection is determined by the rate of placing and compacting the fill.

However, in some cases the economic production of facing units may determine the construction rate, particularly if an original or unique facing is required.

Construction is normally unaffected by weather except in extreme situations.

Damage and Corrosion: Care must be taken that facing elements and reinforcing members are not damaged during construction. Vehicles and tracked plant must not run on top of reinforcement: a depth of fill of 150mm above the reinforcement is frequently specified before plant can be used. Polymer reinforcement should be stored away from ultra violet light.

Compaction: Compaction of the fill in a reinforced soil structure is desirable as it has a beneficial influence on behaviour and increases the efficiency of the structure. Good compaction also reduces internal differential movements, whilst uniform compaction provides the most stable environmental conditions which are important for durability. Uniform compaction of the fill is achieved by placing the fill in layers varying in depth from 100-300mm, and compacting the soil using suitable plant moving parallel to the facing or edge of the structure.

Any normal compaction plant may be used with reinforced soil structures, selection of the most suitable depending upon the properties of the fill being used.

Distortion: Reinforced soil structures are prone to distortion particularly during construction. Many of the construction details adopted in practice are chosen to minimise distortion or the effects of distortion.

Concertina Construction: Structures built from fabrics and geogrids or constructed as temporary structures using the concertina method of construction are particularly prone to distortion of the face. The degree of distortion cannot be predicted. An accepted method of overcoming the problem is to cover the resulting structure either with soil or with some form of facing. An alternative is to provide a rolling block against which the compaction plant can act.

Telescope Construction: An estimate of the internal movements and distortion of the facing can be made from observations of prototype structures. Horizontal movements of the facing are made up of two components:

(i) horizontal movements at the joints
(ii) tilt of the facing units

(With extensible reinforcement some horizontal movement may be expected, in practice this is minor due to the conservative analytical procedures which are normally used.)

Joint movement during construction is not normally significant and is likely to be between 2-5mm depending upon construction details. Tilt of the facing panels can be significant and may have a marked effect upon the facial appearance of the structure, although other elements of serviceability are unlikely to be effected by tilt of the facing. All facing panels in this form of construction tilt, the pivot point depending upon the geometry of the facing.

Sliding Method Construction: When a non-structural facing is used distortion of the facing is likely to occur, the degree being dependent upon compaction. The distortion is accommodated by:

(a) using light plant in the 2m zone. adjacent to the facing
(b) using bold architectural features to mask the distortion.

When a structural facing is used the horizontal movement of the facing will be limited to the joint movement capacity provided by the reinforcement/facing connection. Typical movements are 2-5mm.

Logistics: The speed of construction must be catered for if the full potential of the use of reinforced soil structures is to be realised. Normally this will cause little or no problems with the reinforcing materials, but the production and delivery rate of the facing units may cause problems, particularly if multiple use of a limited number of shutters is expected for economy.

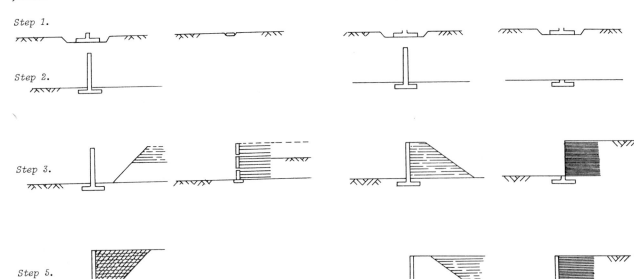

Step 1.

Step 2.

Step 3.

Step 5.

Conventional Structure   Reinforced soil structure

Fig 14.   Construction sequence for conventional
          structures and reinforced soil
          structures.

Conventional Structure   Reinforced Soil
                          Structure

Fig 15.

Transport may cause difficulties and the choice
of structural form and construction technique
may ultimately depend upon the ease and economy
of moving constructional materials. As an
example, the light weight of geogrid materials
with their ability to be transported in rolls
makes them suitable for air freight.

Construction Sequence: Reinforced soil
structures encourage the use of non-conventional
construction sequences. It is possible to
streamline the construction sequence and
eliminate some steps as illustrated in Fig 14.

Alternatively it is possible to invert a
construction sequence as in Fig 15, where the
backfill of a structure is placed before the
structure itself. This is achieved by forming
the backfill into a temporary reinforced soil
structure and using the face of this structure
as the back shutter for the permanent structure.
This technique has been used successfully in
bridgeworks construction.

REFERENCES

Department of Transport (UK) (1978) Reinforced
     Earth Retaining Walls and Bridge Abutments.
     Technical Memorandum (Bridges) BE 3/78,
     London.

Finlay T.W. (1978) Performance of a Reinforced
     Earth Structure at Granton. Ground Engineer-
     ing Vol.11 No.7 p 42-44.

Leece R.B. (1979) Reinforced earth highway
     retaining walls in South Wales. C.R.Coll.
     Int.Reinforcement des Sols, Paris.

Hambley E.C. (1979) Bridge Foundations and
     Substructures. Building Research Establish-
     ment Report, Department of the Environment.

Vidal H. (1963) Diffusion Restpeinte de la
     Terre Armée.

Vidal H. (1978) The Development and Future of
     Reinforced Earth. Keyworth Address ASCE
     Symposium on Earth Reinforcement,
     Pittsburgh.

Jones C.J.F.P. (1978) The York Method of
     Reinforced Earth Construction. ASCE
     Symposium on Earth Reinforcement,
     Pittsburgh.

Schlosser F., and Segresstin P.,(1979)
     Dimensionement des ouvrages en terre armée
     par la méthode d l'équilibre local.
     C.R. Coll.Int. Reinforcement des Sols, Paris.

Economics and construction of blast embankments using Tensar geogrids

J. Paul, *Netlon Ltd*

The paper discusses briefly the various commonly used methods of constructing blast protection and other steep faced embankments.

A typical example is considered from which the economy of the system may be assessed. These savings are generally a combination of total cost and area of land take when compared with traditional systems.

Several structures of this kind have been constructed throughout the world and the various construction techniques are discussed along with surface treatment measures.

Possible future developments are also considered.

Blast Protection embankments and walls are used throughout the world in many different applications.

Obvious examples are ammunition and explosives storage areas, certain types of electrical transformers, some gas and chemicals installations, explosives testing areas and the protection of critical areas from external attack.

Each type of application and each country has its own requirements for blast protection and no attempt will be made to discuss these variations. The example quoted later in this paper is broadly in agreement with the regulations governing the design of embankments around storage buildings containing high explosives ammunition in the United Kingdom. In this case the main aim of the embankment is to provide sufficient mass to contain the horizontal blast and to be sufficiently high to avoid "roll over" of blast and flames into adjoining areas. This latter requirment takes account of the height of the storage stack and the steepness of the face of the containing wall or embankment. With relatively flat sloped embankments the height may need to be increased to avoid the "roll over" problem.

Often, the construction consists of a reinforced concrete wall close to the building with an earth fill behind. The minimum thickness of embankment at the top level of the ammunition stack is 2.4 metres. In normal circumstances the outside face has no retaining wall and is a soil embankment constructed to a face slope suitable for the fill material used.

For lower heights of stack, brickwork or blockwork retaining walls may be constructed rather than reinforced concrete but the general layout is the same.

TYPICAL EXAMPLE

The following example gives an indication of the respective layout and costs for a vertical reinforced concrete faced embankment, a Tensar reinforced embankment and an unreinforced soil embankment using relatively poor quality fill. In this example the top of stack level is taken as underside of roof level and the stack height is 8 metres.

The reinforced concrete wall is constructed using $25N/mm^2$ concrete and the imported backfill has a compacted density of $19kN/m^3$ and a friction angle of 30^0. The narrow crest width ensures light construction traffic therefore no surcharge loading has been included.

In the Tensar reinforced solution a face angle of 60^0 is taken and this uses Tensar SR-2 and SS-1 Geogrid reinforcement. Working load in the SR-2 Geogrid is taken as 18.5kN/metre width which ensures adequate overall factor of safety.

Foundation stability has not been considered in this example. In cases of low bearing capacity foundation material there is often a significant benefit when using reinforced soil techniques which can reduce the maximum toe pressure compared with that from a reinforced concrete wall.

DIMENSIONS

Height to top of stack, Hs	8.0 metres
Building wall thickness	0.3 metres
Minimum footpath width round building	1.5 metres

Minimum height of embankment, H is greater of

a) Hs + 0.6m.
or
b) Hs + 2^0 rise from stack corner.

Minimum width of embankment at Hs level
2.4 metres

Polymer grid reinforcement. Thomas Telford Limited, London, 1985

191

1 - Reinforced Concrete Wall

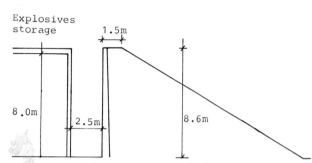

Fig 1 - Reinforced Concrete Wall

Design of Reinforced Concrete wall gives the following dimensions,

top of stem width	= 0.35m
bottom of stem width	= 0.50m
base width	= 4.30m
base thickness	= 0.50m
weight of steel reinforcement	= 0.10t/m³

Per metre run, Volume of concrete in stem
= 3.655m³
Volume of concrete in base
= 2.250m³
Weight of steel reinforcement
= 0.59t

Costs:- using Measured Rates published in "Civil Engineering", October 1983.

Excavation	2.25 x 5.07	=	11.41
Provision of concrete	5.905 x 41.33	=	244.05
Placing in base	2.25 x 5.94	=	13.36
Placing in stem	3.655 x 19.31	=	70.58
Rough formwork, base	2 x .5 x 13.84	=	13.84
Rough formwork, wall	8.6 x 13.20	=	113.52
Fair formwork, wall	8.6 x 15.52	=	133.47
Steel Reinforcement	0.59 x 431.13	=	254.37
Compacted Fill behind wall	73.96 x 7.20	=	532.51
Trim slopes	18.2 x 0.59	=	10.74
Grass seed to slopes	18.2 x 0.32	=	5.82

Total cost per metre run   = £1403.67

Land width required out with buildings = <u>18.9m</u>
(allowing 2.5m footpath for construction purposes)

2 - Tensar Reinforced Embankment

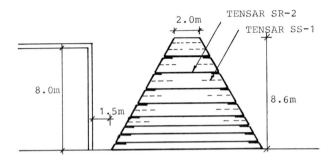

Fig 2 - Tensar Reinforced Embankment

A Tie-Back analysis using the given geometry and soil parameters allows the design of a Tensar reinforcement layout to give the required factor of safety.

Taking the working stress in Tensar SR-2 reinforcement (allowing for all safety factors) as 18.5KN/metre width, 9 layers are required. Vertical spacing varies from 0.6 metres at the base to 1.2 metres near the top.

To avoid possible bulging on the face short intermediate layers of Tensar SS-1 reinforcement are placed as shown in Figure 2.

Therefore, Volume of fill = 60.2m³/m
Area of SR-2   = 111m²/m
Area of SS-1   = 21m²/m

Costs.

a. Compacted fill (allowing for working within edge supports) at £7.70/m³   = 463.54
b. Tensar SR-2 at £3.50/m² (laid) = 388.50
c. Tensar SS-1 at £1.20/m² (laid) = 25.20
d. Edge support at £3.00/m²   = 60.00
e. Grass seed to face at £0.60/m²   12.00

Total cost per metre run   = £949.24

Land width required out with building = <u>13.5m</u>

3 - Unreinforced Embankment

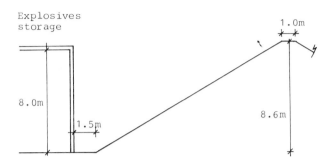

Fig 3 - Unreinforced Embankment

Volume of fill = $137.6 m^3/m$

COSTS

a.	Compacted fill at £7.20/m^3	=	990.72
b.	Trim slopes at £0.59/m^2	=	20.30
c	Grass seed to slopes at £0.32/m^2	=	11.00

Total cost per metre run = £1022.02

Land width required outwith building = 32.5m

SUMMARY

Embankment Type	Land Take	Cost per metre run
Reinforced Concrete Wall	18.9m	£1403.67
Tensar Reinforced Embankment	13.5m	£949.24
Unreinforced	32.5m	£1022.02

Economy of the Tensar Reinforced system is highlighted by this example.

Even greater savings are likely when blast protection is required between existing buildings, or close to boundaries or other obstructions. In these cases reinforced concrete or other vertical walls are required, sometimes on both sides of the structure.

CONSTRUCTION

While a 60° face slope was chosen in the example for the Tensar Reinforced embankment there is no reason why this could not be increased to 90°. Several vertical, or near vertical Tensar Reinforced embankments have now been constructed in various parts of the world and a variety of construction techniques have been developed.

The first major vertical embankment was constructed at Warton, near Preston, England.

Requirements were for a vertical structure 7 metres high, 16.5 metres wide and a minimum of 3.5 metres deep at the crest. The purpose of this sand embankment is to allow the arming of aircraft standing close to the vertical face, ensuring the interception of any accidentally fired weapons.

Similar structures had previously been constructed as three-sided wooden boxes strengthened with a steel frame and filled with sand. In terms of labour requirements and overall cost the Tensar Reinforced solution was very attractive.

A scaffold frame was erected outside the front and side edges of the embankment with boards fixed vertically to support the faces during construction. As each completed layer of reinforcement and fill is self supporting the framework is relatively light using standard scaffold boards with gaps between, as shown in Figure Number 4.

Fig 4 - Face Support During Construction

Figure Number 5 shows a cross-section of the embankment during construction and Figure number 6 shows the completed structure.

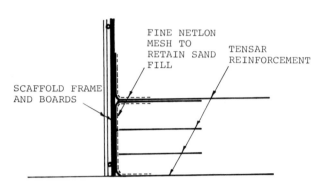

Fig 5

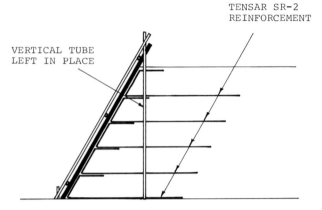

TENSAR SR-2
REINFORCEMENT

VERTICAL TUBE
LEFT IN PLACE

Fig 7 - Edge Support Arrangement

Slopes up to 45° have been constructed without the use of edge supports. The face of each layer of fill is dressed by hand and the Geogrid Reinforcement taken up and secured over the top of each lift.

Above 45°, support is required. Blast walls constructed at Catterick Barracks, Yorkshire, England, were reinforced with Tensar SR-2 Geogrid and had a face slope of 60°. Again a scaffold frame and boards were used with a sacrificial vertical tube at approximately 4 metre centres (Figure Number 7).

As can be seen from Figure Number 8, construction of these steep faced embankments is relatively straight forward and requires no specialised techniques or equipment.

A view of this completed project is given in Figure Number 9.

In higher embankments, a horizontal tube is laid within the top layer of fill in each stage of 3 to 4 metres height. This tube projects through the face to support the sloping tubes for the stage above (Figure Number 10).

A sloping face has in several instances been approximated by a series of vertical sections stepped back at each lift. Light plywood shutters, including the standard modular systems have been used with a leap-frogging erection sequence which minimises the number of forms required. Generally, a row of forms may be removed when construction of the stage above is complete therefore only two rows of shutters are required no matter how high the structure (Figure Number 11).

The Tensar Geocell Mattress may also be used to give a vertical or stepped-vertical face.

Fig 6 - Completed Structure

Fig 8

Fig 9

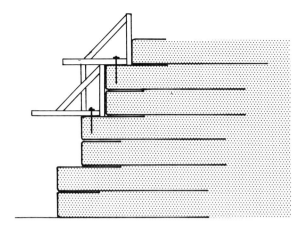

Fig 11

Due to the unique, elongated apertures of
Tensar uniaxial reinforcement a cellular
mattress system is easily erected on site using
pins to secure the joints (Figures 12 and 13).
The 1 metre high cellular structure thus
formed is filled with a granular material and
they may be stacked one above the other to form
an extremely rigid reinforced embankment.
They may be used as a facing to the embankment
(similar to gabions) but do not require to be
filled with large stone which can cause an
additional hazard in blast situations. When
filled with sand or other fine material a
geotextile liner is used to retain the fill.
In narrow, steep sided structures the mattress
extends from face to face and as may be seen
from the diagrammatic view (Figure 14), damage
to any area of the face is contained by the
Tensar diaphragms bounding that cell. This is
of special importance in the case of an
embankment protecting buildings or equipment
from external attack.

FACE TREATMENT

Treatment of the face of the embankment depends
on the requirements of each individual project.

In many cases no surface treatment is specified
and through time, weeds and grasses tend to
establish themselves on the face, the seeds
being trapped in the Tensar apertures.

To encourage this vegetation cover grass seed
may be sown either dry or in hydraulic mulch,
the seeds again being held in the Tensar
apertures.

Some projects have used a different quality
fill (including topsoil) close to the face to
ensure rapid vegetation growth in aesthetically
sensitive areas. Turf, placed against the face
during construction gives even more rapid cover
to the exposed geogrid.

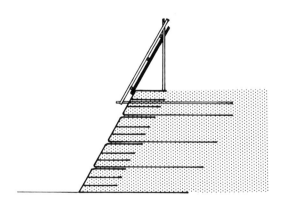

Fig 10

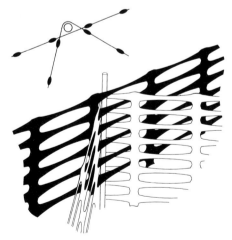

Coupling for diaphragm to a side

Fig 12

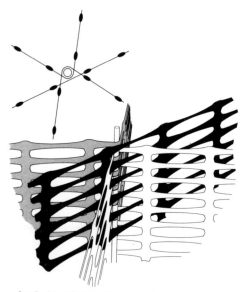

Coupling for two diaphragms to a centre panel

Fig 13

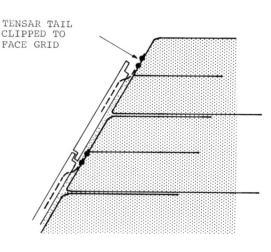

TENSAR TAIL
CLIPPED TO
FACE GRID

Fig 15

FUTURE DEVELOPMENTS USING TENSAR GEOGRIDS

Surface treatment of the steep embankment type
of blast protection has, to date, consisted of
natural vegetation cover. In general this is
the most desirable form of treatment giving the
most natural appearance. There may be some
situations however where a harder finish is
required and there are several possible
alternatives.

A large aperture Tensar Geogrid may be fixed
close to the face and sprayed concrete applied
over the whole surface. This could be a useful
solution in desert areas where no vegetation
cover is possible and some protection is
required for the Tensar against extremely high
levels of ultra-violet radiation.

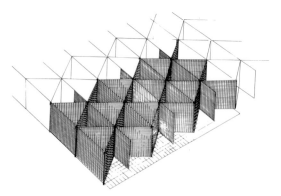

Fig 14

Work has been carried out on assessing the
performance of Tensar reinforced cement
composites under dynamic loading and positive
results are reported in Session 7 of this
Symposium.

Precast, Tensar reinforced concrete slabs may
also be used to cover the face. These may be
constructed with exposed-aggregate or other
featured surface to give a pleasant visual
appearance and a cast-in tail of Tensar would
allow them to be securely fixed to the
reinforced slope (Figure 15).
Heavier concrete panels have been used as
facings to vertical, Tensar reinforced soil
walls as have brickwork and blockwork. These
are almost certainly lower in cost than the
traditional hard surface of a reinforced
concrete wall.

Some low intensity blast situations are
contained by walls of the building housing the
equipment or by relatively thin walls outside.
The development of Tensar reinforced cement
composites will undoubtedly lead to
applications in this area.

CONCLUSIONS

Tensar-reinforced blast protection walls and
embankments are in many cases a feasible
alternative to traditional methods of
construction. Major cost savings are possible
and in some cases there are even greater
benefits.

Landtake may be of primary importance when
working between existing buildings or close to
other obstructions. A steep faced, vegetated
embankment can often be the most acceptable
solution.

In some areas of the world suitable fill for
embankment construction may be difficult to
obtain and concrete plus reinforcement almost

impossible. Tensar reinforcement is light, easily handled and installed and can reduce fill volume to a minimum. The Geocell mattress has proved to be an efficient system to use when a steep wall is required in areas where fine sand is the only fill material available.

The variety of Tensar reinforcement systems suggests that few applications would not warrant investigation of a Tensar alternative.

Construction of a steep sided geogrid retaining wall for an Oregon coastal highway

J. R. Bell, *Oregon State University,* and
T. Szymoniak and G. R. Thommen,
Oregon State Highway Division

The Oregon State Highway Division constructed a near vertical Tensar SR-2 geogrid reinforced wall to stabilize a landslide on the Oregon (USA) Coast. The wall was approximately 10 m high and 50 m long at the top. The site was adjacent to a park and special considerations were given to providing a natural appearance. Sod was placed behind the grid to establish vegetation on the face of the wall.

Initial construction was with forms which were removed and moved up as each segment of the wall was completed. Problems were experienced with the initial form design and a modified system was developed during construction. The grid wall was an economical solution to the special problems of the site.

INTRODUCTION

In December 1981 a slide occurred on an Oregon coastal highway closing the main entrance to the popular Devil's Punch Bowl State Park 25 km north of Newport, Oregon, U.S.A.

The soil profile consisted of 3.5 m of medium to yellow brown sand over a layer of soft gray silty clay varying in thickness from 0 to 3.5 m underlain by gray shale. The failure plane was at the clay-shale interface. Figure 1 shows a typical cross section of the slide and the failure plane.

Several alternatives were considered for stabilizing the slide. A nonwoven geotextile retaining wall and a geogrid reinforced wall had the lowest estimated costs. The geogrid wall was chosen for two reasons: 1) the geogrid retaining wall had the lowest estimated cost, and 2) the open face of the geogrid wall allowed establishment of vegetation on the wall to provide a natural appearance compatible with the state park.

The geogrid wall had the lowest estimated cost because it did not require a facing for protection from ultraviolet (UV) light as did the conventional geotextile wall. Other than the facing, the geogrid wall and the conventional wall had nearly identical estimated costs.

The wall is 21 m long at the bottom and 52 m long at the top. The top is stepped to fit the vertical curve of the roadway. The minimum height of the wall is 9 m. Common backfill is placed over the lower face of the wall to reestablish the natural ground surface. Above the natural ground line sod is placed between the gravel backfill and the geogrid facing. The sod was believed to be the most economical way to establish vegetation on the wall. To accommodate growth, a dirty backfill (Class B) was placed in the first 0.6 m behind the sod.

GEOGRID WALL DESIGN

Tensar® SR-2 was selected for the wall reinforcement grid. The backfill materials were graded crushed basalt with 50 mm maximum size; the general backfill material had a maximum of 10 percent fines, and the Class B material had approximately 20 percent fines. Specifications required at least 95 percent of standard optimum dry unit weight (AASHTO T99). The bulk density and angle of internal friction for the backfill were assumed to be 22 kN/m^3 and 40°, respectively.

To limit possible creep, the working stress for the geogrids was taken as 40 percent of the ultimate strength. Because of the open structure of the grids, the full soil friction was assumed at the soil-geogrid interfaces.

The wall was designed assuming the grids had to resist the active Rankine lateral earth pressures by the portion of the reinforcement extending beyond the theoretical Rankine failure surface. The method of analysis was described by Lee, et. al. (1) and Hausmann (2) for Reinforced Earth® walls and was modified for geotextile walls by Bell and his coworkers at Oregon State University (3,4). This method has been used by the U.S. Forest Service (5,6), New York Department of Transportation (7), Colorado Department of Highways (8), and others to construct geotextile walls in the United States.

Geogrid lengths and vertical spacings were calculated to provide minimum safety factors of 2.0 for dead load only, and 1.15 for dead load plus live load. The reduced factor with live loads was allowed because: 1) after construction, truck traffic would be limited to recreational vehicles and an occasional service vehicle; and, 2) the allowable working load included a safety factor of 2.5 against a short-term grid failure.

198

Polymer grid reinforcement. Thomas Telford Limited, London, 1985

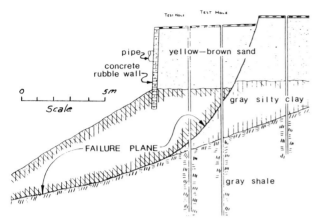

Figure 1. Geological section

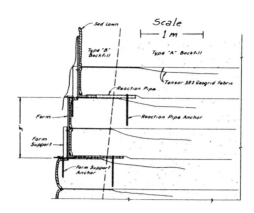

Figure 4. Typical form installation

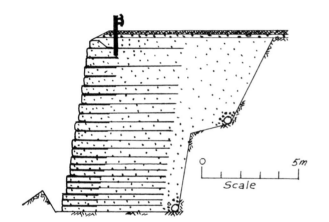

Figure 2. Geogrid wall section

Figure 5. Fabricating geogrid panels

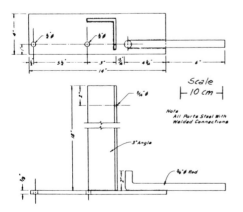

Figure 3. Form support

Figure 6. Original forms in place

For appearance and construction considerations, the wall was detailed with 0.9 m steps. Each step was set back 150 mm from the one below to give the wall an average batter of 1:6. The lower three layers were given reinforcement spacings of 0.3 m, the mid-height layers spacings of 0.45 m, and the top two layers reinforcement spacings of 0.9 m. To give a uniform appearance the geogrids were folded back into the backfill at mid-layer height for the top two layers. This fold was anchored a distance of 1.5 m into the backfill. This anchored distance was the same as the 1.5 m overlap embedment used for each layer.

The geogrid reinforcment lengths were 4.9 m. This length was required at the top for pullout resistance and at the bottom for resistance to horizontal sliding.

CONSTRUCTION

To keep the costs of the geogrid wall competitive, it was necessary to select a simple effective method of supporting the face during construction. Because of the steep site, wall geometry, and the need to operate equipment in front of the wall, scaffolds were not considered practical for this wall. Instead, movable self-supporting forms as used on geotextile walls (5,7,8) were proposed.

This wall required forming 0.9 m steps. Experience on a geotextile wall in Glenwood Canyon, Colorado, indicated that the simple movable forms previously used were not suitable for layers greater than about 0.4 m (9). Therefore, a forming system was suggested that incorporated the concepts of the previously used geotextile forms, but had special features to allow for thicker layers. This forming system is illustrated in Figures 3 and 4.

The suggested form consisted of a 0.9 m by 2.45 m sheet of 19 mm plywood held in place by a form support. The form support was anchored in the backfill. There was concern that if the form support base extended into the backfill far enough to provide stability, friction would make it very difficult to pull the base out at the completion of the layer. Therefore, a sacrificial reaction pipe was anchored in the backfill, and the rod on the form support was inserted into the pipe. Since there was little friction on the form support base, an anchor rod was used to provide lateral resistance.

As shown by the typical installation in Figure 4, it was anticipated that the forms for a completed step would be left in place while the next higher step was constructed. The lower form would add stability to the upper form and help maintain vertical and horizontal alignments. When the upper step was completed, the lower forms would be removed and moved up to form the next step. It was believed this system and procedure would be expedient and stable for the 0.9 m steps.

The general procedure followed by the contractor in the early stages of the construction was:

1. Prefabricate 3 m by 4.9 m geogrid panels.
2. Set the proposed forms at gradeline.
3. Lay out prefabricated grid panels, drape the grid over the forms, and secure the panels with anchor pins.
4. Secure panels to one another at the face.
5. Place backfill in 0.15 m lifts to desired layer thickness.
6. Place sod in position behind the geogrid and place Class B backfill.
7. Pull back and secure grid overlap.
8. Repeat Steps 3-7 until the top of the form is reached, then remove forms and move up for the next layer.

Figures 5 through 10 illustrate this procedure. Figure 5 shows a worker securing the strips of geogrid into panels and splicing the ends of the grid with No. 3 rebar. A masonry circular saw was used to cut the panels to length. Figure 6 illustrates the initial forming system with the grid draped over the form. Figure 7 is an overview of the wall construction which shows the restricted space and the placement of the backfill. Figure 8 shows a worker hanging sod strips on the forms, and shows the space left for the dirty Class B backfill. In Figure 9 workmen are pinning the overlap. The completed wall is shown in Figure 10.

As the wall gained in height, problems began to occur with sagging and bulging of the wall face. This was due to the excessive flexibility of the forms and the loss of Class B backfill through the grid where sod was not used between the geogrid and the backfill. Between the times when the forms were removed and when the face was covered by common backfill significant amounts of the fine Class B backfill fell out from behind the grid. Where sod was placed against the geogrid reinforcement, the fines were retained and the wall face was near vertical. The bulging problems were not deemed important in the lower layers, since the layer would be covered. However, the problem resulted in the contractor modifying the forming method before constructing the higher layers.

The combination of the thin plywood and short form supports on wide centers resulted in deflection of the forms. An even more serious problem resulted from the loss of support from under the forms. It was expected that the forms for a completed 0.9 m step would be left in place until the forms above were set and at least the first lift of that step was in place. The contractor elected, however, not to follow this procedure and moved the forms as each 0.9 m step was completed. Also, the contractor used plastic rather than steel reaction pipes. This resulted in the forms being dependent on the support of the previous layer directly under the metal plate of the form support, see Figure 4. Without the lower form in place, the slight inevitable bulging of the face resulted in tipping of the form support. Loss of the backfill through the grid compounded the problem and with the form support stiffened only

Figure 7. General view of construction

Figure 10. Completed geogrid wall

Figure 8. Hanging sod on forms

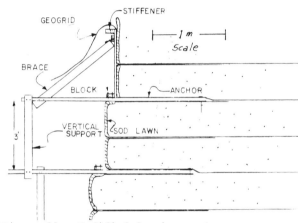

Figure 11. Modified forming system

Figure 9. Pinning geogrid overlap

Figure 12. Construction with modified forms

201

by the plastic reaction pipe, the form tipped even further. Loss of backfill material also reduced the effectiveness of the form support anchor.

The contractor's solution to the forming problem is shown in Figures 11 and 12. The forms were stiffened and braced against a 50 mm by 100 mm member anchored 1.2 m into the backfill. The protruding end of the horizontal anchor was supported by a vertical member. The bottom of the 19 mm plywood form was held in place by a 50 mm by 100 mm block nailed to the anchor support. At least 3 braces were used on each 2.45 m forming unit. The new forming system required considerably more time to construct, but did provide a stable face to build against.

COSTS

The engineers estimated the project cost to be $165,802 (U.S. dollars), the low bid was $166,328 and the actual cost was $183,395. For the wall items of grid, backfill, and sod facing, the corresponding values were $81,710, $97,146, and $85,496, respectively. This cost translates to $255 per m^2 of wall face. The in-place costs for the geogrid including the grid material, forming, and handling were $6.60 per m^2 of grid and $120 per m^2 of wall face.

EVALUATION AND RECOMMENDATIONS

The geogrid wall appears to have stabilized the site. The sod facing was growing and the appearance was very satisfactory at the end of construction; however, lack of irrigation killed much of the sod in a few weeks. Where vegetation is to be established on a wall face, adequate maintenance must be provided.

Geogrid walls are economical and have great potential; however, improvements in construction techniques are necessary to fully utilize their potential. At suitable sites, scaffolding may be the solution to the forming problems. In other situations, a modification of the movable forms suggested for this project are recommended. The following modifications are proposed:

1. Stiffen the plywood along the top edge and secure adjacent plywood sheets to each other with battens.

2. Lengthen the upright on the form supports and use at least three form supports on each 2.45 m form section.

3. Eliminate the reaction pipe and all anchor pins and extend the base plate of the form support 1 m into the backfill.

4. Weld rings on the short end of the form support base plate so mechanical aids can be used, if necessary, for extraction.

5. Use backfill significantly coarser than the grid openings, or use a fine mesh grid or geotextile behind the face of the wall.

With these changes, the forms may be removed and moved up with each layer.

ACKNOWLEDGEMENTS

The project was constructed as an FHWA Experimental Features Project. The wall was built by Dan D. Allsup, Contractor, Eugene, Oregon. Special acknowledgements are due Chuck Elroy, the Project Manager and Claudious Groves, the construction inspector. The authors also wish to acknowledge James Paul of Netlon, Limited for providing advice during the early stages of the construction. Finally, special thanks to Laurie Campbell for her expert and expeditious preparation of the manuscript.

REFERENCES

1. Lee, K.L., Adams, B.D. and Vagneron, J.M.J., "Reinforced Earth Retaining Walls," Journal of the Soil Mechanics and Foundations Division, ASCE, Vol. 99, No. SM10, 1973, pp. 745-763.

2. Hausmann, M.R., "Behavior and Analysis of Reinforced Soil," Ph.D. Thesis, University of New South Wales, Kensington, Australia, 1978.

3. Bell, J.R., Stilley, A.N. and Vandre, B., "Fabric Retained Earth Walls," Proceedings, 13th Annual Engineering Geology and Soils Engineering Symposium, Moscow, Idaho, 1975.

4. Whitcomb, W. and Bell, J.R., "Analysis Techniques for Low Reinforced Soil Retaining Walls and Comparison of Strip and Sheet Reinforcements," Proceedings, 17th Engineering Geology and Soils Engineering Symposium, Moscow, Idaho, 1979.

5. Bell, J.R. and Steward J.E., "Construction and Observations of Fabric Retained Soil Walls," Proceedings, International Conference on the use of Fabrics in Geotechnics, Ecole Nationale Des Ponts et Chaussees, Paris, Vol. 1, April 1977, pp. 123-128.

6. Steward, J. and Mahoney, J., "Trial Use Results and Experience Using Geotextiles for Low-Volume Forest Roads," Proceedings, Second International Conference on Geotextiles, Las Vegas, Vol. III, August 1982, pp. 569-574.

7. Case Histories, Reinforced Earth Walls, Earth-Fabric Wall, New York State Department of Transportation, Soil Mechanics Bureau, Albany, New York, March 1981.

8. Bell, J.R., Barrett, R.K. and Ruckman, A.C., Geotextile Earth Reinforced Retaining Wall Tests: Glenwood Canyon, Colorado, Paper presented at 62nd Annual Meeting, TRB, Washington, D.C. 1983.

The design and construction of a reinforced soil retaining wall at Low Southwick, Sunderland

D. R. Pigg and W. R. McCafferty, *Tyne and Wear County Council*

This paper describes the design and construction of a retaining wall using materials and methods, which at the time, were largely untried.

The wall is 100 metre long, up to 3.5 metres high and supports an access road with associated vehicle parapet. Factors influencing the choice of a polymer grid reinforced soil wall were a variable depth and nature of fill, an adjacent warehouse on 20 metre piles and minimal length availability for the reinforcing elements.

Construction of the wall took eight weeks with inexperienced labour. In the 2½ years since construction recorded movements have not exceeded 30mm horizontally or 50mm vertically.

THE SITE

The subject of this paper was instigated by a design brief which called for a retaining wall to support a new roadway. This would provide access to a future industrial development to the west of the centre of Sunderland on the north bank of the River Wear.

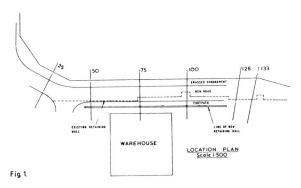

Fig. 1.

A large proportion of the site was occupied by the remains of a concrete products factory which had occupied two separate levels. These levels were separated by an existing reinforced concrete retaining wall, with an access road at the higher level which was to form part of the new roadway. The retaining structure was required to support the new access road, with the existing retaining wall remaining in position. The length of wall required was a little short of 100 metres and the maximum retained height was 3.5 metres. Approximately 2 metres away from the front face of the proposed retaining structure, a new D.I.Y. warehouse was to be constructed before wall construction could commence. This factor would greatly affect site access to the highway scheme and complicate construction of the new wall. In addition to the retaining structure, provision had to be made for the installation of a type

P2 vehicle/pedestrian parapet at the rear of the footway adjacent to the wall, see fig. 3.

GEOTECHNICAL CONSIDERATIONS

At the initial design stage, our own soils information was not available but a soils report for the warehouse project was in our possession. It was decided to undertake an initial reinforced concrete cantilever wall design based upon the available information pending the results of our own site investigation. The lower level of the site was overlain with concrete to a depth of 300mm beneath which variable fill material extended to different depths. In some areas, this fill was in the form of a clay, brick, rubble and concrete, whereas in other areas the fill comprised black ash and gravel. This variable fill extended over the full area of the warehouse site extending to depths between 3.7 metres and 6 metres. Below the fill layer, a fairly thick band of soft to firm clay covered the site which, in some areas overlayed a stiffer clay, and in other areas, a rocky sand layer, above rockhead at some 8 metres below ground level.

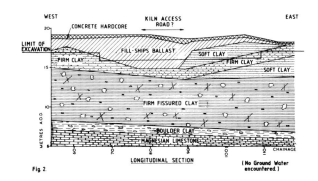

Fig. 2.

Geotechnical section based upon Tyne and Wear County Council Commissioned Survey.

Polymer grid reinforcement. Thomas Telford Limited, London, 1985

Our own soils investigation revealed that maximum formation level for the R.C. wall should be not less than 2 metres below existing ground level, thus founding in the upper region of the soft to firm clay, see Fig. 2. An allowable bearing capacity of 70 kN/m^2 was calculated for this layer, and because of the variation in fill material, it was felt that differential settlement of any retaining structure would be inevitable. We were informed that the piles for the warehouse project had been driven to a depth of some 20 metres before reaching bedrock, a factor confirmed by our own soils investigation. Initial predictions of differential settlement based upon our own survey were in the order of 200mm. As a result of this new information, the reinforced concrete wall design was abandoned at an early stage and alternative methods investigated.

SELECTION OF WALL TYPE

The problems associated with both the low bearing capacity of the soil and the expected differential settlement could have been overcome by the design of a piled reinforced concrete wall. However, due to the excessive lengths of the piles used for the warehouse construction this solution was abandoned for reasons of expense. In addition, access for construction of the wall and the vibration and noise effects of piling on the new D.I.Y. warehouse would have caused severe problems.

The natural choice to deal with the prevailing ground conditions, was a reinforced earth retaining wall, none of which had been constructed on a highway scheme in the County of Tyne and Wear. The main reasons for this choice included low associated bearing capacity of subgrade; a 'flexible construction'; and of prime importance, the finished structure would have very little drawdown effect on the piled foundations of the D.I.Y. warehouse.

The design parameters peculiar to our retaining structure were: a maximum free height of retaining wall of 5.5 metres, comprising difference in level of 3.5m and a depth to subgrade of 2m, supporting a total width of single carriageway and footways of 16 metres. In addition to the retained height of soil, the structure was also required to support full H.A. loading as a working load case. It is general practice to consider this H.A. loading surcharge as an additional 600mm of backfill for the purpose of analysis. A vehicle/pedestrian barrier was required to contain vehicles at the higher level to the rear of the footway and also to protect the adjacent D.I.Y. warehouse. As previously mentioned, the safe bearing capacity of the ground below the reinforced earth wall is 70 kN/M^2 and the new wall was to be constructed between an existing retaining wall and the warehouse property, see Fig. 3. The condition of this existing retaining wall was unknown and it was to be shored during construction.

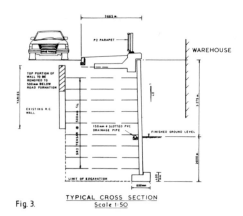

Fig. 3. TYPICAL CROSS SECTION
Scale 1:50

Before the design of the reinforced earth wall could commence, a problem concerning the length of the reinforcing elements had to be resolved. B.E.3/78 which governs the design, states that the minimum length of reinforcing elements should be the greater of 80% of the retained height or 5 metres. As the available space between the existing and new walls was a minimum of 3.1 metres, we could not fulfill this requirement over the full length of the wall. However, after reading an article describing the use of a polymer reinforcement fabric to construct a wall at Newmarket Silkstone Colliery, Ref 1, it was felt that this problem could be overcome. Unlike the normal strip system of reinforcing elements, the Tensar grid system promotes good aggregate interlock between the plastic "fingers" and the possibility of corrosion is completely removed. As BE 3/78 was written before the advent of grid reinforcement and is based upon the "strip" reinforcement principle, it was felt that this minimum length criterion could be relaxed. The design was commenced assuming the use of Tensar SR-2 reinforcing elements and frictional fill material. The top of the existing retaining wall was to be removed where the Rankine active wedge was at its widest in order to accommodate the calculated grip length.

THE DESIGN PROCESS

The general concept of reinforced earth is now well understood. The soil is reinforced using layers of material placed horizontally. These layers are capable of transmitting tensile forces which effectively bind the soil structure to form an integral mass.

The design of a reinforced earth structure consists of the evaluation of lateral stresses within the soil over the full height of the structure and the provision of horizontal reinforcing elements to dissipate these stresses, by friction, to the soil beyond the Rankine active zone. The required length of embedment outside the Rankine active zone, or grip length, is generally shorter at the top of the soil structure than at the bottom. Thus for equal length reinforcing elements there is a balance between a wide Rankine active wedge zone and short grip length at the top and a narrow wedge zone and long grip length at the bottom.

A frictional fill material was chosen for its good internal friction and drainage characteristics. The former enables short grip lengths to be used while good drainage reduces the lateral pressures which have to be accommodated. For the purposes of analysis the following properties were assumed:-
cohesion = O kN/m^2; bulk density = 2038kg/m^3; effective angle of internal friction = 35 degrees.

A vertical spacing of 500mm for the reinforcing elements was chosen as it optimised the grip length properties of the Tensar SR-2 within a narrow site and suited the size of facing units selected. The force in each reinforcing element was calculated and, after summation, the factor of safety against overall tensile failure was computed using the permissible working load supplied by Netlon. The grip length required to resist the tensile force in each reinforcing element was then calculated with a factor of safety of 2.5 against local bond and tensile failure being applied. The factor of safety against overall bond failure was then calculated.

The reinforced "block" of frictional fill and reinforcing elements was then considered as a mass gravity wall supporting not only its own dead and imposed loadings but also the fill material placed behind it. Factors of safety against sliding, bearing capacity failure, overturning and slip circle failure were then calculated. In addition sliding on any horizontal plane within the frictional fill mass and wedge stability above any horizontal failure point were also considered. None of the aforementioned calculations resulted in a factor of safety less than 2.0.

PRACTICAL CONSIDERATIONS

It was decided to use cruciform shaped precast facing units reinforced on each face with a structural mesh fabric and with an exposed aggregate finish on the front face. At this early stage, a decision was taken to cast Tensar "starters" into the rear of the facing units, these "starters" to be lapped to the main reinforcement, in order to obtain a simple connection between facing units and the main Tensar grids. The panel reinforcement was designed as a continuous beam carrying the tensile forces from the reinforcing elements. A plain concrete strip footing was detailed for the bottom facing units to bear upon and a small upstand was cast on the top surface of this concrete footing to aid the location of units in line and level. The front of the facing unit was placed against this upstand and the unit was inclined towards the backfill at a slope of 1:20. This detail was adopted to allow the facing unit to rotate about its lower edge and move outwards during filling and compaction adjacent to its rear face. The joints between facing units were detailed as being a half tongue, half groove arrangement

carrying horizontal shear, with dowels being cast into the "arms" of the cruciform units to aid location and interlock. To enable movement to take place, the joints between facing units were to interlock such that a space of 20mm existed at both the front and rear faces. A cork packing was placed in this space in the horizontal joints only at the rear of the facing units. A foam strip was then inserted to all the rear joints, both horizontal and vertical, the purpose of which was to prevent the wash out of fines from the frictional fill, see Fig. 4. The main layers of Tensar reinforcing grids were fixed to the Tensar starters, cast into the facing units, using galvanised steel rods. Where a step was required in the strip foundation, the footing was made wholly discontinuous to aid differential longitudinal settlement.

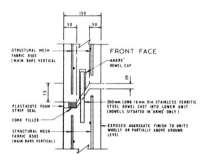

Fig. 4.
TYPICAL HORIZONTAL JOINT TO FACING UNITS
Scale 1:5

An area of poor ground at depth existed over a length of some 40 metres. It was felt that taking the reinforced earth wall facing units to this depth was uneconomic but that the frictional fill should still be reinforced. The solution adopted was to provide Tensar SR-2 running parallel to the wall to form a hammock arrangement which was continued up into the reinforced earth wall by interleaving the layers of Tensar over a depth of 1.5 metres.

Regarding the vehicle/pedestrian parapet, it had been decided at an early stage in the design that the parapet foundation should be totally independent of the wall facing units to avoid any vehicle impact load being transferred to the wall. The parapet base layout was designed to fulfill this criterion and an additional use for the Tensar was adopted to enhance the base stability. It was imperative that the parapet base should be accommodated within the proposed footway and be of reasonable dimensions. A width was thus selected such that the overturning moment due to vehicle impact was resisted by the dead weight of the base providing a factor of safety of unity under parapet impact loading (the wheel loads associated with impact would increase this value). To increase this factor of safety to a more reasonable figure in the order of 1.4, at ultimate, a grid of Tensar was cast into the bottom of the slab and lapped to the main reinforcement. The free end of this Tensar grid was then passed through the frictional backfill and into the carriageway sub-base to prevent pullout. In addition to increasing the factor of safety against overturning, the Tensar was also required to enhance the resistance to horizontal movement during vehicle impact. The parapet base slab was formed such that it acted as a capping beam to the reinforced earth wall and its profile was tailored to overhang the top of the uppermost facing units, thus giving weather protection to both the facing units and the fill material. As the main reason for the selection of a reinforced earth wall had been the expected differential settlement, it was desirable that the aesthetic appearance of the structure should be protected during the service life of the wall. To achieve this it was decided to cast the uppermost Tensar "starter" in the top facing unit into the parapet base slab. The free length of the Tensar and the overhang of the parapet base was calculated such that should the top facing unit be displaced either downwards or outwards, the front face of the unit would not protrude beyond the front face of the parapet base. Therefore, rather than the front face of the units being the focal point to an observer, the front face of the parapet base, which would always remain fixed, would perform this role. It was felt, however, that the construction of the parapet base should be one of the final operations in order that initial wall settlement would have taken place. To allow some differential movement the parapet base/capping units were cast in 3m lengths and separated by transverse dowelled joints while the steel parapet was of bolted construction.

THE CONTRACT

The contract drawings and bills of quantities were prepared in accordance with the Specification for Road and Bridgeworks, the Standard Method of Measurement for Road and Bridgeworks and BE 3/78. The Select List of Tenderers for the scheme comprised five civil engineering contractors in addition to the Borough of Sunderland Public Works Dapartment. The lowest Tender Price for the full scheme of £120,000 submitted by the Borough of Sunderland, included a cost of £57,000 for the construction of the reinforced earth wall and the vehicle/pedestrian parapet. The contract was awarded to the Borough of Sunderland in January 1981 with construction commencing in March 1981.

CONSTRUCTION

The contractor chose to use a burnt colliery shale material which complied with the specification for frictional fill in BE3/78. In order to verify the design assumptions, the University of Newcastle upon Tyne was commissioned to carry out 300mm direct shear box tests on the material and on its interaction with Tensar SR-2. The results of these tests are contained in Table 1.

Table 1

Sample	Dry Density (kg/m³)	Moisture Content (%)	Normal Stress (kN/m²)	Peak Shear Stress (kN/m²)	Angle of Friction (∅) (Deg.)
Burnt Colliery Shale	1909	11	124	266	65
Burnt Colliery Shale + Tensar	1909	11	124	177	55
Design Assumptions	1836	11	—	—	35

These results were better than those assumed for the design and, accordingly, the material was approved. The high sulphate content of the material necessitated the use of Class 3 sulphate resisting concrete throughout and in addition the rear of the facing units were painted with a bitumen emulsion after erection.

As previously mentioned, access available for the construction of the wall was extremely limited and the existing retaining wall had to remain intact wherever possible. Excavation for the strip foundation and wall mass was carried out using a crawler back acter machine. The excavation commenced at the west end of the site and the only means of access at this lower level was from the east, along the proposed line of the new wall. After removing a bucketfull of material from the foundation trench, the excavator had to slew through 180° with the bucket held at full vertical reach, before discharging into a tipper directly behind. The existing retaining wall was supported as the trench and wall excavations progressed eastwards, a factor which added to the access difficulties.

The concrete was placed to the strip foundations, the first "course" of facing units erected, then the placing of the reinforcement and backfilling and compaction of the frictional fill was carried out in 10 metre increments along the full length of the wall. Only when the wall was complete at any one level, was the next level of facing units erected and backfill placed. It was decided to carry out construction in this way for several reasons: to reduce differential settlement between adjacent lengths of wall; to allow vehicle access at low level, (rather than from the top of the existing wall for the purpose of depositing frictional fill while the existing wall was undermined by the excavation) and to ensure, as much as possible, that the facing units interlocked as desired. The complete operation, i.e. the construction of the wall to parapet base formation level took some eight weeks, a time which could have been reduced considerably had possession of the site been obtained before construction of the warehouse project.

Plate 1   Typical construction scene

At the present time, some 2½ years after construction, vertical and horizontal movement of the wall has almost ceased. The total movement recorded to date has not exceeded 50mm vertically or 30mm horizontally.

In conclusion, this wall has proved to be a practical and economic solution for the site conditions. It was constructed by labour with no previous experience of this type of structure in a reasonably efficient manner considering the physical restrictions of the site.

REFERENCES

Heywood, P. (1980) Soil Reinforcement : UK Designers step out of shadows. Contract Journal Aug. 7th 1980.

BE3/78   Reinforced earth retaining walls and bridge abutments for embankments. Department of Transport 1978.

T.R.R.L. Supplementary Report 457.   Reinforced earth and other composite soil techniques.

ACKNOWLEDGEMENTS

1.   P. Morris B.Sc., C.Eng., M.I.C.E., Executive Director of Engineering, Tyne and Wear County Council who was the Engineer for the works.

2.   Dr. C.J.F.P. Jones C.Eng., F.I.C.E., Assistant Director of Structural Engineering Unit, West Yorkshire Metropolitan County Council for his assistance during the planning and design stage.

Constructional details of retaining walls built with Tensar geogrids

Although each reinforced soil retaining wall is unique, elements forming each structure are often similar.

For reasons of economy and serviceability consistent reliable constructional details are required. This paper provides details of some structural systems which have been shown to be efficient when used with retaining walls built with Tensar Geogrids.

C. J. F. P. Jones, *West Yorkshire Metropolitan County Council*

INTRODUCTION

Although the form of any soil structure may vary, the structural elements required to produce different structures are often similar. In keeping with other forms of construction, poor details will produce an inadequate structure, whilst good detailing will ensure success. The difference between good and poor detailing can be subtle and weaknesses and difficiencies may only become apparent during the construction phase or later during the life of the structure. In either event remedial measures may be costly.

Many situations produce common structural problems. The constructional details shown below, although not necessarily the best possible, have been shown to be efficient and effective in general conditions. In some cases the details represent a compromise between constructional efficiency, structural requirements and aesthetics.

FOUNDATIONS AND DRAINAGE

Normally the need for foundations is minimal with reinforced soil, a mass concrete strip footing being sufficient. If the bearing capacity of the subsoil beneath the footing is insufficient additional support may be provided by conventional means such as by the use of piles; alternatively a geogrid foundation mattress may prove suitable, Fig 1. The latter construction has the added advantage of providing subbase drainage to the reinforced structure. An alternative drainage detail used successfully beneath reinforced soil structures is shown in Fig 2. Also shown are details of the sand drainage layer behind the facing units which has been shown to be successful when the bulk fill contains a high percentage of fine material as in the case of pulverised ash. When a well graded fill is used within a reinforced soil structure vertical drainage behind the facing

is not usually required although care must be taken to avoid erosion of fines through joints between facing units.

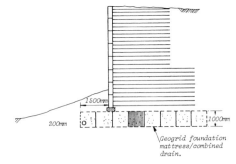

Fig 1. Combined Geogrid Foundation and Drainage Blanket

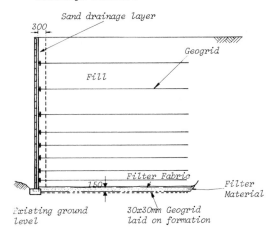

Fig 2. Drainage Beneath Reinforced Structure

Polymer grid reinforcement. Thomas Telford Limited, London, 1985

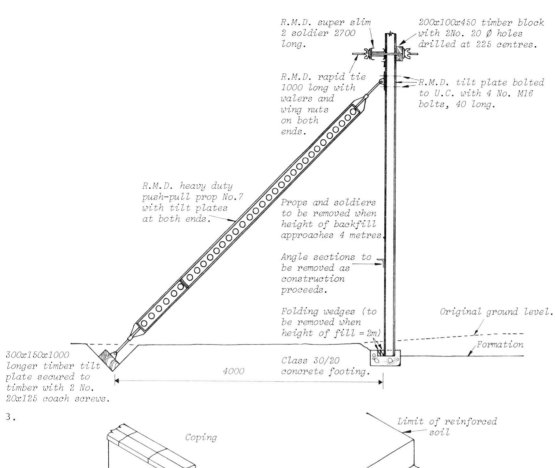

R.M.D. super slim 2 soldier 2700 long.

200x100x450 timber block with 2No. 20 Ø holes drilled at 225 centres.

R.M.D. rapid tie 1000 long with walers and wing nuts on both ends.

R.M.D. tilt plate bolted to U.C. with 4 No. M16 bolts, 40 long.

R.M.D. heavy duty push-pull prop No.7 with tilt plates at both ends.

Props and soldiers to be removed when height of backfill approaches 4 metres.

Angle sections to be removed as construction proceeds.

Folding wedges (to be removed when height of fill = 2m)

Original ground level.

Formation

300x150x1000 longer timber tilt plate secured to timber with 2 No. 20x125 coach screws.

4000

Class 30/20 concrete footing.

Fig 3.

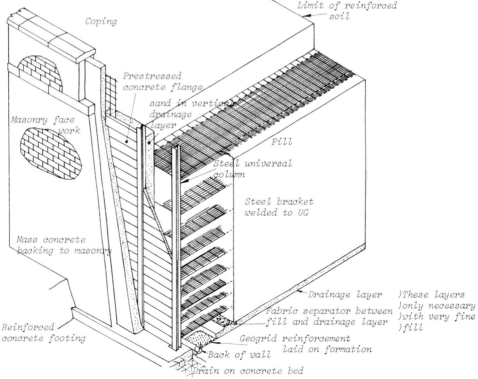

Limit of reinforced soil

Coping

Prestressed concrete flange

sand in vertical drainage layer

Masonry face work

Fill

Steel universal column

Steel bracket welded to UG

Mass concrete backing to masonry

Drainage layer

Fabric separator between fill and drainage layer

Reinforced concrete footing

Geogrid reinforcement laid on formation

)These layers
)only necessary
)with very fine
)fill

Back of wall

Drain on concrete bed

Fig 4.

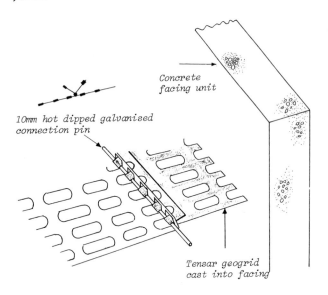

Concrete
facing unit

10mm hot dipped galvanised
connection pin

Tensar geogrid
cast into facing

Fig 5.

FACINGS

The facings used in reinforced soil structures
must conform to the technical needs associated
with the three basic construction systems,
Jones (1984). With both the concertina and the
telescope methods of construction an elemental
form of facing is usual, formed either in steel
or in reinforced concrete. An elemental facing
may also be used with the sliding method of
construction, although a stiff full height
facing formed in prestressed concrete has been
shown to be economic and aesthetically pleasing,
Jewell and Jones (1981). Recently an hybrid
facing has been developed, using full height
structural members and discrete elements, this
facing has been used successfully with Tensar

geogrid. The method of erection and structural
arrangement is shown in Figs 3 and 4.

CONNECTIONS - FACINGS AND GEOGRIDS

The connection between the facing and the
reinforcing material can influence the
distortion developed in the structure during
construction, to reduce distortion to a minimum
there should be no slackness within the
connection. In addition a full strength
connection is required so as to make the most
effective use of the reinforcement. A common
detail between the facing and the reinforcement
is a bolted connection, the bolt passing
through a hole formed in the reinforcement.
Although effective, this method reduces the
total carrying capacity of the reinforcing
element by up to fifty per cent.

An attribute of Tensar Geogrid is that full
strength connections may be obtained easily and
economically. With the telescope method of
construction, short lengths of geogrid can be
cast into the facing element to which the main
geogrid sheet may be attached using a simple
galvanised pin, Fig 5.

The connection used with the sliding method of
construction differs in that the reinforcement
is attached to the facing in such a manner that
movement in the downward vertical plane is
possible, Fig 6.

In the case of the hybrid structure shown in
Fig 4, two alternative connection systems are
possible, Fig 7(a,b). In figure 7(a) the
geogrid is connected to the facing by a
galvanised steel bar passing in front of the
steel columns, the grid passes through the
facing and vertical settlement within the
structure is accommodated on the telescope
principle, Jones (1984).

The alternative method of connection is shown
in figure 7(b) where the grid is attached via

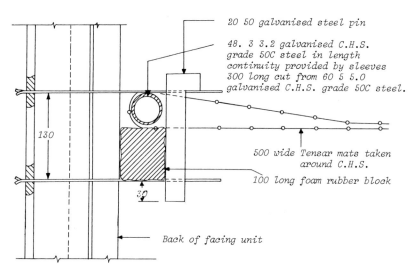

20 50 galvanised steel pin

48. 3 3.2 galvanised C.H.S.
grade 50C steel in length
continuity provided by sleeves
300 long cut from 60 5 5.0
galvanised C.H.S. grade 50C steel.

130

500 wide Tensar mats taken
around C.H.S.

100 long foam rubber block

30

Back of facing unit

Fig 6

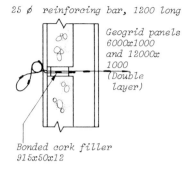

25 ∅ reinforcing bar, 1200 long

Geogrid panels
6000x1000
and 12000x
1000
(Double
layer)

Bonded cork filler
915x50x12

(a) Alternative fixing
     for facing.

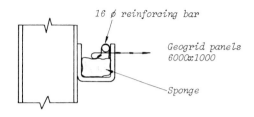

16 ∅ reinforcing bar

Geogrid panels
6000x1000

Sponge

(b) Alternative fixing
     for facing.

Fig 7.

a reinforcing bar held in ∪ lugs welded to the back of the steel columns. Vertical settlement of the connecting bar and geogrid is possible and the connection fulfills the requirement of the sliding method of construction. With this detail the bonded cork gaskets placed between the infill planks shown in figure 7(a) are not used.

CONCLUSION

Geogrid reinforcing material can be adopted for use with any of the current construction systems and may easily be incorporated into new details. Care must be taken not to damage the material during construction, either with construction plant or with sharp fill. The light weight of the material must also be considered and it is particularly important to eliminate slackness from the material and connections during fill placing.

REFERENCES

Jones C.J.F.P.(1984) "Design and Construction of Reinforced Soil Structures". SERC/Netlon Symposium on Polymer Grid Reinforcement in Civil Engineering. London.

Jewell R.A. and Jones C.J.F.P. (1981) "The Reinforcement of Clay Soils and Waste Materials using Grids", p 701-707 Vol 3 X ICSMFE, Stockholm.

The design and construction of polymer grid boulder barriers to protect a large public housing site for the Hong Kong Housing Authority

L. Threadgold, *Leonard Threadgold Geotechnics Ltd,* and D. P. McNicholl, *Hong Kong Housing Authority*

The shortage of sites for housing in Hong Kong has required the construction of estates on hillsides where boulders and rock exposures are a prominent feature. Protection of sites against potential boulder or rock falls requires measures to prevent boulder movement or barriers to intercept falling boulders. This paper concerns the design of protective barriers on a site in which the topography allowed their use, together with preventive measures, to achieve an economic solution.

Barriers constructed using rock-filled polymer grid mattresses which have both mass and flexibility were selected and preventive works comprising boulder removal, buttressing or erosion protection used.

1. INTRODUCTION

Hong Kong Housing Authority administers the largest public housing programme in the world. Since 1953 the Authority has housed over 2,000,000 people and the current construction programme provides for some 35,000 flats per year. Within Hong Kong there is a shortage of suitable housing sites and hence development has to take place in areas which frequently pose significant civil engineering problems. This paper describes one such area where extensive site formation works are required to create platforms in sloping terrain. These platforms are being used for the construction of residential blocks with associated roads and services for a projected 50,000 inhabitants.

The natural slopes which extend above the site are a potential source of boulder fall. In view of the exposure of the housing site and its occupants to any future boulder movements, Hong Kong Housing Authority commissioned a study to investigate the extent of possible boulder instability and movements and to advise on suitable protective and preventive measures.

Throughout the study there was close liaison between the Consultants, the Housing Authority and other Government Departments. This ensured that the work proceeded in a positive manner and took account of technical, practical and administrative requirements.

2. SITE DESCRIPTION

The study area is located on the southern slopes of Lion Rock ridge which dominates the skyline to the North of Kowloon peninsula in Hong Kong, see Figure 1. The area is bounded to the south by the new housing estate. The ground slopes at angles of 50° to 60° adjacent to the ridge line and reduces in a southerly direction to the range 0° to 30° at the foot of the slopes. Exposed rock masses on or close to the ridge

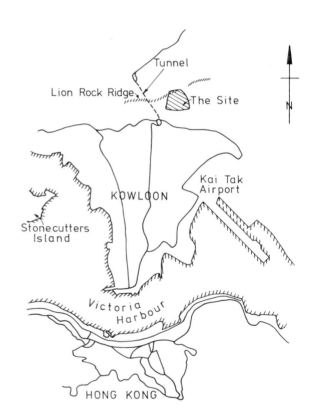

SITE LOCATION PLAN

FIGURE 1

Polymer grid reinforcement. Thomas Telford Limited, London, 1985

line exhibit overhangs and vertical surfaces.

The slopes are divided into a series of North-South trending valleys and ridges resulting from the erosion processes below the main ridge. The site is underlain by Granite, much of which has been weathered to considerable depth, although there are also exposures of fresh rock. Colluvium is present on the ridges and in the valleys. Many boulders exposed as a result of erosion of the soil matrix or derived from rock falls are evident on the slope surfaces. They range in size from less than a metre up to 30m in width although this latter size is exceptional.

At the time of the study, parts of the lower slopes were covered by squatter huts which obscured some boulders.

Grass, trees and shrubs obscured many of the ground features although periodically the vegetation has been removed by man or slope fires, thus allowing the unprotected ground surface to be eroded by wind and water.

3. BOULDER SURVEY

The size of the area from which boulders could fall was such as to render the task of identifying and stabilising all boulders practically unrealistic in terms of time and cost. Furthermore, rock exposures close to the Lion Rock ridge which have been the source of boulders in the past would probably continue as such and the task of stabilising these would also be time consuming, expensive and for this site, probably aesthetically unacceptable. The surveying and preventive option would take so long that it would not be completed ahead of building construction, thus putting the site at risk. It was clearly necessary for the Housing Authority to take positive action to protect the site, even though the source of the threat extended beyond the site boundaries.

For the foregoing reasons consideration turned towards a combination of protection and prevention with protection being of primary importance.

4. LITERATURE REVIEW

Currently available literature on the problem of boulder falls is relatively limited. Work has tended to concentrate on the stability of rock faces, remedial action to maintain stability or measures to protect features such as roads or buildings which would be threatened by any falls from these rock faces.

Work by Ritchie (1) in the United States of America provides a basis for rational design of protective works, with particular reference to highways threatened by rock face instability, but considers slopes no flatter than 38°. Solutions to the problems are presented in terms of the design of rock ditches and the location and height of associated fences. It is largely empirical in character and does not attempt to quantify the loads involved in stopping or retaining the boulders.

In India, problems with roads passing through the Himalayas have been studied by Bhandari (2, 3) of the Central Building Research Institute in Roorkee and protective measures such as fences, netting, trenches etc. designed. Fookes & Sweeney (4) have studied remedial and protective works in relation to rock faces whilst DeFreitas and Watters (5) have studied toppling failure, predominantly in relation to rock exposures. A paper by Mercer (6) deals with measures adopted to deal with rock falls in Tremadog, North Wales.

The papers most appropriate to this study are from a meeting of the I.S.M.E.S. in Bergamo, Italy on 20-21 May 1976, entitled "Rockfall Dynamics and Protective Works Effectiveness" (7). Papers describe experimental studies, model tests, computer models, mathematical work and field observations relating to rock or boulder movements. The papers consider the form of movement, factors affecting the distances of movement and the form of protective works using a site near to Lecco in Italy as the study area.

Camponuovo describes model tests for the slopes of St. Martino. He demonstrated the effectiveness of barriers a few metres high, particularly where the natural ground formed a series of gorges or "forced passages". He also observed that, in the absence of barriers, the finer material tends to form screes, whilst larger and rounder blocks tend to break away and roll further down hill.

Lied of the Norwegian Geotechnical Institute indicated that the maximum reach of a boulder fall could be defined using an angle of reach between the source of boulders on the talus slope and the valley floor. In his paper (8) he indicates an angle of 28° to 30°, but in a private communication to the Author a tentative angle of 23° to 25° was postulated for Norwegian conditions.

Other papers deal with mathematical models for boulder falls and employ coefficients of restitution to deal with the bouncing mode. The rolling mode is less easily dealt with.

It is clear that many factors influence boulder behaviour and the variables of any site would require an extremely large modelling and testing programme. Even then a large number of unknowns would remain.

The overall conclusion from the above papers is that the main factors of practical significance to any design are the path of any falling boulder and its velocity pattern along this path.

5. LOCATION OF PROTECTIVE WORKS

The ridge and valley terrain divides the area into catchments appropriate not only to water, but more importantly, in this context, to boulders. Study of the contoured plans of the area allowed the construction of "flow lines" for boulders perpendicular to the contours. These flow lines enabled protective barriers to be located in the most efficient position towards the bottom of each catchment to intercept the vast majority of boulder trajectories. There remained areas near to the site, however, where the flow lines would not be intercepted by the barriers. For the boulders in these areas simple remedial works such as erosion protection, buttressing or in some cases removal were required.

This approach gave a very practical method of efficiently locating barriers and minimising boulder survey and treatment work. Effort was therefore directed to the location and design of the barriers.

6. BOULDER VELOCITY

A major problem is the determination of the forces involved in stopping a boulder. This in turn is related to the size of the boulder, its velocity and the rate of deceleration which the obstruction imposes. Whilst an estimate of the size and shape of boulder to be dealt with could be made from the ground or aerial survey, determination of its velocity is particularly difficult in analytical terms. After considering several alternatives, a model consisting of a sphere rolling down a slope and hitting a series of steps of varying height and spacing was used. Figure 2(a)

This model appeared to predict reasonable relationships between slope angles, roughness and velocity. At a later stage, the concept of a cubic or hexagonal boulder was incorporated following a similar analogy, to predict velocities. Figure 2(b). The terminal velocity was used as the value for design purposes and typical relationships between step heights and spacings and boulder shapes are presented in Figure 3 (a) (b) (c) (d)

In practice the velocity would be influenced by other factors such as the dynamic modulus of the ground over which the boulder travels, the presence of vegetation, variations in slope and unforseen obstructions etc. It was felt that this work provided an upper bound velocity and a practical basis for design.

There remains an element of doubt in respect of this upper bound velocity, and hence it was felt that any solution adopted should reflect this uncertainty and be able to accommodate variations without failure.

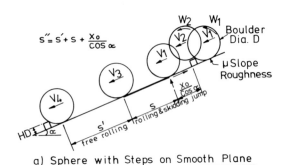

a) Sphere with Steps on Smooth Plane

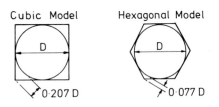

Equivalent Step Height
Cube=0·207D Hexagon=0·077D

b) Cube and Hexagon on Smooth Plane

BOULDER VELOCITY MODELS

FIGURE 2 (a) (b)

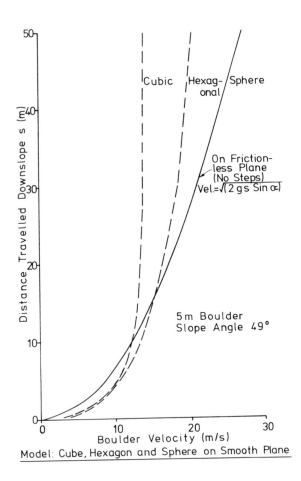

Model: Cube, Hexagon and Sphere on Smooth Plane

FIGURE 3 (a)

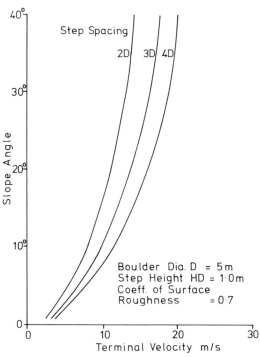

Boulder Dia. D = 5 m
Step Height HD = 1·0 m
Coeff. of Surface
Roughness = 0·7

Model: Sphere with Steps on Smooth Plane

FIGURE 3 (b)

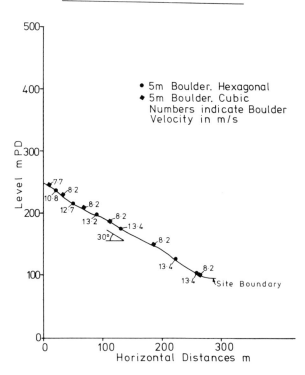

Velocity Model Catchment E2

● 5m Boulder. Hexagonal
♦ 5m Boulder. Cubic
Numbers indicate Boulder
Velocity in m/s

FIGURE 3 (d)

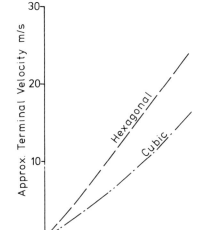

Approx. Terminal Velocity for Non-
Spherical Boulders

FIGURE 3 (c)

7. PROTECTIVE WORKS

In the past the design of remedial or protective works has proceeded in a largely intuitive or empirical manner. A wide range of solutions have been adopted on an ad-hoc basis to deal with the problems of most concern to the feature to be protected. Alternative solutions have included:

Trees or Vegetation
Netting or Wire mesh
Energy absorption areas
Fences
Trenches
Structural walls
Mass barriers

Initially, a toe-of-slope rockfall trap was considered with a geometry designed to bring boulders to rest by a controlled vertical impact. Preliminary estimates indicated that at a cost of some HK$9-12 millions this would be too expensive. Furthermore for this site, the use of traps or trenches had adverse implications for slope stability and drainage for the site below. Hence this solution was not used.

Structural concrete or steel barriers can only tolerate small deflections before failure. The forces required to stop a boulder are inversely proportional to deflection and are proportional to the square of the boulder velocity. Small deflections therefore imply high loads and these loads are sensitive to small changes in velocity. If structural barriers of this type fail the boulder would continue on its path.

Sacrificial fences and netting have been used to achieve some control but this has not been quantified. It is recognised that for sites which are restricted these are frequently the only practical solutions but for this site, space was not a major constraint.

The concept of a barrier having mass as its primary characteristic and deformation without failure as another was therefore favoured for this site. Within fairly broad limits, boulders of a given size travelling at a velocity greater than that estimated would cause greater movement but would not cause "failure".

8. BARRIER DESIGN PHILOSOPHY

The concept embodied in the barrier design was that following impact by a boulder, it would deform, in part within itself and in part by displacement of the base. The momentum would be transferred into the barrier and the whole retarded by the braking action of the friction between the barrier and the base.

Various momentum equations were derived which allowed the relationship between boulder velocity, boulder size relative to the barrier and the distance moved to be determined. From this work, the following relationships were derived.

$$S = \left[\frac{V_B}{1+R}\right]^2 \cdot \frac{1}{2g \tan \phi}$$

Where S = distance moved by the barrier

V_B = Velocity of Boulder

R = Ratio of Mass of Barrier to Mass of Boulder

g = acceleration due to gravity

ϕ = angle of shearing resistance along the base of the barrier.

also $F = \frac{(M_B V_B)^2}{M_B + M_P} \times \frac{1}{2S}$

Where F = Force applied

M_B = Mass of boulder

M_P = Mass of barrier

Examples of these relationships have been plotted in graphical form. Figures 4 (a) to 4 (c).

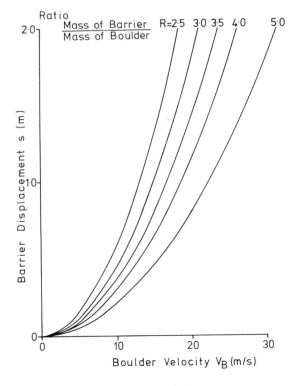

FIGURE 4 (a)

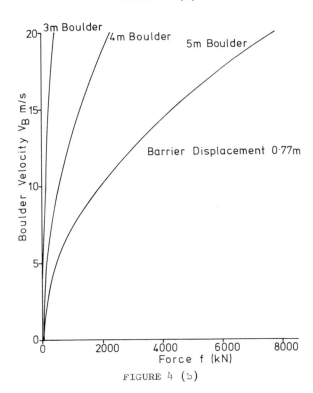

FIGURE 4 (b)

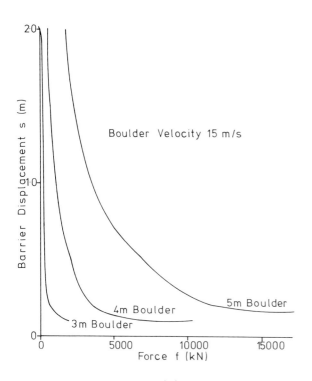

FIGURE 4 (c)

9. BARRIER STRUCTURE

The use of gabions appeared to fulfill the design requirements. The materials used would need to have sufficient strength and elasticity to allow the barrier mass to undergo substantial deformation without failure. The form of fabrication should be such as to facilitate the orientation of the various elements in a manner which would re-distribute loads effectively throughout the mass. The materials should also be durable under the prevailing conditions.

Polymer Grids filled with graded rock appeared to be suitable. Whilst the grids have not been used previously for this purpose it was considered that the test data available and the method of connection and jointing developed by the manufacturer could be adopted to form the barriers. The system of connection allows the interlinked gabions to form cellular mattresses at each level of the barrier. The use of the jointing and diagonal fabrication technique enables efficient transfer of stress throughout the structure which is thus better able to accommodate the imposed loads. It also allows easy adjustment of dimensions to take account of local variations in topography. The barriers were constructed with a vertical face to the uphill side and steps on the downhill side.

This technique of fabrication had been established previously and shown to be relatively straight forward. This was felt to be of particular importance for the terrain in which the barriers are formed. The fabrication does not require sophisticated equipment or very experienced personnel, provided that guidance is given initially and the work supervised throughout.

The use of other materials having characteristics equivalent to those indicated was permitted under the contract, but in the event, the contractor used Tensar SR2, the material on which the design had been based.

Whilst no loading tests of the polymer grid have been performed at the high rate of strain that would be experienced in the event of a boulder strike, tests so far indicate that the strength of the polymer grid increases with increased rate of strain.

Concern was expressed regarding the long term durability of the materials under the heat, humidity and ultra violet radiation to which the material would be subjected by Hong Kong's sub-tropical climate. Available test data shows that any deterioration in strength is unlikely to be significant. Also it should be noted that the structural elements of the material would be mainly within the body of the barrier. External faces would not be subjected to high structural load.

10. DESIGN CONDITIONS

The following potential failure modes were considered:-

1) Punching Shear

The boulder was assumed to cause the displacement of a horizontally orientated truncated "pyramid" of barrier material. This form of movement would be resisted by the polymer grid reinforcement and friction between the base of the pyramid and the barrier platform or ground. It is in this mode that the ability to form the polymer grid in a diagonal pattern was felt to be most advantageous.

2) Overturning

The possibility of overturning about a series of points on the downslope side of the barrier was investigated under a series of loading conditions and boulder impact points. The structure was shown to be stable in this mode.

3) Horizontal Shear

The possibility of horizontal shear displacement within the barrier itself was considered and adequate resistance shown to be present.

4) Water Pressure

The consequence of blockage of drainage provided under the barrier structure and a build up of water pressure for the full height of the barrier was considered and

PLAN ARRANGEMENT OF
POLYMER GRID REINFORCEMENT

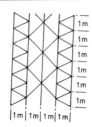

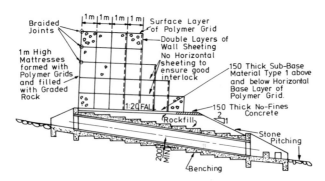

BARRIER TO CATER FOR 5m DIA. BOULDER

FIGURE 5

the structure shown to be stable.

5) **Excessive Strain**

Strain within the grids was predicted from a consideration of the scale and geometry of the anticipated displacements and shown to be well within acceptable limits.

11. DESIGN DETAILS

The final design adopted is shown in Figure 5.

Some particular features are worthy of note.

1) Since the boulder barriers were constructed in the base of valleys it was necessary to provide culverts beneath the barriers to take the predicted flow. In the event, larger pipes than necessary were used to reduce the chances of blockage and facilitate cleaning out. Energy dissipation blocks were included on the down-stream side.

2) Erosion protection both uphill and downhill of the barrier was provided.

3) The base layer of fabric was placed between two 150mm thick layers of well graded

TABLE I : BARRIER CONSTRUCTION TIME AND COSTS

Barrier	Volume m3	Construction Time Weeks	Remarks
D3	468	4	First Barrier Constructed
D1	958	6	Largest Barrier
E2	438	2	Typical Construction Rate

Note: 1) Cost HK$ 480 (Approx £43.5 AT CURRENT EXCHANGE RATE) per cubic metre placed including SR2 polymer grid, rockfill, compaction and erection.

2) The costs and times quoted are for the gabion structure only.

granular material to ensure good frictional interlock between the gabions and the supporting base structure.

4) To prevent the fine material from washing out of the base of the barrier the sides were retained by no-fines concrete.

5) No horizontal sheeting was placed between the "mattresses" in order to ensure good interlock between them. Horizontal layers were only used at the base and at the top surface of the barrier.

6) Lateral reinforcement is provided by the vertical diagonal sheeting which also provided support to the vertical faces.

12. BARRIER CONSTRUCTION

After an initial familiarisation process by the Contractor, fabrication of the gabion structures proved to be extremely rapid. It is to the credit of the main contractors for the work, Kumagi Gumi Hong Kong Ltd., that these unique structures were built quickly and neatly in difficult terrain.

Table I gives an indication of the times taken to fabricate the barriers above culvert structure level, with materials and fabrication costs.

13. FUTURE WORK

It has not been possible to conduct controlled impact testing on the barriers at this site. Such testing would not only aid the prediction of boulder velocity and mode of movement but also lead to refinement of the design in relation to barrier deformation and stress redistribution.

Tests of the materials at the likely ambient temperature and at the rapid rate or strain which is appropriate to boulder impact would be extremely helpful to future design.

14. ACKNOWLEDGEMENTS

This paper is published with the permission of the Hong Kong Housing Authority and their Site

Formation Consultant, Peter Y.S. Pun and Associates. We would like to thank them and in particular acknowledge the contribution of Mr. M.C.Gregson of Hong Kong Housing Authority, Mr. T.A. Rogers of P.Y.S. Pun and Associates and Dr. Y.C.Suen of Leonard Threadgold Geotechnics Limited, Sub-Consultant for the work.

REFERENCES

1. Ritchie, A.M. (1963). Stability of rock slopes. Research Record No. 17, P13-28.

2. Bhandari, R.K. & Sharma, S.K. (1976). Mechanics and Control of Rockfalls. Journ. Inst. of Engineers (India), Vol. 57 P14-24.

3. Bhandari R.K. (1977). Some typical land-slides in the Himalayas, Proc. 2nd Int. Symp. on Landslides and Their Control, Tokyo, P1 24.

4. Fookes P.G., Sweeney M. (1976). Stabili-sation and control of local rock falls and degrading rock slopes. Q. Jnl. Engng. Geol. 9:1:27-35.

5. DeFreitas, M.H. & Watters, R.J. (1973). Some field examples of toppling failure. Geotechnique 23, No. 4, 495-514.

6. Mercer, R. (1982). The stabilisation of Craig-y-Dref, Tremadog. Ground Engineering, January P28-35.

7. Various Contributors (1976). Rockfall dynamics and protective works effectiveness. Proc. meeting Istituto Sperimentale Modelli E. Strutture, Bergamo, 20-21 May 1976.

8. Lied, K. (1976). Rockfill problems in Norway. Proc. meeting ISMES, Bergamo, May 76, P51-54.

Retaining walls: report on discussion

C. J. F. P. Jones, *West Yorkshire Metropolitan County Council*

The Chairman discussed the anchoring effect of different types of reinforcement and described the results of research work at Sheffield University into the performance of 2m long reinforcing elements buried in a tank with a surcharge capacity of 200 kN per square metre.

In the case of steel strip reinforcement, the ultimate pulling resistance was found to be directly proportional to the overburden pressure. With smooth steel strip failure could be induced and the steel strip pulled out after very few cycles. The use of ribbed steel strip increased the number of cycles to failure by a factor of 10. A very marked difference was noted in the performance of Tensar SR2. The Tensar SR2 was subjected to 100,000 load cycles with the pulling load varying from between the limits of 70 per cent of ultimate and zero. The results show that the system became stiffer and stiffer as the number of load cycles increased. Polymer grid reinforcement was therefore believed to provide excellent performance if subjected to severe repeat loadings, as one might well get in bridge abutments.

Referring to paper 6.1 the optimum angle for placing of reinforcement was discussed by a number of contributors. Dr Murray described some model tests at the TRRL where horizontally placed reinforcement had been compared with inclined reinforcement. It had been found that the inclined reinforcement provided twice as much effective reinforcement as the horizontally placed layers.

Reference was made to the fact that ground anchor structures often utilise inclined anchors, and it was felt practical to install reinforcement at an incline by first placing a wedge of soil down at a slope of 1 in 4.

It was reported that research work at the University of Strathclyde had illustrated the effect of the method of construction on the principal strain directions. Having discovered the principal strain direction of the soil, geotextile reinforcement was placed at angles calculated to make the reinforcement most effective, only to discover that changing the mode of construction had in itself changed the principal strain direction.

The conclusions which can be drawn from this work are that the principal tensile strain directions were the

directions in which the reinforcement could be placed but that it is important to take into account the method of construction in order to determine where these will be. Once the principal tensile strain directions had been identified, it was possible to analyse, measure and design structures properly.

In reply Dr Jones recommended that for practical reasons the reinforcement would usually be laid horizontally. In many cases, it was not the cost of the reinforcement which was dominant, but the cost of the fill. A heuristic approach had to be taken. If the installation of the reinforcement at an incline proved more economical, then the contractor would suggest that this mode of construction be undertaken.

Dr Bassett illustrated that in the case of reinforced soil retaining walls, there was a quadrant into which the reinforcement could be inserted with positive effect. With a frictional material this quadrant would have an angle of (45 degrees + $\phi/2$). The effect of surcharge, as described by the Chairman, provided an increase in efficiency.

Dr Bassett commented that Dr Jones was making an important point with regard to the construction when he recommended pulling the reinforcement tight. The result was to provide stress right the way through the structure and the effect was to produce anchored walls with the anchorage zone extending beyond the failure plane.

Dr Bassett disagreed with Dr Jones' comparison of a reinforced soil system with piles, particularly with regard to foundations where the piles pass through the foundations.

He commented that the foundation could be reinforced and used as the classic fan for a 45 degree stress field and the principal stress directions in compression and tension to illustrate the problem.

The same origin of planes and therefore the same rotation of principal stresses was noted, but the piles would have had to be bent to come out horizontally through the fill material if like was to be compared with like. Theoretically the tensile reinforcement should be horizontal under the foundation.

Dr Bassett concluded that piles could be used as reinforcement if provided, for example, at 14" centres to make a wall of vertical timber props with soil packed between them.

Mr Pigg was questioned on the foundation conditions for the reinforced soil retaining wall at Low Southwick, reported in paper 6.4, in view of the very narrow length of Reinforcement to Height of Wall ratio used.

It was pointed out that the Department of Transport Technical Memorandum called for a reinforcement of length 0.8 times the height or a minimum of 5 metres long strips.

The design concepts in the Technical Memorandum recognised the possibility of constructing a reinforced soil wall on poor foundations, whereas with a conventional wall piled foundations would be required. If, however, a narrow wall was built the situation reverted to the condition where the structure creates very high stresses in front or below the toe of the wall.

In reply Mr Pigg explained that when excavation took

Polymer grid reinforcement. Thomas Telford Limited, London, 1985

place for the wall foundation, some fairly poor soil was found, and it was decided to extend the frictional fill to a greater depth. In addition the foundation was reinforced using a hammock construction formed from Tensar running along the length of the wall and tied into the good clay on the inside of the dip.

The longitudinal reinforcement was interleaved with the reinforcement extending back in a direction normal to the wall. The interleaving was carried on up to a height of 2 metres. In view of the fact that the reinforced soil wall was being constructed, the subgrade was made as strong as possible.

In reply to a question on machines operating on top of the reinforcement it was confirmed that no tracked vehicle was ever allowed to run directly on the reinforcement, regardless of the type of reinforcement selected. Good practice demanded that soil was pushed on to the reinforcement.

The use of colliery waste was discussed with particular reference to the problems of pre nationalisation dumps being mixtures of burnt and unburnt shales and the potential problems of unburnt shale being subject to internal combustion, or other forms of degradation.

It was pointed out that the slides shown in the presentation of paper 6.5 had shown that unburnt colliery shale had been used in reinforced soil structures.

In reply Dr Jones stated that colliery waste had been used with both Tensar and fibreglass reinforcement. The mining authorities used unburnt colliery shale, and it appeared to perform extremely well. Structures in which unburnt shale had been used had been designed to Design Memorandum BE 3/78 - as far as it was possible, although from a specification point of view, BE 3/78 did not permit the use of unburnt shale.

The potential dangers of internal combustion were recognised but experience had shown that if the material is properly compacted, in accordance with BE 3/78, that this problem could be eliminated.

Participants: Professor Hanna

Dr Murray

Dr McGown

Dr Jones

Dr Bassett

Dr Merrifield

Mr Pigg

Mr Stebbings

The behaviour of Tensar reinforced cement composites under static loads

R. N. Swamy, R. Jones and P. N. Oldroyd,
University of Sheffield

In concrete construction there are many situations where only nominal conventional steel reinforcement is needed to cater for secondary stresses arising from lifting, transportation or environmental conditions. The development of Tensar high strength polymer grids provides an alternative to steel, and opens up a new set of composite construction materials which are not subject to corrosion and are resistant to various aggressive media. The paper reports the results of extensive tests on the behaviour of Tensar reinforced cement composites under tension, compression and flexure. Results are also presented on the influence of polymer reinforcement on shrinkage, creep and thermal cycling. The paper identifies various areas of application where the substitution of polymer grids for steel can lead to economic and durable construction.

1.0 INTRODUCTION

In concrete construction there are many situations where the nature intensity and incidence of loading are such that only nominal reinforcement is needed for load resistance purposes; such reinforcement also serves to resist shrinkage, temperature and other stresses induced by external or internal restraint. When steel is used as the reinforcement the thicknesses of such members may be dictated by the cover requirements for the steel rather than by structural considerations. If a suitable alternative to steel with adequate bond is available, such members can be made thinner and more slender, the concrete and reinforcement acting compositely together to resist stresses incidental to lifting, transportation and the environment.

The development of Tensar high strength polymer grids has opened up the possibility of using them as reinforcement in cement matrices to produce a new reinforced composite material for these sort of applications in the construction industry. Apart from their high strength, polymer grids have several other advantages such as lightweight, flexibility and safe, easy and convenient handling. Tensar polymer grids are also resistant to most chemicals and under aggressive chemical attack, it will be the cement matrix that will deteriorate first. But the most outstanding advantage of Tensar polymer grids is probably their almost total resistance to water and chemical corrosion (1). Consequently precast or in-situ concrete construction can be carried out with the minimum practical cover possible to the reinforcing elements. This important advantage can not only enable more slender units to be used in practice but also could possibly lead to maintenance-free construction and eliminate the often inconvenient and expensive processes of repair due to steel corrosion and cracking. These are major considerations to the construction industry now, and are also likely to assume greater importance in the future because of financial limitations on maintenance and repair.

Tensar polymer grids also offer other practical attractions. They can be easily folded into shape, which can be an advantage both for storage and to produce shaped sections. They can also be produced in a wide range of grid sizes and shapes which make them highly suitable for an extensive range of applications in the construction industry. Another added bonus of Tensar polymer grids is that their structural performance under dynamic loading is comparable with that of steel reinforcement (2).

This paper reports typical results from an extensive investigation on the use of polymer grids to produce cement composites and subjected to "static" loads. The paper discusses the development of suitable mixes, fabrication techniques, and the behaviour of the composite under tension, compression and flexure. The bond between the Tensar and cement matrix is examined, and this is related to the post-cracking behaviour of the composite material. Results are also presented on the influence of polymer reinforcement on shrinkage, creep and temperature movements. The paper also identifies several areas of application where the substitution of polymer grids for steel can lead to economic and durable construction.

2.0 EXPERIMENTAL DETAILS

2.1 Materials and Manufacture of Test Specimens

In the tests reported here two types of cement matrices were used - a normal weight mortar matrix and a lightweight mortar matrix. The mix details and the basic engineering properties of these mixes are shown in Table 1. The lightweight mortar matrix was obtained by using Leca (expanded clay) fine aggregates. The Leca aggregate consists of hard round particles with a dense skin and a honeycomb interior, and conformed to Zone 2 gradation of the relevant

Polymer grid reinforcement. Thomas Telford Limited, London, 1985

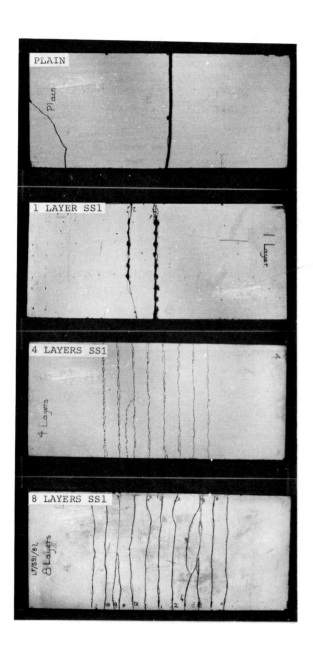

INCREASED FLEXURAL CRACKING WITH
INCREASED POLYMER CONTENT

FIG. 1

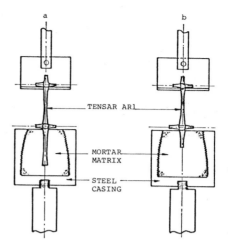

BOND PULL OUT TEST ARRANGEMENT

FIG. 2

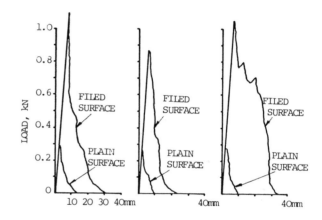

CROSS-HEAD DISPLACEMENT

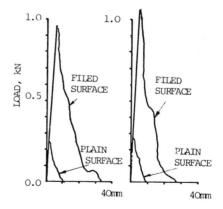

CROSS-HEAD DISPLACEMENT

BOND PULL OUT TEST RESULTS

FIG. 3

223

BS code. The sand for normal mixes was washed and dried sharp sand conforming to Zone 3.

Rapid hardening cement was used throughout. All the test specimens were compacted on a table vibrator, demoulded at one day, cured in a fog room under controlled conditions of temperature ($20^{\circ}C \pm 1^{\circ}C$) and relative humidity (95-100%) and tested at 7 days. The strength and elasticity properties of the two mixes shown in Table 1 are based on the mean of several tests carried out over a long period of time.

Specimens were initially fabricated using traditional steel reinforcement methods, supporting the grid on chairs and casting the concrete around it. This method proved un-suitable since the low density polymer grid tended to float in the denser fluid cement matrix. Several other fixing techniques were investigated, the most successful being to tie the grid to the formwork with tensioned nylon wire. This tendency of Tensar reinforce-ment to float raises careful considerations to be attended to in the fabrication of concrete composites containing Tensar. It is important that the Tensar should, wherever possible, be formed into a rigid cage, and held firmly in position by suitable attachments.

3.0   TEST RESULTS AND DISCUSSION

An extensive range of tests has been carried out on Tensar cement composites containing a normal weight or a lightweight matrix. From these results only typical data are presented here to illustrate the properties and perform-ance of the composite.

3.1   Cement Matrix - Tensar Bond

Since the Tensar reinforcement is in the form of a continuous grid, analogous to a steel mesh, the problem of bond does not strictly arise, unlike in composites made with short discrete fibres. The composite reinforced with such grids must therefore fail by fibre fracture than by fibre pull-out. Nevertheless bond has an important role to play in the transfer of stress between the cement matrix and the reinforcing element and in the development of cracking. In a Tensar reinforced composite, cracks tend to form along the transverse ribs, provided enough volume fraction of the reinfor-cing element is present (Fig. 1).

To investigate further the nature of bond between the Tensar grid and the cement matrix pull-out tests were carried out at 7 days as shown in Fig. 2. When the full length of the rib (between junctions) was embedded in the matrix (Fig. 2a), the grid fractured at the junction just outside the cement matrix at loads between 1.24 kN and 1.35 kN, the main resistive force to pull-out arising from the anchorage of the embedded element. When the whole grid was embedded in the matrix, the pull-out load to fracture amounted to 1.8 kN with a standard deviation of 0.127 kN.

When only half the rib (Fig. 2b) was embedded

failure occurred by pull-out. To examine the influence of the nature of the surface of the Tensar grid the series of tests were repeated with the Tensar surface roughened with a file. The results are shown in Fig. 3 and Table 2. The failure with the roughened surface also occurred by pull-out, but the load to pull-out the specimen increased by about 200%, and approached values similar to that required to fracture the grid. A roughened surface is thus able to produce a pull-out force of a magnitude similar to that to cause fracture of the grid, and any further increase in bond is unlikely to improve the performance of the Tensar composite. These results indicate that a Tensar grid with adequate roughness can be made to fail in a cement composite by fibre fracture than by fibre pull-out. A commercially produced Tensar grid with a roughened surface can thus be designed such that the composite utilises almost the entire tensile strength of the reinforcing element: the full tensile strength of the Tensar could thus be put to economic advantage in a Tensar reinforced cement composite.

3.2   Compression Behaviour of Tensar Cement Composite

To study the effect of Tensar reinforcement in compression, tests were carried out on light-weight mortar cylinders 100mm diameter x 200mm reinforced with various amounts of Tensar grid. Comparative tests were also carried out on specimens with steel mesh reinforcement. The results of the tests are shown in Fig.s 4 and 5 and Tables 3 and 4.

The Tensar reinforced specimens showed results contrary to those reinforced with conventional steel. The latter showed that as the amount of steel was increased, both the peak stress and the secant modulus were increased, but the failure was less ductile than specimens contain-ing polymer reinforcement. With the Tensar reinforced specimens, as the amount of Tensar was increased both the peak load and the secant modulus decreased, but the failure became more ductile. The reduction in gross stress at peak load corresponded to the reduction in the cross-sectional area of the matrix produced by including Tensar. However, the net stresses were higher, and the computed composite stresses are shown in Table 3.

The reduction in the secant modulus was only marginal, and it is therefore fair to conclude that the compressive modulus is not signifi-cantly affected by the presence of Tensar. The strain at maximum load is also reduced in relation to the reduction in load capacity, but the unique feature of the Tensar is the enormous increase in ductility as indicated by the ultimate strain at failure, the compress-ibility of the specimens (Fig. 5) and the area of the stress-strain diagram (Table 4). The ultimate strain was more than seventy times that of the plain concrete specimen and the energy absorption capacity, as indicated by the areas under the stress-strain curve was about 600 times for 1 layer of Tensar and 1500 times for 3 layers compared with that of the plain unreinforced specimen. The post-cracking load

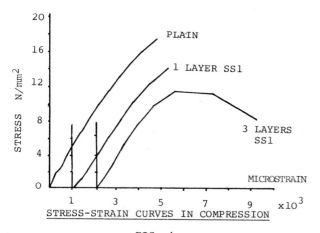

STRESS-STRAIN CURVES IN COMPRESSION

FIG. 4

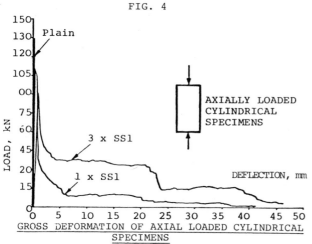

GROSS DEFORMATION OF AXIAL LOADED CYLINDRICAL
SPECIMENS

FIG. 5

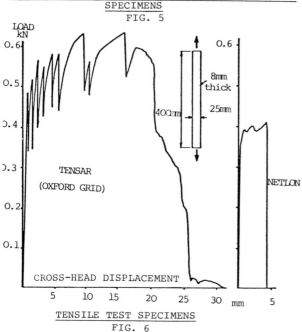

TENSILE TEST SPECIMENS

FIG. 6

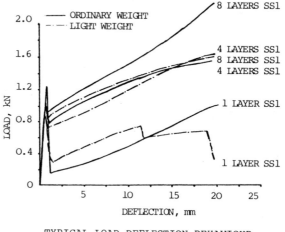

TYPICAL LOAD DEFLECTION BEHAVIOUR

FIG. 7

NW - STEEL - 3.30 J
PLAIN - 0.48 J
LW - STEEL - 2.77 J
PLAIN - 0.41 J

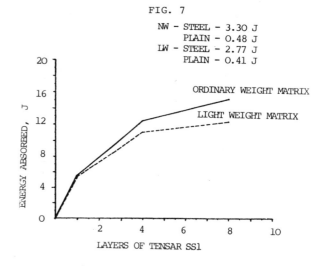

STATIC FLEXURAL ENERGY ABSORPTION

FIG. 8

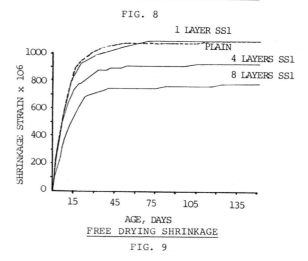

FREE DRYING SHRINKAGE

FIG. 9

225

capacity was further illustrated by the fact that the specimen with 3 layers of Tensar, although considerably deformed at 50mm compression, began once again to take load.

These results indicate the enormous ductility and energy absorption capability of Tensar when embedded in concrete and suggest that Tensar could find applications where these two parameters are the major factors in design. For example, Tensar could thus provide an alternative to timber supports presently used in mine workings. Other situations where ductility and energy absorption are major design parameters under static and dynamic loading are shatter proof construction, earthquake resistant construction, and as protective linings for piers and columns against collision and impact in car parks, highway bridges and off-shore structures.

3.3 Behaviour in Direct Tension

These tests were initially carried out on mortar sheets containing 3% by volume of "Netlon" (an unoriented polymer grid) or 3% by volume of Tensar (an oriented polymer grid). The reinforcement grid consisted of 1mm thick filaments at about 10mm centres in both directions. Specimens 400 x 25 x 8mm were cut from these sheets and tested in direct tension at a cross-head speed of 20mm per minute. The results are shown in Fib. 6 and Table 5.

The results show that the Tensar grid composite was generally superior to the "Netlon" grid composite both in crack distribution and load carrying capacity, although the latter showed higher elastic modulus. The presence of Tensar reduced the first crack load, but unlike the plain, unreinforced sample, the Tensar composite displayed increasing load capacity after first crack. The modulus itself was only slightly affected by the presence of Tensar. However, the Tensar composite showed again, as in compression, a large capability for energy absorption, and the composite remained in one piece although it was extensively cracked.

Tensile tests on specimens 300 x 100 x 20mm reinforced with two layers of SS1 grid positioned at 5mm from each face show similar results i.e. increasing post-cracking load capacity after first crack, extensive ductility and energy absorption capacity of the composite. Tensile strains in excess of 1000×10^{-6} have been measured on the surface of the Tensar grid itself.

3.4 Behaviour in Flexure

Behaviour in flexure is important, since most concrete elements are subjected to largely flexural loads in practice. The specimen size for this series of tests was 700 x 300 x 20mm and the specimens were loaded at third points over an effective span of 600mm. The specimens were reinforced with increasing amounts of Tensar, and for comparison purposes, specimens reinforced with a welded steel mesh of 1mm diameter wire at 50mm centres were also tested. The test results are shown in Table 6 and

Figs. 7 and 8.

The results shown in Table 6 indicate the unique properties of Tensar cement composites. The first crack load remains almost unaffected when the composite is reinforced with only one layer of Tensar. The steel composite also showed similar behaviour. However, as the number of Tensar layers was increased, the first crack appeared at progressively earlier loads; however, the reduction in the first crack load was less with the lightweight matrix. These results show that with a matrix of elastic modulus comparable to that of the reinforcing element, a better composite action can be achieved than when the moduli of matrix and reinforcing element are wide apart.

All the specimens showed a sudden drop in load immediately on first cracking (Fig. 7). The decrease in load was greatest with only one layer of Tensar, and became progressively less with increasing volume contents of Tensar. However, inspite of the fall in load, all the specimens showed an increasing post-cracking load capacity, and this was evident even with one layer of Tensar.

To examine the post-cracking ductility of the test specimens, all the specimens were loaded, beyond first cracking, to a deflection of 20mm. The plain, unreinforced specimens, and the specimen with one layer of steel mesh were unable to reach this deflection, and failed to show any post-cracking ductility. On the other hand, Tensar composites with only a single layer of reinforcement were able to show remarkable post-cracking ductility, although the post-cracking peak load in this case was less than the first crack load (Table 6).

Composites with four and eight layers of Tensar, on the other hand, demonstrated not only post-cracking load capacity, but also higher loads at 20mm deflection than first crack loads, thus both the post-cracking load capacity and the ultimate strength of 20mm deflection increased with increasing volume content of Tensar. In many cases, it was impossible to fail the specimen completely, increasing load producing further large deflections, and maintaining the integrity of the test specimens.

3.4.1 Cracking behaviour

The initial crack spacing in flexure was dominated by the transverse rib spacing in the outermost layer of the grid, and this soon became evident even on initial loading (Fig. 1). Increasing amounts of Tensar also increased the number of cracks. One important implication of this crack formation and crack control is that concrete elements could be designed to achieve systematic and uniform cracking, and that any crack spacing can theoretically be achieved by suitably sized grids. When cracking occurs, the tensile force is carried largely by the Tensar reinforcing element at the crack, and with adequate volume fraction, further cracking can be developed by bond stress transfer between the matrix and the reinforcement. With further increases in load, cracking will

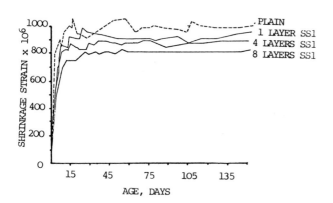

RESULTS OF THERMAL CYCLING

FIG. 10

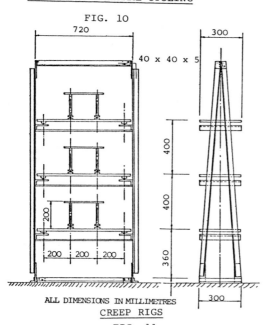

ALL DIMENSIONS IN MILLIMETRES
CREEP RIGS

FIG. 11

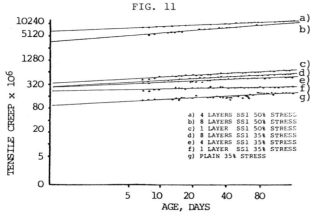

VARIATION OF TENSILE CREEP

FIG. 12

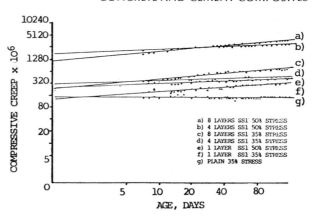

VARIATION OF COMPRESSIVE CREEP

FIG. 13

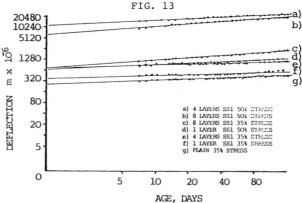

VARIATION OF DEFLECTION

FIG. 14

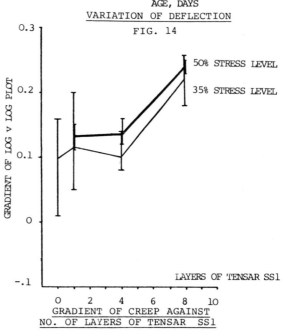

GRADIENT OF CREEP AGAINST
NO. OF LAYERS OF TENSAR SS1

FIG. 15

again occur at the transverse ribs, which is in effect act as stress raisers.

A regular uniform crack spacing implies smaller crack widths, and a more homogeneous behaviour of the composite. The fact that the crack spacing can be controlled by the geometry of the reinforcing grid offers a unique opportunity to the designer to produce elements such as cladding panels and partition walls that can be more effectively assessed in their cracking and post-cracking performance.

Two further important flexural characteristics need to be emphasized. Firstly Tensar composites on reloading, after the first loading and unloading stage, retained significant proportions of the first loading behaviour. Secondly, Tensar composites showed almost total recovery on unloading after repeated loading and extensive cracking. Both these characteristics are important in the performance of concrete elements; Tensar composite elements would not only recover, even after extensive cracking and deformation, but would also retain their shape and integrity in a structural system.

Composites reinforced with one steel mesh only, on the other hand, showed little recovery particularly if the steel mesh had yielded in the first loading stage. Further reloading continued to produce plastic behaviour until failure.

3.4.2 Energy absorption in flexure

The area under the load deflection curve up to 20mm deflection was taken as the static flexural energy absorption capacity of the composite. This is shown in Fig. 8 which again confirms the behaviour in tension and compression i.e. that the energy absorption capability increases with increasing volume content of Tensar, although at a decreasing rate. The lightweight matrix composites gave a reduced energy absorption compared to the normal weight matrix, probably due to the higher compressive strength of the latter. The data shown in Fig. 8 compare with the following energy absorption values:

Normal weight matrix:

Unreinforced specimens - 0.48 J
Specimen with steel mesh - 3.30 J

Lightweight matrix:

Unreinforced specimen - 0.41 J
Specimen with steel mesh - 2.77 J

3.5   Shrinkage of Tensar Composites

The shrinkage tests were also carried out on 700 x 300 x 20mm thick specimens. All the specimens were demoulded at one day and then stored in a controlled environment at 15°C and 55% R.H. Fig. 9 shows the shrinkage behaviour of specimens with one, four and eight layers of Tensar together with that of the plain un-reinforced specimen. No allowance of shrinkage due to carbonation has been made in this data,

and the shrinkage values shown are total shrinkage values. The effect of carbonation is thought to be small.

The results show two important characteristics. The free shrinkage decreases as the volume of Tensar increases. With 4 (1.4% min. CSA at rib) and 8 layers (2.8% min. CSA at rib) of Tensar respectively the free shrinkage of the mortar matrix of 1100×10^{-6} m/m at 150 days is reduced by about 15 percent and 28 percent. The figure also shows that the shrinkage of the composite stabilizes fairly early in its life, and then continues to remain at a more or less constant value, unless of course the ambient environment changes.

3.6   Effect of Thermal Cycling

Many concrete elements, particularly thin non-structural units such as cladding panels, may undergo thermal cycling in their everyday life. To examine the performance of Tensar cement composites under these conditions, a number of specimens has been subjected to thermal cycling. To simulate the temperature conditions to which concrete members are subjected to in practice, the specimens were subjected to two cycles in every 24 hrs. - 8 hrs. at 50°C and 16 hrs. at room temperature following the heating period. Fig. 10 shows the shrinkage behaviour under thermal cycling for a period up to 150 days.

The results indicate that whilst the plain unreinforced sample shows considerable variation in its movements, and thus considerable instability, the Tensar reinforced composites showed a more uniform as well as reduced deformational behaviour. This ability to control movements and display less variability in movements increased as the number of layers of Tensar increased. In fact with 8 layers of Tensar in the 20mm thick specimen, the movements became not only consistent but did so at an early stage of the thermal cycling. With this number of layers of Tensar (2.8% cross-sectional area), the movements were reduced by about 15% (Fig.10).

The time for visual cracks to appear with the thermal cycling also varied with the type of specimen. The plain unreinforced specimens remained uncracked at 310 days whereas the composites with 1, 4 and 8 layers showed the first visible cracks at 150 days, 141 days and 124 days respectively. The Tensar reinforcement thus appears to have a negative effect, in the sense that the cracks due to thermal movements appear earlier, but this is probably due to the difference in their coefficients of expansion, and also the fact that the presence of Tensar reduces the effective area of the matrix.

As the thermal cycling proceeded, the Tensar composites began to exhibit surface cracking, but again, emphasizing the unique nature of Tensar, these cracks tended to close once the thermal effects had been dissipated. This aspect again is important in the serviceability of structures, and Tensar reinforced concrete members are shown to have the ability to return to their apparently uncracked state once the effects of adverse thermal cycling cease.

3.7 Flexural Creep Behaviour

The flexural creep tests were carried out on 700 x 300 x 20mm specimens loaded at third points. Each test rig had three samples under test (Fig. 11) with companion unloaded specimens. All the specimens were kept under controlled conditions of temperature and humidity (15°C and 55% R.H.). The results, shown in Figs. 12, 13 and 14, are plotted on a log-log basis.

The specimens had 1, 4 or 8 layers of Tensar together with unreinforced samples, and were stressed to 35% (uncracked) or 50% (cracked) of their respective flexural strength. In general, both tensile and compressive creep as well as the deflection under sustained loading, increased as the volume content of Tensar and the applied stress increased. The plain un-reinforced specimens showed in all cases the least amount of creep or sustained deflection. With the cracked specimens loaded at 50% of flexural strength, the composite with 8 layers of Tensar tended to show less creep and deflection than the one with 4 layers.

The results also show that tensile creep in bending is generally greater than the compressive creep except for uncracked specimens containing Tensar in the compression zone. This is probably due to the movement of the neutral axis towards the tension zone and the consequent increased compressive creep.

The creep results have been analysed on the basis of a power law, $y = C X^m$. Applying this law to the creep behaviour up to 150 days, the gradients and the y intercepts have been calculated by linear regression analysis and these values are shown in Table 7. The gradients plotted against Tensar quantity is shown in Fig. 15. These results seem to indicate that as the amount of Tensar is increased, the gradient of the log-log plot also increases indicating a greater rate of creep.

The very large creep strains shown in Fig. 12 and 13 (which incidentally also include the crack widths in the case of tensile strains) probably indicate that Tensar reinforced composites are likely to undergo large deformations if they are subjected to sustained loads over a long period of time. These deformations are inherent in the nature of the material, in which a stiffer matrix is reinforced with a relatively less stiff component. It is obvious then that such composites are unlikely to find application in areas where substantial loads have to be carried over a long period of time. With elements such as cladding panels, and partition walls, however, the load carrying capacity is much more restricted both in magnitude and duration, and such large deformations are therefore unlikely to occur.

The tests so far seem to indicate that Tensar composites should be able to carry loads of short duration without exessive deformation. In any case, the ability to return to their original state is one of the unique qualities of Tensar composites which is unlikely to be impaired by creep deformation for short periods.

4.0 PRACTICAL APPLICATIONS

The data presented so far show that concrete composites reinforced with Tensar polymer grids have a number of desirable characteristics which can be exploited in the construction industry. The type and nature of application will obviously depend on the properties and performance characteristics specified by the engineer, but the following list, whilst not being exhaustive, is intended to be indicative of the uses where Tensar grid reinforcement could be used to advantage:

1) cladding panels

2) sprayed concrete construction and repairs such as sea wall repairs and rock stabilization

3) alternative to timber supports in mine workings

4) tunnel protective linings

5) pavement/foot path slabs

6) shatter proof construction

7) earthquake construction

8) protective linings against collision and impact in bridge structures, car parks, sea structures etc.

Several trial tests have been carried out to examine the practical difficulties that are likely to be met with in using Tensar grids in concrete construction, and in the following two specific areas namely that of cladding panels (Fig. 16) and sprayed concrete construction/repair (Fig. 17) are briefly discussed.

4.1 Cladding Panel Tests

A commercially available cladding panel, 3.3m x 0.9m, used as a non-load bearing unit and designed to resist wind load only was tested with part of conventional steel reinforcement replaced by a Tensar grid (Fig. 16). The panel was tested with a horizontal transverse load, the specimen being held vertically as in the structure. The panel was initially loaded to first crack, unloaded and the recovery monitored. The panel was reloaded again until the central deflection was about equal to that on first loading, unloaded and allowed to recover.

At the end of the first loading-unloading stage, the panel recovered 50% of the central deflection, and the crack width of 1.5mm at maximum deflection was reduced to 0.5mm. On reloading, the loading was continued until the central deflection was about 25mm at 35% of the peak load of the undamaged slab. On unloading, the residual deflection was reduced to about 10% of that under load and with time, the panel was restored to almost the original state. The load carried by the panel in the cracked

	MORTAR	LECA
AGG/CEMENT	3.0	2.0
WATER/CEMENT	0.6	0.45
CUBE STRENGTH N/mm^2	33.4	19.2
ELASTIC MODULUS GPa	29	8.5
DENSITY kg/m^3	2100	1450
MOR N/mm^2	2.9	2.1

Table 1

MIX PROPORTIONS AND PROPERTIES

	MEAN NOTIONAL BOND STRESS N/mm^2	STD DEVIATION N/mm^2
PLAIN SURFACE	0.92	0.14
FILED SURFACE	2.82	0.23

Table 2

PULL OUT BOND TEST RESULTS

LAYERS OF SS1	O	1	3
PEAK LOAD kN	135	110	91
GROSS STRESS N/mm^2	17.2	14.0	11.6
NET STRESS (NODES) N/mm^2	17.2	14.1	11.7
NET STRESS (RIBS) N/mm^2	17.2	15.0	14.1
COMPOSITE STRESS	17.2	19.8	23.1

Table 3

COMPRESSION TEST RESULTS

LAYERS OF SS1	O	1	3
STRAIN AT MAX. LOAD $\mu\epsilon$	2383	2134	1737
STRAIN AT ULTIMATE LOAD $\mu\epsilon$	3000	218000	225000
SECANT MODULUS GPa	9.3	8.0	8.9
ENERGY ABSORPTION J	0.7	411	1152

Table 4

DEFORMATION PROPERTIES IN COMPRESSION

	TENSAR	NETLON
1ST CRACK LOAD kN	0.48	0.37
Ultimate load kN	0.63	0.41
1ST CRACK gross stress N/mm^2	2.40	1.90
net stress N/mm^2	2.47	1.91
composite stress N/mm^2	8.22	8.22
Ultimate gross stress N/mm^2	3.15	2.05
net stress N/mm^2	3.24	2.11
composite stress N/mm^2	8.91	8.52
Initial tangent modulus GPa	1.02	1.40

Table 5

PROPERTIES OF TENSAR CEMENT COMPOSITES IN TENSION

SPECIMEN	PLAIN		25/25 SWG		1 LAYER SS1		4 LAYERS SS1		8 LAYERS SS1	
	OW	LW	OW	LW	OW	LW	OW	LW	OW	LW
LOAD TO 1ST CRACK kN	0.64	0.48	0.63	0.48	0.61	0.48	0.52	0.47	0.49	0.46
LOAD TO 20mm DEFL kN	-	-	-	-	0.50	0.26	0.76	0.81	1.11	0.80
GROSS STRESS AT 1ST CRACK N/mm^2	3.20	2.40	3.15	2.40	3.05	2.40	2.60	2.35	2.45	2.30
GROSS STRESS AT 20mm DEFLN N/mm^2	-	-	-	-	3.50	1.30	3.80	4.05	5.55	4.00

Table 6

FLEXURAL LOAD TEST RESULTS ON TENSAR COMPOSITES

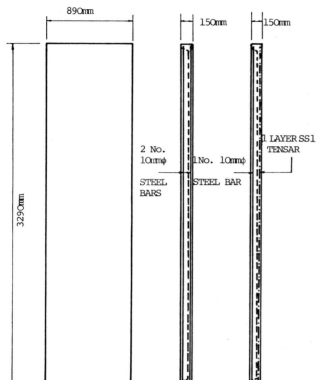

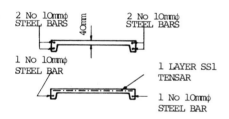

PRECAST CONCRETE PANEL WITH 2 10mm DIAMETER
BARS REPLACED BY 1 LAYER OF SS1 TENSAR

FIG. 16

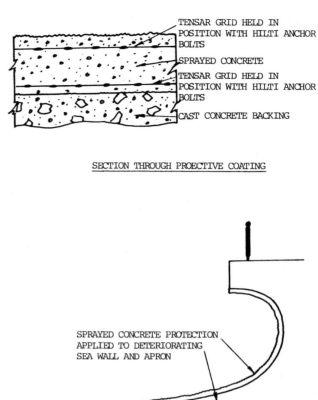

SECTION THROUGH PROECTIVE COATING

SCHEMATIC REPRESENTATION OF PROPOSED SEA WALL REPAIR

FIG. 17

condition was approximately equivalent to a
wind loading of 1.5 kN/m^2.

4.2 Sprayed Concrete Construction/Repairs

Several sprayed concrete panels were fabricated
reinforced with SS1 and AR1 grids and using the
dry mix spray process. The grids were held in
position in the mould by various methods, and
the thickness of both the cast concrete backing
and of the sprayed concrete were varied. In
addition panels were also made by spraying
directly into the mould without any concrete
backing (Fig. 17).

After curing, the test panels and various test
samples cut from them, were subjected to the
following tests: visual examination, compaction
tests through density and pulse velocity on
cores, interface bond tests through four point
flexural tests on 700 x 300mm panels and pull-
out tests on cores, repeated impact tests on
300 x 300mm panels and permeability tests on
100mm diameter, 50mm thick cylindrical specimens
under a steady water pressure of 3.45 x 10^5
Pascals.

These tests showed that the most important
factors to be considered in using sprayed
concrete together with Tensar grids are the
correct and firm location of the grids and
ensuring a well compacted, dense matrix covering of the Tensar. The shadowing behind the
grids is minimised if the grid closest to the
back of the mould or nearest to the cast
concrete is positioned at least 25mm away from
the interface.

The core tests showed matrix densities of 2050
to 2250 kg/m^3 and matrix pulse velocities of
4.23 to 4.57 km/sec. The interfacial bond

tests showed the tensile strength of the sprayed concrete to be greater than that of the bond between the sprayed and cast concrete. The pull-out tests on 50mm diameter cores carried out to examine the tensile strength of the sprayed concrete containing Tensar gave tensile stress values varying from 1.00 to 1.90 N/mm^2. The failure often occurred at the concrete interface and rarely within the matrix or at the Tensar grid. The impact and permeability tests also indicated that the sprayed concrete can be combined with the polymer grids to obtain all the desired properties of the composite.

5.0 CONCLUSIONS

The results presented in this paper show that Tensar high strength polymer grids can offer an alternative to steel reinforcement particularly in situations where such reinforcement is used for non-structural purposes to resist stresses arising from lifting, transportation or environmental conditions. Tests in tension, flexure and compression show that Tensar reinforced cement composites possess considerable post-cracking load capacity, extensive ductility and substantial energy absorption capability. In many cases the composite, in spite of extensive cracking and deformation, was able to recover, on unloading, almost completely, and maintained its integrity.

The crack spacing in the composite was dominated by the spacing of the ribs. A systematic and uniform crack spacing could therefore be obtained by suitably sized grids. Roughening the surface of the Tensar grid also produced failures approaching fibre fracture rather than fibre pull-out.

The presence of Tensar polymer grids reduced shrinkage movements, and tended to stabilize such movements earlier than in the unreinforced specimens. Under thermal cycling also, Tensar cement composites showed a more uniform as well as reduced deformational behaviour. Although the presence of Tensar grids tended to crack the composites earlier than the unreinforced specimens, the composites were able to return to their apparently uncracked state once the effects of thermal cycling ceased. The large creep deformations sustained by the Tensar cement composites indicate that such composites are unlikely to find application when sustained loads have to be carried over a long period of time.

Several areas of application are identified where Tensar composites can be used to advantage in practice.

ACKNOWLEDGEMENTS

This project is being carried out in the Department of Civil and Structural Engineering, University of Sheffield and is supported by the Science and Engineering Council and Netlon Ltd. The authors express their thanks for all the technical and administrative help and advice given by the staff of these organisations. Thanks are also due to their colleagues involved in the project.

REFERENCES

1. B.P. Chemicals Ltd., Technigram T19/1, Chemical resistance of rigidex.

2. Watson, A.J., Hobbs, B. and Oldroyd, P.N. The behaviour of Tensar reinforced cement composites under dynamic loads, Symposium on Polymer Grid Reinforcement in Civil Engineering, I.C.E., London, March 1984.

The behaviour of Tensar reinforced cement composites under dynamic loads

A. J. Watson, B. Hobbs and P. N. Oldroyd,
University of Sheffield

Details are given of experiments involving impulsive loading from contact explosive charges, high velocity, low mass impacts from armour piercing bullets and low velocity, high mass impacts from a drop hammer. Results for the permanent damage created are presented in terms of crater and spall volumes and diameters, total crack lengths and bullet penetration depths. The transient phenomena leading to this damage are discussed and results are presented for the transient flexural response of the drop hammer specimens. Data from static re-loading tests on some of the damaged slabs is presented. It is concluded that the structural performance of Tensar grids under dynamic loading is comparable with that of steel reinforcement but that Tensar has potential advantages in many circumstances due to its inherent resistance to water and chemical corrosion.

1. INTRODUCTION

There are many semi-structural units such as wall and roof cladding panels which are very lightly loaded under normal service conditions, but may receive impact or impulsive loads caused by vandalism, accident or mis-handling during construction. These exceptional but occasional loads can be so much greater than the static loads that it would be unrealistic to design so that they cause no significant damage. In concrete or mortar units large deformation and crushing can be used to absorb most of the energy, providing total collapse can be avoided. The damaged unit can then be either repaired or replaced. At present many such units are reinforced with steel. In many cases the thickness of the unit is dictated by the cover requirements for the steel, rather than by structural considerations, and attempts at repair often leave a risk of corrosion. Polymer grids offer a suitable alternative because the material is highly resistant to corrosion. The extensive cracking which occurs under high rates of loading can be repaired without the necessity to completely seal the cracks. In addition Tensar polymer grids have a high ultimate strain and high ultimate tensile strength. Strength and stiffness both increase significantly with rate of strain.

The experiments described in this paper have been carried out using concrete and mortar slabs which were either reinforced with Tensar polymer grids or a steel mesh or were unreinforced. The purpose of these experiments was to determine the resistance of the slabs to explosive pressure, low mass high velocity projectiles and larger mass low velocity projectiles.

The maximum energy available was 150kJ from 25g of high explosive, 3.2kJ from a 9.75g armour piercing bullet, 100J from a 5kg steel weight and 40J from a 2kg steel weight. The rates of straining in the composite varied between approximately 750/s, and 8×10^{-3}/s.

The independent variables in these experiments were input energy, aggregate type and matrix properties, percentage cross-sectional area and type of reinforcement. The dependent variables were energy absorption, cracking, deflection, cratering and residual load capacity.

2. EXPERIMENTAL DETAILS

2.1 Materials and manufacture of specimens

The materials and mix proportions used are listed in table 1. The properties of the different types of reinforcement are listed in table 2 and the test specimen types are illustrated in Fig. 1. All tables are given in the Appendix.

The concrete constituents were mixed in the standard way, placed in the forms and vibrated using table vibration or a 'Kango hammer'. The Tensar reinforcement was held in position by tying it to the sides of the form with nylon thread. All specimens were stripped from the mould at one day and cured at a temperature of 20°C $\pm$ 1°C and a relative humidity of 95% to 100% until they were tested; minimum 7 days.

2.2 Equipment

The equipment and arrangements used for the dynamic tests are illustrated in Fig. 1.

2.3 Instrumentation

The instruments and techniques listed in table 3 were used to record transient phenomena and post test damage parameters. Not all the measurements were made on each specimen. The residual strength of flexural specimens was determined in subsequent static four point flexural loading tests.

3. RESULTS

The results from these experiments are illustrated in Figs. 3 to 20 and they show how the following phenomena produced by dynamic forces

Polymer grid reinforcement. Thomas Telford Limited, London, 1985

233

were influenced by the type and quantity of reinforcement in mortar and concrete slabs:

(i) Crack formation and development.
(ii) Deflection.
(iii) Crater formation and dimensions.
(iv) Projectile penetration.

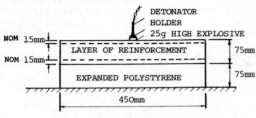

EXPLOSIVE IMPULSE EXPERIMENT

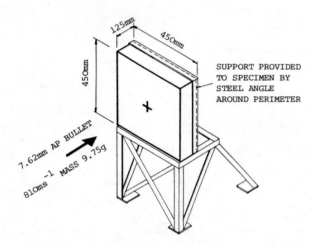

BULLET IMPACT EXPERIMENT

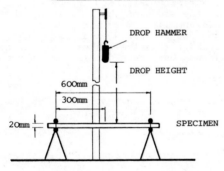

DROP HAMMER IMPACT EXPERIMENT

FIG. 1

4. DISCUSSION

4.1 General

The objectives of these experiments were to determine the resistance of concrete slabs to impact and impulsive forces where the associated energy levels were 150kJ, 3.2kJ and 50J.

For each energy level, the principal independent variables were the type and percentage of reinforcement and the density of the concrete. The cross sectional area of the Tensar was measured on the rib, midway between the nodes.

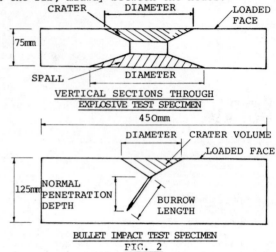

VERTICAL SECTIONS THROUGH
EXPLOSIVE TEST SPECIMEN

BULLET IMPACT TEST SPECIMEN
FIG. 2

The main purpose in altering the reinforcement was
(i) to compare the dynamic resistance of slabs reinforced with Tensar grids to that of steel reinforced slabs.
(ii) to compare the various types of Tensar grids available.
(iii) to compare the response of reinforced with unreinforced specimens.

The ultimate static tensile resistance of the steel mesh reinforcement was 1.75 kN or 70 kN. By using different types of Tensar grid and by altering the number of layers, the ultimate static tensile resistance of the Tensar was varied from about 20 kN to 220 kN.

The dependent variables measured, where appropriate in these dynamic experiments, were

(i) the dimensions of the crater and spall.
(ii) the depth of projectile penetration.
(iii) the length of the cracks.
(iv) the midspan maximum transient deflection.
(v) the period of forced oscillation.

After the drop hammer experiments these slabs were loaded in flexure as in the static load experiments[1] in order to determine the static mid span load required to produce a 20mm deflection at mid span. This measured residual load capacity is a good indicator of whether it is feasible to repair the cracks in an impact damaged slab.

4.2 Crater and spall dimensions, depth of projectile penetration

4.2.1 Introduction

The explosive impulse and the bullet impact experiments both produced significant cratering on the loaded face and sometimes spalls on the parallel but unloaded face, fig. 2. The projectile also produced a hole at the base of the crater, Fig. 2. The drop hammer did not produce a crater or spall but only a small indentation at the hammer impact point.

The volume, maximum diameter and depth were used to quantify the crater. The normal penetration depth was used to measure the hole produced by the bullet. These have been plotted against the principal independent variables in Figs. 3-9. From these figures the following observation can be made, although in many cases these are tentative results requiring more experiments to establish the random variability present in each parameter.

4.2.2 Explosive impulse experiments

The most obvious result from these experiments with 25g of high explosive at the centre of 450mm square x 75mm concrete slabs, was that an unreinforced slab was fractured and the parts propelled some distance from each other but a slab containing any of the Tensar grids or the steel mesh, was held together even though it was cratered and cracked, Fig. 10. Within the crater the reinforcement was fractured.

The volume of the crater and spall were influenced by the principal independent variables as follows:

(a) The crater volume was marginally less and the spall volume much less, in normal weight concrete (NWC) slabs doubly reinforced with Tensar than with steel, even though the steel provided a larger tensile resistance than any of the Tensar grids. Fig. 3.

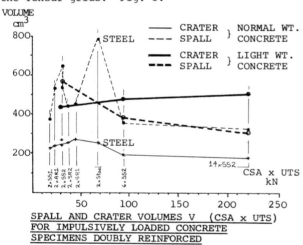

SPALL AND CRATER VOLUMES V (CSA x UTS) FOR IMPULSIVELY LOADED CONCRETE SPECIMENS DOUBLY REINFORCED

FIG. 3

(b) In lightweight concrete (LWC) slabs the volume of the crater was almost double that in NWC slabs with the slabs doubly reinforced with the same percentage of SS2 Tensar grid but the spall volumes were approximately equal. Fig. 3.

(c) The crater volume varied inversely with the percentage of SS2 Tensar in singly reinforced NWC slabs with the SS2 placed on the far face without concrete cover, Fig. 4.

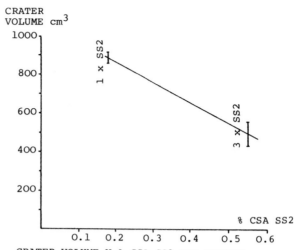

CRATER VOLUME V % CSA SS2 FOR SINGLY REINFORCED IMPULSIVELY LOADED NORMAL WEIGHT SPECIMENS

FIG. 4

(d) The crater volume was less in doubly reinforced NWC slabs with 15mm cover to the Tensar than in singly reinforced NWC slabs with the same total percentage of SS2 Tensar. Figs. 3 & 4.

(e) The crater volume reduced as the number of layers of SS2 Tensar in NWC slabs, doubly reinforced with 15mm cover, increased from 2 to 6 but above this it was unchanged. Fig. 3.

(f) The spall volume varied inversely with the percentage of SS2 in doubly reinforced LWC and NWC slabs. Fig. 3.

(g) The spall volume was greater than the crater volume in NWC slabs irrespective of the percentage or type of reinforcement. Fig. 3.

The crater and spall diameters were influenced by the independent variables as follows:

(a) The spall diameter was always greater than the crater diameter irrespective of the type of concrete or percentage and type of reinforcement. Fig. 5.

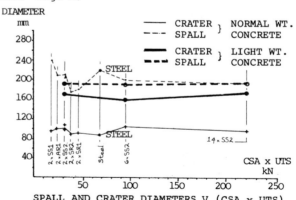

SPALL AND CRATER DIAMETERS V (CSA x UTS) FOR IMPULSIVELY LOADED CONCRETE SPECIMENS DOUBLY REINFORCED

FIG. 5

(b) The spall diameter in many of the doubly reinforced NWC slabs was less using Tensar than using steel reinforcement even when the steel provides a greater tensile resistance. Fig. 5.

(c) The crater diameters in LWC slabs were approximately double those in NWC slabs with the same percentage of SS2. Fig. 5.

(d) The crater diameter was almost the same in all NWC slabs doubly reinforced, irrespective of the type or percentage of reinforcement. Fig. 5.

The mechanism of crater formation is by both compressive crushing and tensile cracking of the concrete, Fig. 6. The expanding stress pulse produces trajectories of principal stress which are confocal orthogonal ellipsoids for tension and hyperboloids for compression[2].

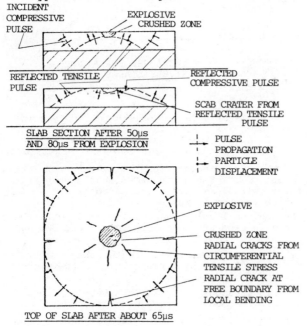

INCIDENT COMPRESSIVE PULSE
EXPLOSIVE CRUSHED ZONE
REFLECTED TENSILE PULSE
REFLECTED COMPRESSIVE PULSE
SCAB CRATER FROM REFLECTED TENSILE PULSE

SLAB SECTION AFTER 50µs AND 80µs FROM EXPLOSION

PULSE PROPAGATION
PARTICLE DISPLACEMENT

EXPLOSIVE

CRUSHED ZONE
RADIAL CRACKS FROM CIRCUMFERENTIAL TENSILE STRESS
RADIAL CRACK AT FREE BOUNDARY FROM LOCAL BENDING

TOP OF SLAB AFTER ABOUT 65µs

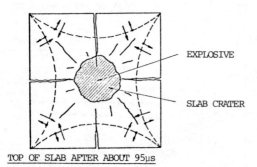

EXPLOSIVE

SLAB CRATER

TOP OF SLAB AFTER ABOUT 95µs

DIAGRAMS OF SEQUENTIAL CRATER AND CRACK DEVELOPMENT IN AN UNREINFORCED CONCRETE SLAB

FIG. 6

The volume of the crater will depend partly upon the extent of the compression fracture zone, and is therefore larger in the LWC than the NWC slabs which have a greater compressive strength.

It also depends on the length of the tensile cracks and it is these tensile cracks which can be controlled by the reinforcement. The results above have shown that a Tensar grid close to the cratered surface reduces the crater volume and is at least as effective as the steel mesh in limiting the size of the crater, even though the steel, on the basis of the static ultimate strength, can produce a greater tensile resistance.

Although reinforcement is most effective in reducing the volume of the crater when placed close to the cratered surface, the results above have shown that the volume can also be reduced by increasing the percentage of Tensar in singly reinforced specimens when the Tensar is on the face opposite to the cratered surface. In this position the Tensar is well below the bottom of the crater but can influence the reflected stress pulse.

Because the bottom of the slab is an interface with the lower density polystyrene, Fig. 1; then the compressive pulse is reflected as a tensile pulse and on reaching the, by now, cratered surface, is again reflected but as a compressive pulse. If the amplitude is sufficient, it will then expand the crater which has already been formed by the mechanism described above. It appears from the results that Tensar on the bottom of the slab will reduce this amplitude.

The spall, which is the crater formed on the concrete-polystyrene interface, is produced when the incident compressive pulse and reflected tensile pulse interact to produce a net tensile stress which exceeds the tensile strength of the slab. This will occur over the ellipsoidal surface of the reflected wave front and the spall forms a shallow dome of concrete. The mechanism of spall formation then depends only on tensile cracking of the concrete and not on compressive crushing. A further difference between crater and spall formation is that the tensile stresses produce cracks propagating obliquely into the slab for crater formation, but more parallel to the slab boundary in spall formation.

For these reasons the spall is expected to have a larger volume than the crater and its volume depends much more on the reinforcement near to the spalled surface.

4.2.3 Bullet impact experiments

The 90° impact of a 9.75g AP 7.62mm bullet at the centre of a 450mm square x 125mm thick concrete slab, Fig. 1, produced a surface crater and hole, Fig. 2. The hole diameter was equal to that of the bullet and although it was often straight it was not often perpendicular to the impacted face.

The diameter and depth were used to quantify the crater and the most important parameter of the hole was considered to be its depth measured perpendicular to the impact surface. These results have been plotted against the principal independent variables in Figs. 7-9.

The crater diameter in unreinforced NWC slabs was not significantly different from that in reinforced NWC slabs, whether one of the Tensar grids or a steel mesh was used as the reinforce-

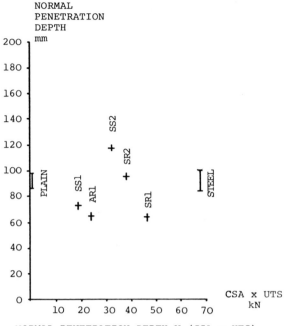

NORMAL PENETRATION DEPTH V (CSA x UTS)
FOR SPECIMENS TESTED WITH HIGH VELOCITY
ARMOUR PIERCING BULLET

FIG. 7

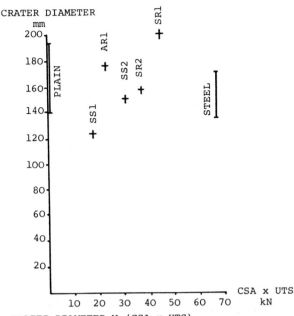

CRATER DIAMETER V (CSA x UTS)
FOR SPECIMENS TESTED WITH HIGH VELOCITY
ARMOUR PIERCING BULLET

FIG. 8

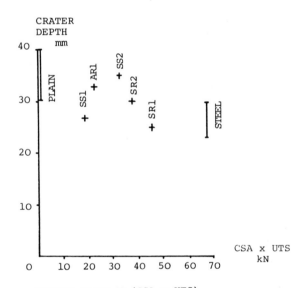

CRATER DEPTH V (CSA x UTS)
FOR SPECIMENS TESTED WITH HIGH VELOCITY
IMPACT FROM ARMOUR PIERCING BULLET

FIG. 9

ment. The variation in the maximum diameter of the crater for two unreinforced NWC slabs of equal strength was almost as much as the variation for the slabs with different types of reinforcement. Fig. 8.

The crater depth was smallest when the NWC slab was reinforced with a Tensar grid having a high tensile resistance or with the steel mesh. Fig.9.

The normal penetration depth in unreinforced NWC slabs was about the same as that in slabs reinforced with the steel mesh. The variation between the normal penetration in two impacts in equal strength, unreinforced NWC slabs and between two impacts in equal strength, NWC slabs reinforced with steel, was less in both cases than the difference between that in equal strength NWC slabs, one reinforced with SR1 Tensar and the other with SR2 Tensar. Fig. 7.

The simple mechanism of crater and hole formation by projectile impact is different from that by explosive impulse. If the projectile penetrates the concrete then it produces high compressive pressures perpendicular to the projectile path. This radial compressive pressure produces perpendicular tensile stresses which near to the impact surface are perpendicular to that surface and the subsequent cracking cannot be controlled by reinforcement placed parallel to the impact surface. In fact the reinforcement could produce a plane of reduced tensile strength in the concrete, and those grids with the greatest area of material in plan and the weakest bond to the concrete, will cause the greatest weakness. The crater depth would then

be determined mainly by the plan area of material in the grid and the cover to the reinforcement. The results have shown that the crater diameter is not dependent upon the type or tensile resistance of the reinforcement but the crater depth is less with steel or SR1 Tensar reinforcement.

4.3 Crack formation and transient deflection

4.3.1 Introduction

The explosive impulse and the drop hammer impact experiments both produced distinctive cracking in the concrete slabs, Figs. 10, 11. In contrast very few cracks were produced by the bullet impact, Fig. 12.

2 LAYERS SS2

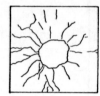

2 LAYERS SS2

2 LAYERS 12/16 GAUGE STEEL

6 LAYERS SS2

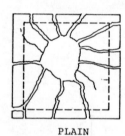

PLAIN

NORMAL WEIGHT
CONCRETE (NWC) SPECIMENS

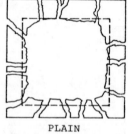

PLAIN

LIGHTWEIGHT CONCRETE (LWC)
SPECIMENS

TYPICAL SPECIMENS AFTER IMPULSIVE LOADING
PRODUCED BY 25g OF HIGH EXPLOSIVE

FIG. 10

The 450mm square x 75mm thick concrete slabs used in the explosive impulse experiments were continuously supported on polystyrene, Fig. 1, and most of the cracks were caused by transient longitudinal stress waves. In contrast, the 700mm x 300mm x 20mm thick concrete slabs used in the drop hammer impact experiments were supported on two line supports, Fig. 1, and the cracks were produced partly by transient longitudinal stress waves and partly by transient bending stresses.

The length of the permanent cracks was measured in all the experiments, and the transient deflection at mid span during the drop hammer impact experiments. These results are plotted against the principal independent variables in Figs. 13, 14 & 15.

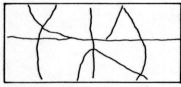

NWC SPECIMEN WITH 1 LAYER
OF SS1 TENSAR

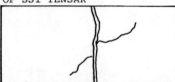

NWC SPECIMENS WITH 1 LAYER
OF 25/25 GAUGE STEEL

LWC SPECIMEN WITH 1 LAYER
OF SS1 TENSAR

LWC SPECIMEN WITH 1 LAYER
OF   25/25 GAUGE STEEL

IMPACTED FACE OF SPECIMENS AFTER DROP
HAMMER TEST

FIG. 11

4.3.2 Explosive impulse experiments

Typical crack patterns in the 450mm square by 75mm thick concrete slabs are shown in Fig. 10 for Tensar and steel reinforced NWC and LWC slabs and for reassembled unreinforced slabs. The reinforced slabs were cracked but held together whatever the type of reinforcement, but, the total crack length was not particularly influenced by the percentage area, or type of reinforcement. The observations from the experiments are as follows:

(a) The total crack length is about the same in doubly reinforced NWC slabs whether two layers of any of the Tensar grids or the steel mesh were used, even though the steel provides the largest tensile resistance. Fig. 13.

(b) The total crack length is not significantly influenced by the percentage of SS2 Tensar in either singly or doubly reinforced NWC slabs. Fig. 14.

The typical crack patterns in Fig. 10 show cracks which are all radial from the point of application of the explosive impulse. Although they are all radial they are not all produced by the same tensile conditions and don't all propagate outwards from the centre of the slab, Fig. 6. Those radial cracks which reach the corners

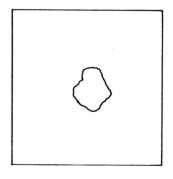

2 LAYERS SS1

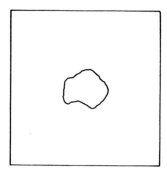

2 LAYERS 12/16 GAUGE STEEL

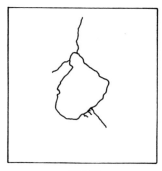

PLAIN

FRONT FACE OF SPECIMENS SUBJECTED TO HIGH
VELOCITY IMPACT FROM ARMOUR PIERCING BULLET

FIG. 12

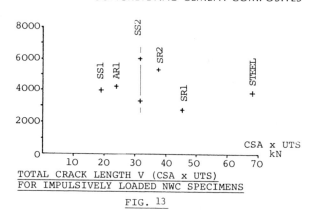

TOTAL CRACK LENGTH V (CSA x UTS)
FOR IMPULSIVELY LOADED NWC SPECIMENS

FIG. 13

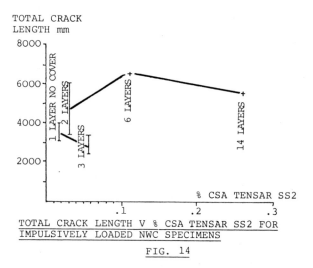

TOTAL CRACK LENGTH V % CSA TENSAR SS2 FOR
IMPULSIVELY LOADED NWC SPECIMENS

FIG. 14

RC-type, or the mid points of the sides of the
slabs, RM-type, depend upon the edge conditions
and were initiated at the edges of the slab.
The other radial cracks, R-type are formed by
the radially expanding compressive stress wave
before it reaches the edges of the slab. Some
of these may have connected with RM or RC type
cracks.

Fig. 16 shows that although any of the Tensar
grids or the steel mesh were sufficient to control the radial cracks, the RM-type cracks were

extremely wide in slabs with low percentages of
SS2 Tensar. If the reinforcement was to fracture along these cracks then the slab would
disintegrate.

4.3.3 Drop hammer impact experiments

Typical crack patterns in the 700mm x 300mm x
20mm slabs spanning 600mm are shown in Fig. 11
for NWC and LWC slabs singly reinforced with
Tensar or steel. The hammer produced a point
impact at mid span, Fig. 1.

The most obvious results were for the unreinforced slabs where the impact produced a
single crack near mid span, and across the full
width of the slab producing total collapse.
In the LWC and NWC slabs reinforced with steel,
only a few cracks were formed, Fig. 11, and
the crack near mid span across the full
width of the slab, developed as the steel yielded and
ultimately fractured, producing total collapse
of the slab.

In contrast to steel, Tensar **rein**forced slabs
had a larger tensile resistance; none of these
collapsed. The influence on total crack length
of the principal independent variables in LWC
and NWC slabs reinforced with different percen-

239

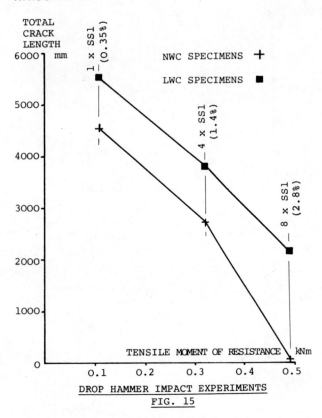

TOTAL CRACK LENGTH
6000 mm

1 x SS1 (0.35%)

NWC SPECIMENS +

LWC SPECIMENS ■

5000

4 x SS1 (1.4%)

4000

8 x SS1 (2.8%)

3000

2000

1000

TENSILE MOMENT OF RESISTANCE kNm

0

0.1   0.2   0.3   0.4   0.5

DROP HAMMER IMPACT EXPERIMENTS
FIG. 15

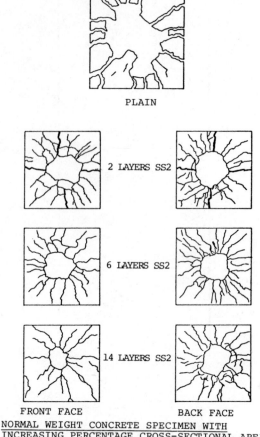

PLAIN

2 LAYERS SS2

6 LAYERS SS2

14 LAYERS SS2

FRONT FACE   BACK FACE

NORMAL WEIGHT CONCRETE SPECIMEN WITH
INCREASING PERCENTAGE CROSS-SECTIONAL AREA
OF TENSAR AFTER IMPULSIVE LOADING WITH 25g
OF HIGH EXPLOSIVE

FIG. 16

tage areas of SS1 Tensar grid is shown in Fig.
15 and the following observations are made:

(a) The total crack length decreases considerably
in both LWC and NWC slabs singly reinforced with
SS1 Tensar, as the percentage area of SS1
increases from 0.35% to 2.8%, Fig. 15. In the
slabs with 0.35% area of SS1, tensile moment of
resistance was 0.1kNm and considerable cracking
occurred but collapse was prevented. In contrast, the slabs with steel had a tensile
moment of resistance of 0.03kNm and collapsed.

(b) The total crack length was greater in LWC
than in NWC slabs with the same percentage area
of SS1 Tensar reinforcement. Fig. 15.

Of the cracks which typically formed in the
Tensar reinforced slabs, some, like the orthogonal cracks through the impact point and parallel to the sides of the slab were formed by
flexural deformation. These cracks were usually
through the full depth of the slab. Other
cracks, such as those which were radial from
the impact point are formed by the tensile
stresses associated with the longitudinal compressive stress pulse propagating radially from
the impact point, producing hoop tensile stress.

The periodic oscillations were recorded photo-
graphically and typical records are given in
Fig. 17 for NWC and LWC slabs singly reinforced
with different percentage areas of SS1 Tensar
after equal energy impacts from a single blow of
the drop hammer.

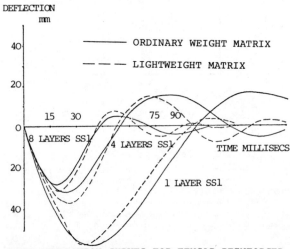

DEFLECTION mm

40

——— ORDINARY WEIGHT MATRIX

- - - LIGHTWEIGHT MATRIX

20

15   30   75 90

0

8 LAYERS SS1   4 LAYERS SS1   TIME MILLISECS

20

1 LAYER SS1

40

TIME DEFLECTION CURVES FOR TENSAR REINFORCED
SLABS IN THE DROP HAMMER IMPACT EXPERIMENTS

FIG. 17

The obvious results from these experiments are that the steel reinforced slabs did not oscillate but collapsed on impact. The oscillations of the Tensar Reinforced slabs damped out to leave practically no permanent deflection of the slab. The maximum transient deflections at mid span and the period of forced oscillation obtained from curves such as Fig. 17 are plotted in Figs. 18 and 19.

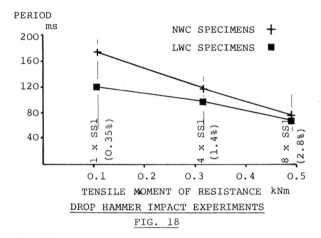

DROP HAMMER IMPACT EXPERIMENTS

FIG. 18

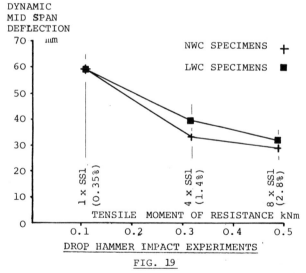

DROP HAMMER IMPACT EXPERIMENTS

FIG. 19

The following observations are made:

(a) The period of forced oscillation is less in LWC than NWC slabs singly reinforced with equal percentage areas of SS1 Tensar, particularly when the area of SS1 was below 1%. Fig. 18.

(b) The period of forced oscillation and the maximum midspan transient deflection, were almost halved in both LWC and NWC slabs singly reinforced with SS1 Tensar, as the percentage area of SS1 was increased from 0.35% to 2.8%, Figs. 18 and 19.

(c) The maximum transient mid span deflection

is slightly greater in LWC than in NWC slabs, singly reinforced with equal percentage areas of Tensar, Fig. 19.

These results have indicated that all Tensar reinforced slabs have a pseudo-elastic response to a single impact of 50J energy. Increasing the percentage area of SS1 Tensar increases the stiffness and decreases the total length of cracks in the slab.

4.4 Residual load carrying capacity of impact damaged slabs

The results plotted in Fig. 20 show the influence of the principal independent variables on the residual strength of impact damaged slabs, i.e. the static load required for 20mm deflection at mid span. A comparison is also made with the load required to produce the same deflection in identical slabs, but which had not previously been loaded.

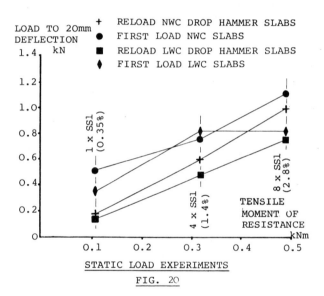

STATIC LOAD EXPERIMENTS

FIG. 20

The following observations are made:

(a) A greater load was required to produce 20mm deflection of impact damaged NWC slabs than for LWC slabs, singly reinforced with equal percentage areas of SS1 Tensar, Fig. 20.

(b) The residual strength of impact damaged NWC and LWC slabs was about 40% of the strength of identical undamaged slabs with 0.38% of SS1 Tensar, but about 90% with a 2.8% SS1 Tensar. Fig. 20.
The Tensar reinforced slabs retained a considerable proportion of their load carrying capacity after an impact of 50J. Because there is no risk of corrosion from water penetrating to the Tensar, it is quite feasible to repair the cracks and leave the slab in place on the structure.

5. CONCLUSIONS

These results show that Tensar can be effective as reinforcement in concrete components subjected to dynamic loading and its structural performance is comparable with steel reinforcement in many instances but has potential advantages arising from its inherent resistance to water and chemical products[3].

ACKNOWLEDGEMENTS

This research is being carried out in the Department of Civil and Structural Engineering, University of Sheffield and is supported by the Science and Engineering Research Council and Netlon Ltd. Thanks are given for all the technical and administrative help and advice given by the staff of all these organisations.

REFERENCES

1. Swamy, R.N., Jones, R., Oldroyd, P.N. "The behaviour of Tensar reinforced cement composites under static loads", Symposium on Polymer grid reinforcement in Civil Engineering, ICE, London, March 1984.

2. Watson, A.J., Sanderson, A.J. "The resistance of concrete to explosive shock" pp. 307-317, Proceedings of the Second Conference on the Mechanical Properties of Materials at High Rates of Strain, Oxford, 28-30, March 1979.

3. B.P. Chemicals Ltd. Technigram T19/1 "Chemical Resistance of Rigidex."

APPENDIX

TEST	AGGREGATE FINE			AGGREGATE COARSE			AGG/CEMENT BY WEIGHT	WATER/CEMENT BY WEIGHT
	TYPE	GRADE	%	TYPE	GRADE	%		
SURFACE EXPLOSIVE ORDINARY WEIGHT	LIMESTONE	ZONE 2	50%	LIMESTONE	10mm DOWN	50%	4.72	0.45
SURFACE EXPLOSIVE LIGHTWEIGHT	LYTAG	ZONE 2	50%	LYTAG	10mm DOWN	50%	2.2	0.6
BULLET IMPACT	LIMESTONE	ZONE 2	50%	LIMESTONE	10mm DOWN	50%	4.72	0.45
DROP HAMMER ORDINARY WEIGHT	SHARP SAND	ZONE 3	100%	-	-	-	3.0	0.6
DROP HAMMER LIGHTWEIGHT	LECA	ZONE 2	100%	-	-	-	2.0	0.45

(ALL MIXES USED RAPID HARDENING PORTLAND CEMENT)
MATERIALS AND MIX PROPORTIONS

TABLE 1

TYPE	SHAPE	C.S.A.		STATIC PROPERTIES				ρ
		MAIN mm^2/M	SECOND mm^2/M	TENSILE		SECANT MODULUS		
				MAIN N/mm^2	SEC N/mm^2	MAIN GPa	SEC GPa	kg/m^3
SS1	BOP	71	44	288	151	1.9	0.6	900
SS2	BOP	141	86	248	136	1.2	0.7	900
AR1	BOP	182	154	145	113	1.0	0.8	900
SR1	UOP	239	-	310	-	1.7	-	900
SR2	UOP	312	-	260	-	1.5	-	900
SR3	UOP	230	-	480	-	5.6	-	900
12/16 STEEL	RECTANGULAR	179	51	420	420	210	210	7800
25/25 STEEL	SQUARE	16.5	16.5	350	350	210	210	7800

PROPERTIES OF TENSAR AND STEEL REINFORCEMENT

TABLE 2

LOAD	MEASUREMENT	INSTRUMENTATION	TECHNIQUE
25g HE	Crater volume	-	Sand replacement
	Crater diameter	Planimeter	-
	Crater depth	Ruler & straight edge	Depth on sawn C/S
	Crack lengths		Line interaction
bullet 9.75g	Crater diameter	Ruler ± 0.5mm	-
	Crater depth	Ruler & straight edge	Depth on sawn C/S
	Hole dia & length	Divider	On sawn C/S
5kg mass 2kg mass	Transient cracks	High speed camera	-
		Silver paint crack detectors	
	Transient strains on concrete	E.R.S. gauges	
	Cracking	Pundit	U.P.V.
		-	Line interaction
	Deflection	LVDT	
	Residual strength		Static reload test

MEASUREMENTS OBTAINED FROM DYNAMIC EXPERIMENTS

TABLE 3

Concrete and cement composites: report on discussion

G. Somerville, *Cement and Concrete Association*

Discussion on papers 7.1 and 7.2 followed two themes:-

1) A series of questions seeking clarification on details of the experimental work and asking for information on some aspects of performance apparently not covered implicitly in the papers.

2) Possible applications for the material, and the procedures to be followed in bringing these to fruition.

On the first of these, it was noted that fairly small specimens had been used, and the relevance of the results to full size elements was queried. In response, the authors correctly stated that the use of small scale models was a recognized and well-established research technique, and that ample correlation had been established to check on the laws of similitude between specimens of different size and scale. Additionally, for the impact work in particular, there were practical limitations, if an extensive series of tests covering a wide range of variables was to be completed in a reasonable period.

This first theme was developed further in assessing the possibility of using Tensar grid materials in very thin sections; one speaker asked about the possibility of drawing the material into a very fine mesh or even into fibre-type reinforcement. Thin sections were seen as being important, since the properties established in the papers indicated that the most promising applications were those involving relatively light loads (such as cladding panels) where weight reduction and good durability were especially important. However, it was felt that the interlocking characteristics of Tensar grids should not be given up lightly, and hence material development in fibre form was perhaps not the best approach.

In developing the second (major) theme, it was felt that real progress had been made along the learning curve in terms of establishing material characteristics and performance under a range of loading types. Some possible applications had been identified in the two papers, but these had not been pursued in any depth. It was considered that this was the next essential step in the development process. This would help define more precise performance requirements which could be matched against established material characteristics. Essential ingredients in this process were:

- fitness for purpose in technical terms

- favourable costs compared with other solutions

- ease of manufacture and/or fabrication.

Making progress in this interactive way could in turn lead to further research to modify the material characteristics to suit specific applications or indeed to adapt to the more promising fabrication techniques.

In the discussion, one particular civil engineering application was illustrated where the ability of the composite material to distribute cracking was the key performance requirement. In general, however, it was felt that the properties identified so far showed particular promise in thin sections where the high resistance to corrosion could be used to advantage. This could lead to production in sheeting form - possibly using a range of fabricating techniques - and serious consideration in applications outside the traditional building and civil engineering fields.

Participants: Mr Bird

Professor Hannah

Dr Jones

Dr Mercer

Dr Somerville

Dr Watson

Polymer grid reinforcement. Thomas Telford Limited, London, 1985

A new method of soil stabilization

F. B. Mercer, *Netlon Ltd,* and K. Z.
Andrawes, A. McGown and N. Hytiris,
University of Strathclyde

The paper describes a novel method of soil
stabilisation which involves mixing into the soil
molecularly oriented mesh elements in the form of
squares, rectangles or ribbons. Laboratory compaction,
C B R, triaxial and model footing tests are detailed
in which 40 mm square mesh elements are mixed into
sand in order to identify the important properties of
the mesh and the effect of the mesh element content
on the behaviour of the stabilised soil. The results
indicate that the basic operating mechanism is that
each mesh interlocks with the adjacent soil particles
to form an aggregation and these aggregations are
locked together by the surrounding mesh elements to
form a coherent matrix with improved stress resistant
properties and increased ductility. These benefits
are obtained even when the mesh element content is
small.

INTRODUCTION

The stress resistant properties of soils can be
improved in a variety of ways. For example,
sheets, strips or rods of metal or polymeric
materials can be placed in a soil to create a
composite material with improved stress
resistant properties. This is known as
"Reinforced Soil" and can be regarded as soil
strengthening at the macro-scale. Alternatively,
the matrix of a soil can be modified by
physically or chemically binding the particles
together in such a way as to increase the
overall stress resistant properties of the
soil. Binding agents and processes used include
lime, cement, bitumen, heating and freezing.
All of these are termed "Soil Stabilisation
Techniques" and depend on soil strengthening
at the micro-scale.

A possible disadvantage of reinforced soil is
that strengthening is confined to specific
directions or surfaces and with some
reinforcing materials, weakening of the soil
may occur in some other specific directions or
surfaces, McGown et al (1978). Possible
disadvantages with the presently available
soil stabilisation techniques derive from
improvements in stress resistance often being
accompanied by reductions in permeability and
ductility.

The new soil strengthening technique discussed
in this paper involves mixing soil with small
squares, rectangles or ribbons of molecularly
oriented mesh. These mesh elements interlock
with groups of particles and provide tensile
resistance to the soil matrix. Their action
thus lies somewhere between existing soil
reinforcement and soil stabilisation techniques
and may thus be described as soil strengthening
at the meso-scale. By ensuring that the mesh
elements are randomly distributed within the
soil matrix, the anisotropy of soil reinforcement is largely overcome. As will be indicated
later, no reduction in void ratio occurs when
using mesh elements thus no reduction in
permeability is likely. Also from tests it is
found that soil ductility is not reduced, hence
this technique does not suffer the possible
disadvantages of available soil stabilisation
techniques.

It is envisaged that this innovative method of
soil strengthening has the potential of being
a most valuable additional means of increasing
the stress resistant properties of soils. It
is likely that it will be used in its own right
in many situations; however, it is also
considered possible that it will be employed
in conjunction with soil reinforcement or soil
stabilisation techniques in other situations.
Development and evaluation studies of the
technique are still under way and include an
intensive laboratory test programme and a study
of the practical problems of efficiently mixing
the soil and the mesh elements on-site. In
this paper, the early development work on the
identification of the soil-mesh interaction
mechanism, the evaluation of important mesh
properties and the laboratory scale testing of
soil-mesh mixtures are described.

SOIL-MESH INTERACTION MECHANISM

When a soil mass is subjected to a stress
system, it strains. Even when the externally
applied stress system is entirely compressive,
tensile strains may develop within the soil
mass. If tension resistant inclusions are
present within the soil in the zones and
directions of tensile strains, they will be
strained and so develop tensile resistance.
This will induce a transfer of stress from the
soil to the inclusions, thereby reducing the
strains induced by the externally applied
stress system, eventually leading to an increase
in the overall load carrying capacity of the
soil mass.

The mechanism of this load-transfer depends
upon the surface characteristics of the

Polymer grid reinforcement. Thomas Telford Limited, London, 1985

reinforcing materials. For the inclusion types currently in use, two main processes can be identified. Where flat sheets, strips or rods are used, the load-transfer depends on the frictional resistance developed at the soil-inclusion interface, but for grids and meshes it depends on their structural elements interlocking with the soil particles. This latter mechanism can be much more efficient than surface friction as it does not require relative movement of the soil and inclusion to mobilise it and it is not limited by the frictional properties of reinforcement material, depending mainly on the number and size of the openings in the grid or mesh and the size and shape of the cross members forming the structure.

It can readily be appreciated that the interlock principle applies to small mesh elements mixed in soil. In fact, it applies at two levels; each mesh element first of all interlocks with a small group of particles to form an aggregation, as in Fig.1(a), then these aggregations are locked together by adjacent mesh elements to form a coherent matrix, as shown in Fig.1(b).

Fig.1(a)   Aggregation of mesh element and soil

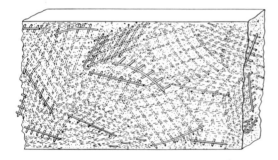

Fig.1(b)   Soil - mesh matrix

Attempts have been made to employ staple (short lengths) and continuous filaments to strengthen soil, Andersland & Khattak (1979), Hoare (1979) and Leflaive (1982). The load-transfer mechanism in these cases is essentially surface friction between the fibres and the individual particles. However, where large amounts of continuous filaments are used, the filaments tend to wrap themselves around particles or even small groups of particles and this can give an additional binder effect. It is not, however, directly comparable to the two level interlock mechanism operating with the mesh elements.

IMPORTANT PROPERTIES OF THE MESH

In order to optimise the interlock mechanism, the opening sizes and shapes of the mesh elements are important and must be related to the size of the soil particles in which they are placed. Equally so, the sizes and shapes of the filaments comprising the mesh are important. For best results, the filaments should have a high profile available for containment of the soil particles. In contrast, in order not to weaken the soil, the volume of the void space must not be significantly increased by the mixing of mesh elements into the soil. Equally, the mesh elements must not significantly reduce void space otherwise soil permeability will reduce. Thus restrictions must be placed on the sizes and shape of the filaments in the mesh and on the gross amount of mesh that may be added. Thus the size and shape of the filaments and mesh openings must be balanced for maximum efficiency and the amount of the mesh added to the soil minimised.

As stated previously, the mesh elements interlock with soil particles to form aggregations and these are in turn locked together by adjacent meshes to form a coherent matrix. When this matrix is stressed, tensile strains may develop and the tensile resistance of the mesh mobilised. In order to minimise the amount of mesh required to achieve any specific improvement, the mesh should possess as much tensile strength as possible over the range of operational tensile strains and sustain this throughout the operational lifetime of the structure in which it is placed. Thus the entire load-extension-time behaviour of the mesh elements is important.

Mixing of the mesh elements into the soil must, however, be easily and efficiently achieved. Generally the requirement will be that the mesh elements are evenly distributed and randomly oriented throughout the soil matrix. Preliminary tests using a wide variety of mesh types have shown that the crucial factors in achieving this are the size and shape of the elements and their flexural stiffness and recovery. For different methods of mixing in different end uses, it is likely that the elements will vary in shape from squares to rectangles to continuous ribbons. For ease of mixing and maintenance of their geometrical stability during this stage and during subsequent stressing, the flexural stiffness and recovery of the elements has been found to be absolutely vital. Figure 2 shows the form of some very flexible meshes removed from soil after only hand mixing. It is clear that these would form loose bundles and large voids within the soil and not interlock as intended. At the opposite end of the stiffness range, rigid elements were found to form bridges and so void spaces within the soil, which is highly undesirable. Thus the flexural stiffness and recovery properties of the mesh elements are important and must be carefully selected.

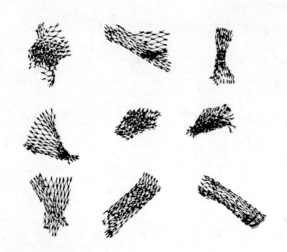

Fig.2   Flexible meshes after mixing

PRELIMINARY LABORATORY TESTING
MATERIALS USED

Prior to testing soil-mesh mixtures, it was necessary to choose a suitable soil and match this with suitable mesh elements.

Soil: The soil chosen was a readily available processed fluvial glacial sand. It has sub-angular particles with the gradation shown in Fig.3 and is known as Mid-Ross sand.

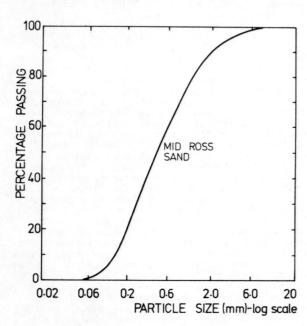

Fig.3   Particle size distribution of Mid-Ross sand

Mesh Elements: Many possible mesh types and element shapes were subjected to some exploratory tests and from these 40 mm square elements of Netlon Mesh Type 7 constructed with filaments of the shape shown in Fig.4, were chosen as the most suitable.

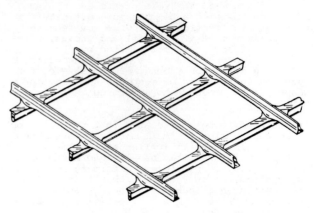

Fig.4   Structure of mesh used in tests

The physical properties of the mesh are given in Table 1 together with the tensile strength obtained from tests conducted on samples 200 mm wide by 100 mm long at 2% per minute constant rate of strain and temperature of 20°C.

TABLE 1   PROPERTIES OF MESH ELEMENTS TESTED

Production process	Netlon
Type	7
Polymer	Polypropylene
Overall size	40 mm x 40 mm
Mass/Unit area	52 g/m^2
Opening size	6.1 x 7.1 mm
Filament thickness	0.5 mm M.D.
	0.48 mm X.M.D.
Maximum Tensile Strength	3.5 kN/m M.D.
(2%/min. at 20°C)	3.80 kN/m X.M.D.

TESTING UNDERTAKEN

Four sets of laboratory tests were undertaken during the preliminary test programme using the Mid-Ross sand and the 40 mm square Netlon Type 7 mesh elements.

Compaction Tests: The tests were carried out in a standard CBR mould. This is 6" (152 mm) diameter and 7" (177 mm) high. The soil-mesh mixture is placed into this in three equal layers and each layer is then subjected to 55 blows using a 5.5 lb (2.5 kg) hammer dropping through a height of 12" (300 mm) onto the soil. First of all several tests were conducted on the soil alone using different water contents to establish the relationship between dry density and water content for this particular soil. As shown in Fig.5, the optimum moisture content for the Mid-Ross sand was 7.5% and the maximum dry density was 1818 kg/m^3 for this

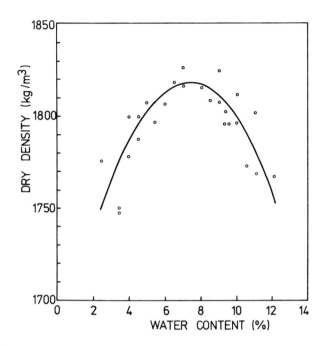

Fig.5   Compaction curve for Mid-Ross sand

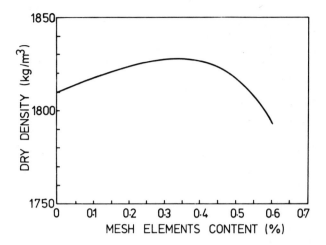

Fig.6   Effect of mesh element content on
         the dry density of compacted samples

level and method of compaction.

Further to establishing these data, a soil
moisture content of 9.3% was chosen for use in
all subsequent compaction and CBR tests on soil
alone and soil-mesh mixtures. This was chosen
as it ensured that the air void space available
for occupation by the mesh filaments, without
causing increase in void space, was at a
minimum. Thus tests at this moisture content
provided a "worst case" condition.

Following this, various percentages of mesh
(by dry weight of soil) were mixed into the
soil at a water content of 9.3% and subjected
to exactly the same compaction test as the soil
alone. The data obtained are presented in
Fig.6 and indicate that up to a mesh content of
0.6%, the dry density of the mixture is the
same or slightly greater than that for the soil
alone. Therefore the mesh elements are not
causing an increase in void space in the soil.
At percentages of mesh content in excess of
0.6%, the dry density of the mixture rapidly
decreased, indicating that the meshes were
forming additional void space in the soil,
which is not desirable.

C.B.R. Tests: The compacted soil and soil-mesh
mixtures were all subjected to CBR tests. In
this test, the soil in the compaction mould is
placed in a testing machine, an annular
surcharge load of 5 lb (2.27 kg) is placed on
the trimmed top surface of the soil and a
2" (50 mm) diameter plunger is pushed into the
soil at a constant rate of 1.25 mm/min. The
test specimen is then turned upside down, the
base of the mould removed, the annular ring

positioned and the test repeated as before on
the bottom of the specimen. The relationships
between penetration and load applied are
plotted and corrected where necessary for
bedding errors, and the CBR values calculated
for both top and bottom of the specimen as
follows:

For 0.1" (2.5 mm)
penetration $\qquad$ $CBR = \dfrac{Test\ Load}{3000\ (lb)} \times 100\%$
$\qquad\qquad\qquad\qquad\qquad\quad (13.3\ kN)$
and

For 0.2" (5.0 mm)
penetration $\qquad$ $CBR = \dfrac{Test\ Load}{4500\ (lb)} \times 100\%$
$\qquad\qquad\qquad\qquad\qquad\quad (20\ kN)$

Figure 7 shows the effect of mixing various
percentages of mesh, by dry weight of soil,
into the Mid-Ross sand, in terms of CBR values
at both 0.1" and 0.2" penetration. It also
shows data from both the top and bottom of the
test specimens but it must be pointed out that
the results obtained using the top of the
specimen should be discounted due to unavoid-
able disturbance of the specimen during
trimming of the surface of the soil-mesh
mixture. The results from the bottom of the
test specimen were undisturbed and are
therefore more indicative of the CBR value of
the soil-mesh mixtures. These bottom values
show that there is a steady improvement in
CBR values as the percentage of mesh is
increased up to almost 0.6% where some 400%
improvement over soil alone is evidenced. It
is envisaged that smaller mesh contents would
be used in practice; however, it is clearly
demonstrated in this test that such smaller
mesh contents would still provide substantial
improvement in soil properties.

Triaxial Tests: Dry samples of sand alone and
sand mixed with various percentages (by weight)
of mesh elements were subjected to drained
triaxial tests. The specimens were 6" (152 mm)

247

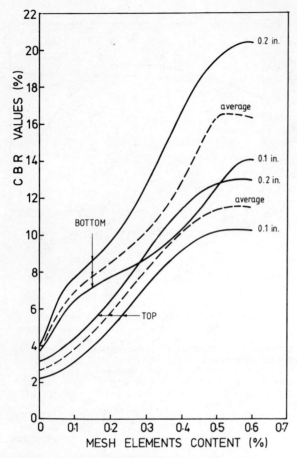

Fig.7   Effect of mesh element
         content on CBR values

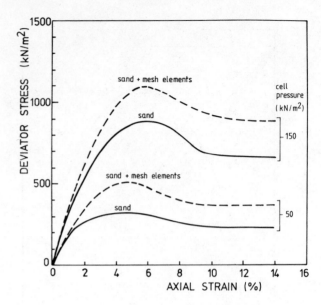

Fig.8   Triaxial test results for sand
         and sand containing 0.19% by
         weight of mesh elements

diameter and 7" (177 mm) high, the same size as the CBR specimens. Indeed they were prepared by the same compaction procedure in split CBR moulds, the only difference being that the soil was dry. The samples were then placed in the triaxial test apparatus using lubricated end platens and subjected to cell pressures in the range 50 to 300 kN/m^2, after which they were sheared under a constant rate of strain of 5.6×10^{-4}%/min. Figure 8 shows the results of tests on the sand with and without 0.19% by weight of mesh elements at cell pressures of 50 and 150 kN/m^2. These data show that the peak deviator stresses, at these cell pressures, increased by 60% and 25% respectively, which confirms the ability of the mesh elements to strengthen soils. The data also indicate that soil improvements can be generated at low strains and low stress levels using these mesh elements and that the strength of the soil-mesh mixture is maintained over a larger strain range than with the soil alone. Thus the soil-mesh mixture is more ductile than the soil alone.

Footing Tests: In order to ensure that the previously described improvements in soil

strength were not limited to these tests, plane strain model footing tests were undertaken. The test apparatus consisted of a rigid glass sided tank 640 mm long x 300 mm deep x 75 mm wide. The tank was filled with dense, dry Mid-Ross sand with and without mesh elements and a smooth metal footing 75 x 75 mm pushed down into it at a constant rate of penetration of 1 mm/min. The average data obtained from five tests conducted on the sand alone is shown in Fig. 9 which is typical of the load-settlement behaviour of this type of soil. A series of tests were then conducted with a layer containing 0.19% (by weight) of mesh elements over the dense sand. The depth of the soil-mesh layer (D) was varied from 0.5 to 4 times the breadth of the footing (B). Each test was repeated at least twice and the data from these are presented in Figs. 9 and 10.

The data from the tests illustrate that the bearing pressure increases, for any given settlement, when the soil-mesh layer is present. Also from these figures it can be seen that for a soil-mesh layer depth of 1.5B or greater no significant further increase in load at any given settlement is achieved. This implies that deep treatment of soils in such a situation is not required. Further, as the load bearing capacity at any settlement increases with increase in settlement, the soil-mesh layer has increased the ductility of the soil system, a factor previously noted in the triaxial test data.

CONCLUSIONS

The early development work on the use of small amounts of molecularly oriented mesh elements to increase the stress resistant properties of

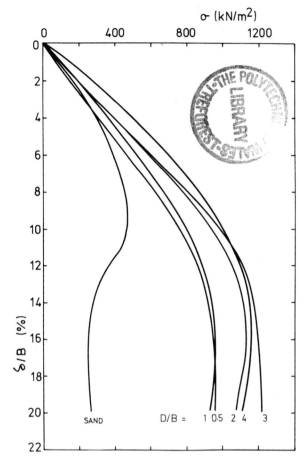

Fig.9   Footing tests – effect of the
         depth of the stabilised layer
         on the load-settlement behaviour;
         mesh element content = 0.19%

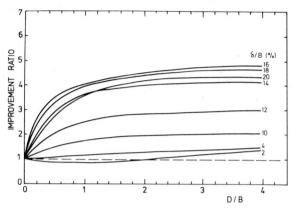

Fig.10   Footing tests – effect of the
         depth of the stabilised layer on
         the load carrying capacity at
         different settlements;
         mesh element content = 0.19%

that the rapid development of this new concept
into a practical soil improvement will be
possible before very long.

ACKNOWLEDGMENTS

This research work is being carried out at the
University of Strathclyde with the financial
support of Netlon Limited.

It is based on fundamental research conducted
by F. Brian Mercer.

Patent applications have been filed in a large
number of countries.

soils has shown that improvements can be
achieved with this method without any reduction
in soil density or ductility. Indeed increased
ductility was a feature of the mixture
identified in both triaxial and model testing.
The basic responses of the mixture in the
various tests were also essentially similar
to those of the soil alone, thus a new design
technology will not need to be developed for
this technique to be employed; conventional
designs using modified (improved) soil
parameters should be adequate.

Further laboratory testing is now under way to
further identify and optimise the important
properties of the mesh elements and to relate
these to different soil types. In parallel to
this, field mixing equipment is being developed
and mixing trials are planned. Numerous end
uses for this new method of soil strengthening
are also being identified and it is envisaged

REFERENCES

Andersland, O.B. and Khattak, A.S. (1979).
    Shear strength of kaolinite/fiber soil
    mixtures. Int. Conf. on Soil Reinforcement:
    Reinforced Earth and other Techniques,
    (1), 11-16, Paris.

Hoare, D.J. (1979). Laboratory study of
    granular soils reinforced with randomly
    oriented discrete fibres - A laboratory
    study. Int. Conf. on Soil Reinforcement:
    Reinforced Earth and other Techniques,
    (1), 47-52, Paris.

Leflaive, E. (1982). The reinforcement of
    granular materials with continuous fibers.
    Proc. 2nd Int. Conf. on Geotextiles,
    (3), 721-726, Las Vegas.

McGown, A., Andrawes, K.Z. and Al-Hasani, M.M.
    (1978). Effect of inclusion properties
    on the behaviour of sand.
    Geotechnique, (28), 327-346.